The Biology of Social Insects

Also from Westview Press

Insect Behavior: A Sourcebook of Laboratory and Field Exercises, J. R. Matthews and Robert W. Matthews

Metabolic Aspects of Lipid Nutrition in Insects, edited by T. E. Mittler and R. H. Dadd

Orthopteran Mating Systems: Sexual Competition in a Diverse Group of Insects, edited by Darryl T. Gwynne and Glenn K. Morris

Pest Control: Cultural and Environmental Aspects, edited by David Pimentel and John H. Perkins

World Food, Pest Losses, and the Environment, edited by David Pimentel

The Biology of Social Insects

Proceedings of the Ninth Congress of the
International Union for the Study of
Social Insects, Boulder, Colorado, August 1982

edited by Michael D. Breed,
Charles D. Michener,
and Howard E. Evans

Routledge
Taylor & Francis Group

NEW YORK AND LONDON

First published in paperback 2024

First published 1982 by Westview Press

Published 2019 by Routledge
605 Third Avenue, New York, NY 10158

and by Routledge
4 Park Square, Milton Park, Abingdon, Oxon OX14 4RN

Routledge is an imprint of the Taylor & Francis Group, an informa business

Copyright © 1982, 2019, 2024 by International Union for the Study of Social Insects

Library of Congress Catalog Card Number 82-72267

Publisher's Note
The publisher has gone to great lengths to ensure the quality of this reprint but points out that some imperfections in the original copies may be apparent.

ISBN: 978-0-367-30582-6 (pbk)
ISBN: 978-0-367-29036-8 (hbk)
ISBN: 978-0-429-30911-3 (ebk)

DOI: 10.1201/9780429309113

Contents

SYMPOSIUM VI: THE EVOLUTION AND ONTOGENY OF EUSOCIALITY

SYMPOSIUM VII: CASTE AND ERGONOMICS

SYMPOSIUM VIII: PREDATION, SOCIAL PARASITISM, AND DEFENSE

SYMPOSIUM IX: COMMUNICATION

SYMPOSIUM X: NEUROBIOLOGY AND BEHAVIOR OF SOCIAL INSECTS

Preface

    Social insects (ants, termites, some wasps, and some bees) are the organisms par excellence for the experimental, observational, and theoretical study of the evolution and ecology of altruism and related biological phenomena such as kin selection. The adaptations of workers for foraging and defense give them vital and often dominant roles in ecosystems, particularly in the tropics. High levels of social organization require complex mechanisms of social integration that are expressed, for example, as castes and as communication. Social insects provide significant model systems for studies in such fields as (1) the endocrine control of development, (2) the evolution of communication systems (particularly those involving pheromones), and (3) neurobiology and behavior. Because their neurobiology is amenable to investigation (electrophysiologically and otherwise) and because their behavior is complex and readily studied, social insects are major experimental subjects for neurobiologists.

    Certain social insects exhibit the most complex behavior of any invertebrates. They are therefore ideal for discerning the potential accomplishments of small-brained creatures. The famous works of von Frisch, Lindauer, and others on honey bee orientation and communication are examples of such studies. Social insects have advantages over other potential research subjects for many investigations. They are small, cheap (compared, for example, to primates), abundant, and exhibit a great variety of social organization so that they invite comparative studies.

    While no one today believes that by watching social insects we can learn anything directly applicable to improvement of human societies, there are meaningful similarities among all social organisms. Considered along with the great differences, these similarities can give us a wealth of ideas and information about the prerequisites and moving forces relevant to the evolution of societies. Thus the elucidation of such issues as altruism in the social insects has a bearing on our understanding of evolution generally, including social evolution of primates. Insights thus gained into our own social evolution give us a better understanding of human potentials.

Many social insects have major economic impact as household, structural, agricultural or forest pests or as stinging creatures of medical significance. The following names bring to mind some of the destructive forms: subterranean and dry wood termites, grazing termites, leafcutting ants, fire ants, Africanized (= "killer") bees, pharoah's ant. Most groups of destructive social insects are particularly abundant in the tropics, hence in the "third world" countries. For example, nearly one third of the animal biomass of the Amazon forest is said to be ants and termites. It is in tropical countries, therefore, that biological understanding of such insects is most important.

Other social insects are major benefactors of man. The honey bee, largely because of its pollination activities, is probably the world's most useful insect. (It is also the best known insect and some say the best known animal other than man.) Other useful social insects include wild bees as pollinators, ants as biological control agents for forest or agricultural pests, soil-inhabiting termites as turners of tropical soils, etc.

This book consists of the Proceedings of the Ninth Congress of the International Union for the Study of Social Insects (IUSSI), Boulder, Colorado, August 9-13, 1982. It provides information on the current status of knowledge and ideas about many facets of social insect biology. In recent years there has been a rapidly increasing number of persons studying social insects. This is partly a result of intensification of interest in animal behavior generally, but is largely a response to the stimulus provided by W. D. Hamilton (1972) and others interested in the evolution of altruism and by E. O. Wilson through his books "The Insect Societies" (1971) and "Sociobiology" (1975). The present collection of papers should be a useful adjunct to "The Insect Societies." Coming eleven years after the publication of that monumental work, the papers herein give a measure of the gains made in our understanding of social insects in the last dozen years.

IUSSI is an international organization (established in 1952) devoted to the study of social insects—not only the eusocial forms included among the termites, ants, bees, and wasps, but also various other taxa that live in less highly organized social units. At four to five year intervals IUSSI holds an International Congress. At irregular intervals it organizes International Symposia on special topics. And finally, IUSSI publishes the journal Insectes Sociaux/Social Insects, now in its twenty-ninth volume.

In preparation for the Ninth Congress, circulars were sent to specialists on social insects all over the world asking for suggestions as to symposium topics. The response was almost overwhelming; after eliminating duplication, there remained approximately 100 suggested topics, varying from extremely specialized to broad and of very general interest. Our Organizing Committee had to reduce this number to ten.

The noteworthy feature of the suggested topics was the large number from all parts of the world in the areas of ecology, ecological physiology, and significance of social insects in natural ecosystems and agroecosystems. The field of neurobiology and behavior was also stressed. Emphases in prior congresses were in such areas as caste determination, physiology, and evolution of social behavior; these areas are not omitted here but are less predominant than in past Congress Proceedings.

In ending this Preface, I wish to thank the many persons who have helped in the preparations for the Congress--the many IUSSI members who have made suggestions, the symposium organizers (listed elsewhere), the authors, the other members of the editorial committee (M. D. Breed, H. E. Evans) and especially Dr. Michael D. Breed of the University of Colorado who not only is the Secretary-Treasurer of the Congress but has also handled all the local arrangements including those for publication of this book.

Charles D. Michener
President, IUSSI,
and
Watkins Professor of Entomology and
of Systematics and Ecology
University of Kansas
Lawrence, Kansas

Acknowledgments

The IUSSI thanks the following organizations and
individuals for financial support of the Congress:

Dr. Caryl P. Haskins
Dr. Charles D. Michener
Mobay Chemical Corporation
Monsanto
The National Science Foundation (USA)
The University of Colorado
The University of Kansas Endowment Association

Editors' Note

Most of the contributions to this book were prepared in camera
ready form by the authors and are published as received. The
editors and at least one individual familiar with the specific field
reviewed each paper. We wish to thank these reviewers, the authors,
and the symposium organizers. Any errors are, of course, the
responsibility of the editors or of the individual authors.

Of Insects and Man

Edward O. Wilson, Harvard University

The history of ideas has witnessed wide cycles in the perceived relevance of the social insects to the study of man. Since the time of the Greeks and Hebrews, the industry of the insect workers, the superb organization of their colonies, and the absence of internal strife have been a cause of admiration and recommended emulation. On the other hand, the too-close comparison of humanity to the social insects has been construed as a sinister exercise. We are fearful of the "myrmidons" of the totalitarian state, who might seem to have attained a new order of life that supplants our old independent mammalian individuality. The perfect organization of the honeybee hive becomes the metaphor of social stagnation in the human community. To a few critics, especially radical political groups such as Science for the People, it follows that the very use of the common language in entomology and human affairs legitimates the status quo and argues against humanity and social progress: ant slavery and human slavery, termite royalty and human royalty, colony territory and tribal territory, and so on down the long list of familiar expressions.

All of this is nonsense, the result of too much emphasis on analogy in reasoning. There are no lessons to be taken home from the direct observation of colonial life in the insects. On the other hand, undoubted similarities exist between insects and human beings. Properly evaluated in combination with the even more striking differences, they offer us an entrée into a much deeper analysis of social evolution. The social insects comprise a separate high pinnacle of evolution, the closest approach we may ever see to colonial existence as it could emerge on another planet. Ants and termites originated from protostome ancestors that separated from our own deuterostome line as much as a billion years ago, and they achieved full sociality a hundred million years before Homo appeared on the scene. Each of the more than 20,000 species of social insects represents an independent evolutionary experiment by which the general theories of sociobiology can be tested and extended.

Consider the overwhelming ecological dominance of the social insects. Ants and termites are among the most abundant of all arthropods. They constitute, for example, about 30 percent of the animal biomass of the Amazonian rain forest. There and in other parts of the tropics they far exceed earthworms in the amount of soil and humus

1

they turn. Species for species, the leaf-cutting ants of the genus
Atta are the prevalent herbivores in Central and South America. The
frequency of all kinds of ants relative to other arthropods in the
Baltic and Dominican amber indicates that they have been numerically
dominant during a large part of their geological existence. In all
probability ants stung dinosaurs, at the same time that termites were
reducing dinosaur dung to powder.

We are naturally inclined to think that this great success was
and remains the outcome of social organization. But it is well to
remember that some of the most social of vertebrates--the communally
nesting parrots, for example, and the pack-hunting wild dogs--are
relatively rare and limited in biogeographic distribution. In fact,
no theory has yet served to predict and hence persuasively explain
the prevalence of social insects over nonsocial ones, any more than
an explanation exists of why certain eusocial genera (e.g., Nasutitermes, Vespa, Camponotus, Pheidole) are overwhelmingly more common
and widespread than most others. Certain features of social existence seem likely candidates, including the capacity for parallel
action of many tasks instead of sequential action, persistent nest
sites, perennial life cycles, and others; but none can be advanced as
contributing decisively, at least not on the basis of more than a
purely intuitive argument. When progress is made on this problem,
new light will be shed on the meaning of the success and failure of
certain kinds of vertebrate social organizations, including those of
human beings.

The relative transparency of the details of insect social organization, enhanced by the large number of species available for study,
has led to other kinds of important conceptual advances in general
sociobiology. The most dramatic individual salient has been in the
theory of kin selection. It is important to note that W. D. Hamilton's 1964 formulation was influential not simply because of priority;
others, including J. B. S. Haldane, G. C. Williams, and even Darwin,
had understood the role of kin selection in broad terms. The success
was due to the fact that Hamilton produced a version that explained a
great deal of the peculiarities of social life across the entire pan-
orama of the social insects: the concentration of eusociality in the
Hymenoptera, the all-female worker force of hymenopteran societies,
the peculiar behavior of drones, and so on. The fit was close enough
to persuade other investigators, including those working on human
behavior, that the evolutionary-genetic analysis of social organization has merit.

The social insects also offer superb opportunities to develop
the concept of optimization in the evolutionary studies of behavior.
Few modern biologists doubt the widespread occurrence of natural
selection, which causes species to evolve in one direction or another.
On the other hand few accept the idea of global optima--the concept
of a "perfect" honeybee or ant toward which species of Apis or for-
micids might be moving. The more reasonable conception is that of
local optima for particular species evolving within specific environments. There conceivably exist many small adaptive peaks on which
such populations may or may not come to rest for lengthy periods of
evolutionary time. To measure the nearness of the approach to local
optima is technically very difficult in large animals with complex,

flexible behavior. But it is much more readily achievable in social insects, which have rigid behavioral repertories and well defined ecological niches and can be easily manipulated through many experimental replications in the laboratory. Plausible estimates of the approach to optimization have recently been made for several social phenomena, including division of labor and foraging in Atta leaf-cutting ants and number of queens in incipient colonies in Myrmecocystus honeypot ants (see the special symposium on Caste and Ergonomics in this volume).

Another striking contribution to general biology has been a deeper understanding of pheromones. Ants and termites communicate primarily by chemical means, in contrast to human beings, who rely principally on audiovisual signals. As a consequence, the social insects provide a view of how a higher degree of social organization can be reached in a radically different sensory world. The colony of a typical ant species utilizes as many as ten distinct kinds of pheromones, including mixtures of alarm and trail substances that serve multiple functions of attraction, excitement, orientation, and identification. Detailed studies of the chemistry and physical properties of pheromones in ants and honeybees provided the first optimization assessment of this form of communication and have served as a stimulus and guide for similar studies of vertebrate pheromones.

The social sciences and humanities still suffer dramatically from a blinkered concentration on the single species Homo sapiens. The most effective social theory will eventually be a comparative social theory, in which other forms of organization are contrasted with the human form in a conception of all possible social worlds. Human beings are characterized by a fundamentally mammalian social plan, overlain by an enormous linguistic capacity and a moderate (but easily overestimated) degree of flexibility in decision making. Human behavior has evidently evolved during approximately the past two million years through gene-culture coevolution, in which the genetic rules of mental development and cultural innovation and choice change in a coupled manner. The social insects occupy a second, very interesting envelope within the space of all possible worlds. In their rigidly programmed repertories, sharp reflection of kin selection, close approach to local evolutionary optima, reliance on pheromones, and other distinctive properties, they can be profitably compared to human beings in ways that transcend mere analogy and contribute to the formation of a truly scientific body of social theory.

On Insects and Insects: Twists and Turns in Our Understanding of the Evolution of Eusociality

Ross H. Crozier,
The University of New South Wales

We generally credit Hamilton (1963, 1964, 1964b) with ushering in the present era of theoretical exploration of the problem of eusociality, or the existence of the worker caste. But no train of thought springs full-grown from nowhere, as Athena is supposed to have sprung from the head of Zeus. Relatively immediate predescessors of Hamilton include Williams & Williams and Haldane, and Charles Darwin wrestled with the problem, and has been interpreted as taking a parental-manipulationist stance. Before Darwin, Shakespeare proved himself no biologist with his description of kingly rule in honey bee life, Marcus Aurelius could be interpreted as an early group selectionist, and Solomon noted both the female-dominated and democratic (anarchic?) characteristics of ant society.

INCLUSIVE FITNESS

Hamilton (1964) introduced the concept of <u>inclusive fitness</u>, which can be defined as:

$$J_1 = S_1 + \Sigma G_{Y1} . \delta S_{1Y}$$

where J_1 = inclusive fitness of individual 1
S_1 = personal reproductive success of individual 1
G_{Y1} = relatedness of individual Y to individual 1
δS_{1Y} = effect of individual 1 on individual Y's reproductive succes
$\Sigma G_{Y1} . \delta S_{1Y}$ = the <u>inclusive fitness effect</u>, or <u>extended fitness</u>, of individual 1.

Inclusive fitness is clearly not just a sociobiological concept: every organism will have an inclusive fitness. It also leads us to <u>kin selection</u>, which occurs when the inclusive fitness effect is consistently greater than zero (while not generally discussed, <u>negative</u> kin selection is also clearly possible).

4

The choice of a relatedness estimator has proved to be more difficult than seems appreciated in many textbook accounts, and has received much recent attention (e. g., Michod and Hamilton, 1980; Uyenoyama and Feldman, 1980; Pamilo and Crozier, 1982). Whereas the pedigree-based concept of identity by descent has been regarded as the basis of most theory, regression measures are more useful empirically (Craig and Crozier, 1979), and apparently more predictive sociobiologically as well (Uyenoyama and Feldman, 1980). In addition, identity by descent can only vary between zero and one, whereas the -1 to +1 range of regression coefficient values makes easier the formulation of situations favoring spite etc.

Despite the undoubtedly great heuristic importance of the concept of inclusive fitness, it has proved very difficult to enter explicitly into population genetic models, although it has figured prominently in games-theoretic approaches. But progress has been made towards formal integration of inclusive fitness in population genetic models (Michod and Abugov, 1980; Abugov and Michod, 1981).

THE RISE OF KIN SELECTION
AS AN EXPLANATION FOR EUSOCIALITY

As has been widely discussed (e. g., Wilson, 1971; Crozier, 1979), kin selection was seen at once as a powerful general explanation, not only for the relatively profuse flowering of eusociality in the Hymenoptera, but also for the female-dominance of hymenopteran (but not termite) societies. Although other explanatory models were proposed, the most powerful being that of parental manipulation (Alexander, 1974; Michener and Brothers, 1974), kin selection remained essentially unchallenged by modellers as the dominant approach until the late '70s. The same period saw an early exclusive reliance on games-theoretic approaches largely replaced by (more) exact allele-frequency modelling, which does not always give the same answers (although they may -- see Craig, 1980). The '70s closed with theoretical studies severely shaking the predominance of kin selection contra parental manipulation as an explanation for the evolution of workers.

KIN SELECTION IN CRISIS

The high point of acceptance of kin selection was probably the influential and pivotal games-theoretic paper of Trivers and Hare (1976), which linked the evolution of eusociality with that of sex-ratios. Unfortunately, this paper was quickly found to be flawed, with Alexander and Sherman (1977) pointing out statistical weaknesses and Crozier (1977) noting that the model required protoeusocial insects to possess the eusocial ability of being able to tell the sex of larvae before such species would need it (because solitary females have other cues as to the sex of larvae, such as which cell they are in). But the explanation of 3:1 sex ratios in established eusocial species has withstood the test of time, although it now appears that female-biassed sex ratios do not favor the evolution of eusociality (Craig, 1979; cf. Wade, 1979), and hence male-haploidy does not either.

The wrinkles that appeared on the ageing face of the kin selection paradigm can be understood after considering the real "fundamental theorem of kin selection", usually called "Hamilton's Rule":

$$b_B . G_{BA} > c_A$$

where, b_B = benefit to beneficiary
c_A = cost to altruist from an altruistic act
G_{BA} = relatedness of beneficiary to altruist

Allele-frequency modelling has brought out unexpected complexities of kin selection not readily apparent from Hamilton's insightful rule, such as that it is a form of frequency-dependent selection, that polymorphisms for altruism alleles are possible, that the dominance relationships between alleles affect the outcome of (kin) selection, and that it also matters whether fitness components are assumed to be additive or multiplicative.

Such complexities, however, do not constitute as much of a crisis as the direct comparison of the required benefit/cost ratios for altruism to be selected for under the parental manipulation and kin selection models (Charlesworth, 1978; Charnov, 1978; Craig, 1979). It turns out that the threshold is lower for parental manipulation than for kin selection.

But don't discard kin selection in favor of parental manipulation. Firstly, there will be a range of benefit/cost ratios over which kin selection will oppose the effects of parental manipulation in promoting altruism (Craig, 1979; Starr, 1979), so that the significance of the lower threshold under parental manipulation is unclear. Secondly, there are severe problems with the mode of action of parental manipulation (Crozier, 1979), although these problems can be substantially alleviated by postulating prior selection under kin selection for genotypes susceptible to manipulation (Craig, unpublished). Of course, the necessity for such a preparation phase restores kin selection to a central role in the evolution of eusociality!

MORE ADVANCES AND RETREATS IN IMPLICATING RELATEDNESS
IN THE EVOLUTION OF EUSOCIALITY

Sherman (1979) noted that present-day eusocial insects tend to have higher chromosome numbers than their non-eusocial relatives, and suggested that a reduced ability to discriminate between sibs on the basis of relatedness would favor indiscriminate altruism to all. Templeton (1979) pointed out that higher chromosome numbers would reduce the variance of the inclusive fitness effect, which should facilitate the evolution of social behavior under kin selection.

If the previous lack (now filled -- Aoki, 1979) of social aphids was embarassing for kin-selection theory, the presence of eusocial termites seemed to be even more so, necessitating the classical suggestion of initially-forced aggregations and inbreeding (see Hamilton, 1972) due to the need to transfer internal symbionts after molts. Bartz (1979) strengthened this argument by pointing out that under plausibly-hypothesized conditions the prototermite workers would be more related to siblings than to offspring, favoring social behavior. Lacy (1980) added an additional genetic dimension by noting that the frequently-observed sex-linked translocation complexes in termites would boost relatedness between sisters in the same way as does the total sex-linkage of Hymenoptera. Of course, this argument runs into the same problem as that invoking male-haploidy in the case of the Hymenoptera, namely that from current theoretical work, relatedness effects stemming from male-haploidy have probably had nothing to do with the evolution of eusociality! In fact, it now appears that, genetically speaking, male workers should evolve as readily as female ones in male-haploids (Craig, 1981), supporting an earlier suggestion by Alexander (1974) that hymenopteran males haven't become workers because they make poor ones. Thus the lack of stings makes males of primitive species less effective as foragers and defenders of the nest. One possible genetic factor remains: new mutations will make males more expensive to produce than females (Smith and Shaw, 1980), but it is uncertain by how much.

SYNERGISTIC FACTORS PROMOTING EUSOCIALITY

Various authors (e. g., Crozier, 1979) have stressed that other conditions than genetic system must be conducive for the evolution of eusociality, both because of the arguments given above and because of the occurrence of other male-haploid, but non-social, animals. Such factors include the possession of mandibles or other means of manipulating the environment, and a suitable life cycle bringing together potential altruists and beneficiaries.

A further factor is a preadaptation allowing group cohesion. This could initially be little more than the ability to return to a common site, but a kin-recognition system would be selectively advantageous to prevent "robbing" etc. Site-recognition substances might furnish raw materials for both goals, and are widely known among lower aculateas and many Parasitica, but are as yet unreported for non-hymenopterous parasitoids such as tachinids. An alternative route to development of kin-recognition systems might be from mate-recognition systems such as reported in Lasioglossum zephyrum and Drosophila melanogaster, and the subsocial isopod Hemilepistus reaumuri seems to possess an analog of a hymenopterous colony odor system (Linsenmair, 1972).

WHITHER COMPLETE EUSOCIALITY?

Many aspects of the biology of forms such as ants, honey bees, and termites seem scarcely relevant to evaluating theories on the origin of eusociality, because for them (especially those ants with sterile workers) there is no option open to workers for selfish behavior -- the benefit/cost ratio is intrinsically extremely large. Thus, the spate of findings of low levels of relatedness in many ants (Crozier, 1980) does not mitigate against a key role for relatedness in the ancestors of ants. The focu for most of these highly eusocial forms will be on the dynamics of established eusociality, especially with respect to the effect of the multiple levels at which selection can act in social insects, and on such problems as internest "drifting" and sex ratios. The nature(s?) of the kin recognition system will affect these phenomena; while the study of kin recognition is now showing real progress (Hoelldobler and Michener, 1980), strong implications cannot yet be drawn about selection in natural populations, especially for higher forms.

But not all ants are necessarily unsuitable for studies on the evolution of eusociality: the Australian genus _Rhytidoponera_ includes many species without normally-differentiated queens whose place is taken by a coterie of mated workers (Haskins and Whelden, 1965). Intra-colony relatedness is very low where it has been examined in this genus (Crozier, unpublished): what holds these colonies together, or, alternatively, why don't these species re-evolve queens?

References

ABUGOV R., MICHOD R.E. 1981. - On the relation of family structured models and inclusive fitness models for kin selection. _J. Theoret. Biol., 88_, (in press).

ALEXANDER R.D. 1974. - The evolution of social behavior. _Annu. Rev. Ecol. Syst., 5_, 325-383.

AOKI S. 1979. - Further observations on _Astegopteryx styracicola_ (Homoptera: Pemphigidae), and aphid species with soldiers biting man. _Kontyu, 47_, 99-104.

BARTZ S.H. 1979. - Evolution of eusociality in termites. _Proc. Nat. Acad. Sci. U. S. A., 76_, 5764-5768.

CHARLESWORTH B. 1978. - Some models of the evolution of altruistic behaviour between siblings. _J. Theor. Biol., 72_, 297-319.

CHARNOV E.L. 1978. - Evolution of eusocial behavior: offspring choice or parental parasitism? _J. Theor. Biol.,75_, 451-465.

CRAIG R. 1979. - Parental manipulation, kin selection, and the evolution of altruism. _Evolution, 33_, 319-334.

CRAIG R. 1980. - Sex ratio changes and the evolution of eusociality in the Hymenoptera: simulation and games theory studies. _J. Theor. Biol., 87_, 55-70.

CRAIG R. 1981. - Evolution of male workers in the Hymenoptera. _J. Theor. Biol., 93_. (in press).

CRAIG R., CROZIER R.H. 1979. Relatedness in the polygynous ant _Myrmecia pilosula_. _Evolution, 33_, 335-341.

CROZIER R.H. 1979. - Genetics of sociality. pp 223-286 in: HERMANN H.R. (ed.). Social insects. Volume I. Academic Press. 437p.
CROZIER R.H. 1980. - Genetical structure of social insect populations. pp. 129-146 in: MARKL H. (ed.). Evolution of social behavior: hypotheses and empirical tests. Dahlem Konferenzen, Verlag Chemie GmbH.
HAMILTON W.D. 1963. - The evolution of altruistic behavior. Amer. Nat., 97, 354-356.
HAMILTON W.D. 1964. - The genetical evolution of social behaviour. I. J. Theor. Biol., 7, 1-16.
HAMILTON W.D. 1964B. - The genetical evolution of social behaviour. II. J. Theor. Biol., 7, 17-52.
HAMILTON, W.D. 1972. - Altruism and related phenomena, especially in social insects. Aanu. Rev. Ecol. Syst., 3, 193-232.
HASKINS C.P., WHELDEN R.M. 1965. - "Queenlessness", worker sibship, and colony versus population structure in the formicid genus Rhytidoponera. Psyche, 72, 87-112.
HOELLDOBLER B., MICHENER C.D. 1980. - Mechanisms of identification and discrimination in social Hymenoptera. pp 35-58 in: MARKL H. (ed.) Evolution of social behavior: hypotheses and empirical tests. Dahlem Konferenzen, Verlag Chemie GmbH.
LACY R.C. 1980. - The evolution of eusociality in termites: a haplodiploid analogy? Amer. Nat., 116, 449-451.
LINSENMAIR K.E. 1972. - Die Bedeutung familienspezifischer "Abzeichen" fuer den Familienzussamenhalt bei der sozialen Wuestenassel Hemilepistus reaumuri Audouin u. Sauvigny. (Crustacea, Isoonda, Oniscoidea). Z. Tierpsychol., 31, 131-162.
MICHENER C.D., BROTHERS D.J. 1974. - Were workers of eusocial Hymenoptera initially altruistic or oppressed? Proc. Nat. Acad. Sci., U.S.A., 71, 671-674.
MICHOD R.E., ABUGOV R. - 1980. Adaptive topography in family-structured models of kin selection. Science, 210, 667-669.
MICHOD R.E., HAMILTON W.D. 1980. - Coefficients of relatedness in sociobiology. Nature, 288, 694-697.
PAMILO P., CROZIER R.H. 1982. - Estimating relatedness in natural populations: methodology. Theor. Popul. Biol., (in press).
SHERMAN P.W. 1979. - Insect chromosome numbers and eusociality. Amer. Nat., 113, 925-935.
SMITH R.H., SHAW M.R. 1980. - Haplodiploid sex ratios and the mutation rate. Nature, 287, 728-729.
STARR C.K. 1979. - Origin and evolution of insect sociality: a review of modern theory. pp 35-79 in: HERMANN H.R. (ed.). Social insects. Volume I. Academic Press, 437pp.
TEMPLETON A.R. 1979. - Chromosome number, quantitiative genetics and eusociality. Amer. Nat., 113, 937-954.
UYENOYAMA M.K., FELDMAN M.W. - 1980. On relatdness and adaptive topography in kin selection. Theor. Popul. Biol., 18. (in press).
WILSON E.O. 1971. - The insect societies. Harvard Univ. Press.

FORAGING BEHAVIOR AND POLLINATION
Organized by Gordon D. Waller

Introduction

Gordon D. Waller, USDA-ARS,
Carl Hayden Bee Research Center

Historically, the study of foraging behavior by pollinators originated with Aristotle who recorded in his History of Animals that the honey bee (Apis mellifera L.) generally restricted its visits to a single plant species during a given foraging trip. Pollination, as a science, was also a part of ancient Greece and Rome, where date palms were pollinated by bringing the "male" to the "female." However, the assignment of sexuality to plants was not firmly established until the middle of the 18th century, at which time the role of. insects in distributing pollen was first recognized. Nevertheless, early naturalists who contributed to this body of knowledge from the 18th century through the early years of the 20th century tended to emphasize the botanical aspects of pollination. These include Kölreuter, Sprengel, Darwin, Fritz Müller, Hermann Müller, Knuth, Loew, and others.

During the 20th century we've seen the development of ethology as a science and its application to plant-pollinator interactions, viz., the analysis of form and function to elucidate the coadaptation of flowering plants and their pollen vectors. Although flowers had been properly described by early botanists, they could only be interpreted properly after the behavior of pollinators was studied and a new discipline called pollination ecology had developed. This historical note can best be summarized with a quote from the book by Proctor and Yeo (1972) as follows: "The pollination of a flower by an insect (or a bird or bat) takes place not in isolation, but in a habitat often of bewildering complexity, populated by numerous species of plants and animals, competing and interdependent - in short, an ecosystem."

A body of knowledge about olfactory, gustatory, and visual perception by honey bees has accumulated over the past sixty years, owing largely to the pioneering efforts of von Frisch and his colleagues. Information about the sensory perception of other pollinators is comparatively incomplete, as is our understanding of the stimuli (search image) offered by the flowers.

To properly understand how pollinators locate and exploit nectar and pollen sources requires that one apply all the basic sciences. Recent technological advances and the resultant output of newly developed analytical equipment have produced precise

11

measurements of petal colors, and have separated, quantified, and identified a whole spectrum of chemicals that comprise floral aroma. Color and aroma must be considered merely markers or cues that enhance a pollinator's ability to orient to flowers and facilitate recruitment. The relative importance of color and aroma to foraging honey bees has been elucidated primarily by the work of Koltermann and Kriston.

Pollinators visit flowers to obtain a reward, usually nectar and/or pollen. Nectar quantity and quality have been the subject of much research activity, possibly because nectar is relatively easy to remove, measure, and analyze and appears to stimulate honey bee foragers to maximum foraging activity. Nectars, once considered to be dilute solutions of sugars, have been shown by Herbert Baker and others to be complex mixtures of many compounds that affect both their acceptance by and their nutritional value to pollinators. Numerous workers have studied honey bee responses to nectar sugars and the minor components of nectar.

Pollen is more difficult than nectar to remove, measure, and analyze and therefore its role in stimulating foragers is less well understood. Isolation and identification of attractant factors in pollen is a task that has not been completed. Also, we do not know whether a bee can distinguish between those pollens that are nutritionally superior and those that tend to be nutritionally inadequate.

Although a vast literature has been accumulated on the behavior of one bee species, Apis mellifera, there is relatively little known about other pollinators. Of particular interest are the interactions among bee species competing for one or more nectar sources. Only three species of wild bees are currently managed as pollinators of crops on a commercial basis. The alfalfa leafcutter bee (Megachile rotundata Fabr.) and the alkali bee (Nomia melanderi Ckll.) are used to pollinate alfalfa, mostly in the U.S.A. and Canada, and Osmia cornifrons Rad. is used in fruit pollination in Japan.

We assume that our main goal as pollination ecologists is to understand the interface between flowering plants and their pollinators and thereby help to improve plant-pollinator relationships. Given this broad objective, we must search for answers to unresolved questions and evaluate areas for further improvement. With this acquired knowledge and appropriate efforts, we should be able to aid the preservation of natural ecosystems and improve the productivity of agroecosystems.

Our attempts to analyze foraging behavior usually fit one of two categories: 1) observe and record foraging under "natural" conditions and develop a mathematical model that describes this activity or 2) construct an artificial foraging arrangement (bioassay) that permits the altering of individual stimuli to measure their effects on foraging behavior. A comprehensive analysis can be developed if we employ elements of both techniques in our research.

Another dichotomy occurs because some scientists are found working only in natural, undisturbed habitats while others prefer to work in agricultural ecosystems that offer vast areas of a single plant species. Some might even insist that exotic plants and pollinators must be excluded from their study area. We need the best efforts by both the basic scientists, viz., the ethologists, the evolutionary ecologists and the mathematical modelers, and the pragmatists, viz., the apiculturists and plant breeders.

New knowledge about pollination ecology should assist plant breeders in selection of crop plants that are attractive to the intended pollinator. Alternatively, bee breeders might select for genotypes that are especially adapted to pollinate certain crops. Such considerations are important when hybrid cultivars are developed and their success or failure is dependent upon the economics of seed production. Thus, the fascinating field of foraging behavior and pollination extends from the analytical chemist's and animal psychologist's laboratory settings to the farmlands where one product of this scientific activity is increased production of seed and fruit.

This symposium presents the results of scientific efforts to describe plant strategies to assure pollination and insect strategies to obtain food. We will hear about foraging activity as it occurs in a natural plant community, in an agricultural field, and in a test arena composed of artificial flowers. We will hear how to build a mathematical model to explain foraging behavior and we will hear a challenge to the theory of optimal foraging - the basis for much of this modeling. The direction of future research on foraging behavior and pollination may be profoundly affected by what transpires at today's symposium.

Competition Among Social Species

*William M. Schaffer, David W. Zeh,
and M. Valentine Schaffer,*
The University of Arizona

INTRODUCTION

Traditional accounts of competition have focused either on the ways in which consumer species partition an array of available resources (e.g., Brown and Lieberman, 1973), or on the role of overt interference by which dominant species are able to exclude competitors from the most productive food species or habitats (e.g., Miller, 1967; Johnson and Hubbell, 1975). Recently, Schaffer et al (1979) considered a third kind of interaction. In this case, three species of bees (Table 1) compete exploitatively for the same resources, pollen and nectar produced by the "shin-dagger," *Agave schottii* Engelm. As is often the case in situations involving aggression, the species sort out along a gradient of increasing productivity. Interestingly, however, behavioral dominance is not involved. Instead, the outcome

Table 1

SPECIES	COLONY SIZE*	INDIVIDUAL WT
Xylocopa arizonensis	Solitary	170–320 mg
Bombus sonorus	10 - 100	70–130
Apis mellifera	4000 - 64000	30–50

*Number of individuals

of competition appears to be determined by the costs of foraging by individual workers (which scale with forager body size) and by the costs of colonial existence (which scale with colony biomass). In this regard, it is useful to distinquish the rate at which resources are produced (PRODUCTION) and the levels of resources actually available to consumers at any particular time (STANDING CROP). Consumer species can then be viewed as operators (in the mathematical sense) that map production into standing crop. Figure 1 illustrates this view. Two points deserve attention. First, there is some minimum level of production, P_m, such that the consumer population can maintain its numbers. For production levels below P_m, the consumer species cannot become established, and the equilibrium standing crop, SC*, equals the production. If production exceeds P_m, the consumer species can become established, and as production levels increase, the equilibrium standing crop will decline. The limit to which the consumers can

14

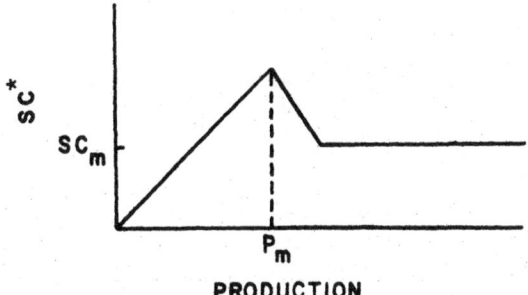

Fig. 1

depress SC* is the level SC_m, at which individual consumers can forage at a profit. The difference

$$D = P_m - SC_m \qquad (1)$$

represents the cost of reproduction and maintenance in excess of the cost of acquiring food. D will be positive for all organisms, but will be especially large, in social species, where the production of numerous workers is necessary for colony reproduction.

Returning to the relationship of SC* to P reduction (Fig. 2), we see that species with large colonies will have a larger P_m than species with small colonies. Concomitantly, species with small workers can depress SC* to lower values than species with larger workers. This is because a small animal can can forage at a profit at lower standing crops than can a larger animal. Thus, in the case of competition between *Apis mellifera* and species of *Bombus*, for example, we expect that bumblebees should be able to exploit less productive habitats because they have smaller colonies and therefore lower values of P_m. On the other hand, in very productive habitats, *Apis* should be able to outcompete *Bombus* by reducing the standing crop of available nectar to levels below the breakeven point for bumblebees. At intermediate levels of environmental productivity, the two species should coexist. In the present paper, we report on the results of experiments designed to test the validity of these ideas.

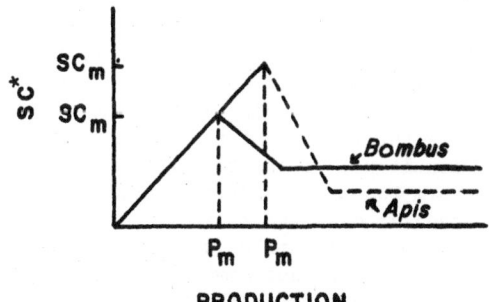

Fig. 2

METHODS

The observations were carried out at an elevation of 1500 m in the Santa Catalina Mountains north of Tucson Arizona. Approximately 400 flower stalks of *Agave schottii* were tagged and measured as to the number of flowers in bloom. Nectar production was measured by bagging selected flower stalks and draining the flowers six times daily. The bases of these stalks were treated with Tree Tanglefoot to exclude crawling insects. Bee activity was measured by censusing flower stalks seven times daily by an observer who walked the study plot and counted the number of bees on each stalk. Standing crops were measured by draining flowers to which the bees had access.

Ants (principally *Forelius pruinosis* and *Crematogaster dentinodus*) were excluded first from 10 of the flower stalks, and later from all of the stalks on and surrounding the site. The object of the manipulations was to enrich the site, so far as the bees were concerned, by reducing the amount of nectar consumed by ants at night. The 10 tanglefooted stalks were considered to represent a less productive resource "patch" than the entire plot when treated with Tanglefoot. It was predicted that progressive site enrichment would result in: a progressive increase in numbers of honey bees; an initial increase in *Bombus* numbers followed by a decline with further enrichment; and an initial increase followed by a decline in standing crops.

During the course of the experiments, first two and then four hives of Cordovan *Apis mellifera* (a light colored strain of honey bee) were placed on the study site. Toward the end of the study, the hives were removed. The object of the removal was to reduce the number of honey bees using the flowers and to determine whether or not the numbers of native bees would increase as a result.

RESULTS

Agave schottii produces approximately 90% of its nectar at night. In previous years, the pre-dawn standing crop was indistinguishable from the calculated nocturnal accumulation. In 1980, the present study, pre-dawn standing crops were significantly less than the nightly accumulation (2.7 µl/fl vs. 17.6 µl). However, when stalks were treated with Tanglefoot, the difference disappeared. Since the flowers on these stalks were open to flying visitors, and since no other crawling insects were observed consuming nectar, we conclude that the ants were taking 85% of the nightly accumulation.

Table 2 gives the results of the bee censuses, reported as the number of individuals per stalk, for the various treatment classes. The results conform to our predictions. Thus both *Apis* and *Bombus* visited the 10 tanglefooted stalks in greater numbers than the controls. However, when all stalks were treated with Tanglefoot, *Apis* continued to increase, but *Bombus* activity declined back to the control level. An increase in small solitary bees was also observed, primarily in the middle of the day after *Apis* and *Bombus* ceased visiting the flowers. Since the ants had been removed from the system and since some small production (about 10% of the daily total) continues throughout the day, an increase in these midday foragers is to be expected.

The effect of removing the Cordovan hives at the conclusion of the experiment is shown in Figure 3. During an interval of about two weeks, the number of feral honey bees increased until they approximated the previous combined total of ferals plus Cordovans. During this period, the numbers of bumblebees and small solitary bees and wasps (natives) first increased and then declined.

Table 2

Dates	6/18-6/24	6/20-6/24	6/28-7/2
Treatment	Control	10 Stalks TF	All Stalks TF
Number of Observ.	4056	278	5981
Apis *	.22	.32	.63
Bombus	.18	.49	.18
Xylocopa	.005	.00	.01
Solitary Bees**	.06	.02	.20
Standing Crop	.46 µl	5.60 µl	3.45 µl

*Feral plus Cordovans. Prior to removal of the Cordovan hives, Cordovan *Apis* consistently accounted for 75% of the total honey bee observations

**Principally *Halictus tripartitus*. Also *Augochlora sp.*, *Dialictus* (2 sp.), *Ceratina* (2 sp.), *Ancistrocerus sp.*, *Scolia ardens*, and an unidentified pompillid wasp.

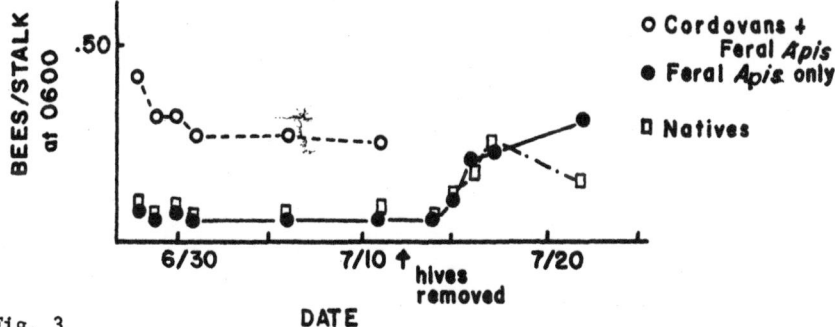

Fig. 3

CONCLUSIONS

The results support the hypothesis that honey bees are able to outcompete native North American bees at productive resource patches. As patch quality increased, the levels of available nectar (standing crop) first increased, and this was paralleled by an increase in the number of native bees. Further increases in production resulted in still greater visitation by honey bees, and as a result, standing crops declined along with the number of native bumblebees. Removing the honey bees from the system resulted in increased activity of native species.

ACKNOWLEDGMENTS

Suzanne Kleinhans, Dana Maxfield, and Jeb Antrim helped in the field. Bee identifications are due to Stephen L. Buchman; ant identifications to Roy L. Snelling. The Cordovan bees were provided and maintained by Stephen L. Buchmann and Charles Shipman of the USDA Bee Research Center, Tucson, Arizona. The work was supported by NSF Grant #DEB-8003783 to the senior author.

LITERATURE CITED

Brown, J.H., and G.A. Lieberman. 1973. Resource utilization and coexistence of seed-eating rodents in sand dune habitats. Ecology 54:788-797.

Ginsberg, H.S. 1981. Foraging ecology of bees on an old field. Ecology (In press).

Johnson, L.K., and S.P. Hubbell. 1975. Contrasting foraging strategies and coexistence of two bee species on a single resource. Ecology 56:1398-1406.

Schaffer, W.M., D.B. Jensen, D.E. Hobbs, J. Gurevitch, J.R. Todd, and M.V. Schaffer. 1979. Competition, foraging energetics, and the cost of sociality in three species of bees. Ecology 60:976-987.

Nectar Biology and Pollinator Attraction in the North Temperate Climate

Edward E. Southwick, State University of New York College at Brockport

The standing crop of nectar is the primary attractant to pollinators in many species of flowering plants in north temperate habitats. However, basic biological data on nectar and its secretion are needed (Baker and Baker, 1975; Robinson and Oertel, 1975; Shuel, 1970). Although nectar tissue structure has been examined in a few species (eg., Fahn, 1979), nectar constituents including sugar and non-sugar fractions, nectar secretion rates and timing, and standing crop, in most species are unknown. Yet many features of production and specific constituents of nectars influence the preference of foragers (Baker and Baker, unpub.; Waller, Carpenter and Ziehl,1972).

In this paper, I elucidate some of the factors of nectar biology important in analyses of pollinator attraction to plants and ultimately in pollination success of the plants.

MATERIALS AND METHODS

Blossoming flowers and their nectar pollinators were studied in northern Michigan (45°33'N 84°41'W) and northwestern New York (43° 12'N 77° 58'W). Nectar samples were collected utilizing aspirated glass micropipettes. Samples were analyzed for sugar concentration on optical refractometers. Quantitative fractions of sucrose, fructose and glucose were obtained from high precision liquid chromatography (HPLC). Sample volumes were determined from the length and diameter of the capillary columns. Visits of nectarivores were observed and recorded during 15 min counts over defined areas of known numbers of blossoms. Further methodological details are described in Southwick, Loper and Sadwick (1981).

RESULTS

Nearly 3000 nectar samples were collected from nine plant families. Twenty-one species of nectar feeders representing 14 families were observed. The standing crop of nectar of each of the plant species was analyzed and plants were grouped according to nectar sugar ratios (sucrose/hexose, S/H) as defined by Baker and Baker (1979). Sugar ratios and mean values of data on nectar are presented in Table 1.

19

Table 1.—Characteristics and pollinator attraction of nectar standing crop in north temperate flowers. Data presented are mean values.

Plant Species (Family)	S/H	volume (ul/flwr)	conc (g/100 ml)	sugar (mg/blos)	Principal Nectarivore
			Sucrose-poor		
Crataegus phaenopyrum (Rosaceae)	.4	.2	58.1	0.1	Bombyliidae
Malus Spp. (Rosaceae)	.4	1.1	20.8	0.2	—
Catalpa bignonioides (Bignoniaceae)	.4	5.0	70.7	3.5	—
			Balanced		
Philadelphus hyb. (Saxifragaceae)	0.5	1.2	91.0	1.1	Halictidae
			Sucrose-rich		
Trifoliumrepens (Leguminosae)	.9	.7	39.6	0.3	Apidae (Apis)
Tilia europaea (Tiliaceae)	.99	.1	78.2	0.1	—
			Sucrose-dominant		
Glechoma hederacea (Labiatae)	1.7	.3	51.8	0.2	Halictidae
Lonicera maackii (Caprifoliaceae)	2.8	.5	39.9	0.2	Apidae (Apis)
Echium vulgare (Boraginaceae)	5.0	.2	37.4	0.1	—
Asclepias syriaca (Asclepiadaceae) (NY)	4.9	1.9	36.7	0.7	Apidae (Apis)
(Michigan)	11.9	1.3	27.1	0.3	Sphecidae

All three sugars were detected in the nectars from all the plant families. Two plant families contained nectar that was sucrose-poor (S/H < 0.5); one was balanced (S/H = 0.5); two were sucrose-rich (S/H >0.5<1.0); and four were sucrose dominant (S/H > 1.0). Nectar volumes and concentrations were not correlated with sugar-ratio subgroups. The largest flowers produced the most nectar and the most sugar.

Available nectar varied through the day and was affected by many factors including microclimatic conditions and previous feeding by nectarivores. Except for rare encounters, blossoms of most species sampled contained progressively smaller amounts of nectar through the day. Field data on bagged blossoms (protected from foragers) show time dependence of nectar production, both through the day and during the life of the blossom (Fig. 1 and Southwick and Southwick, 1982).

Among the insects feeding at the blossoms of the six plant species where I have counts, definite preferences were seen (Table 1). Honey bees (Apis mellifera) were the principal nectar feeders at Trifolium repens (80% of nectar feeders counted were Apis), Lonicera maackii (41% were Apis), and Asclepias syriaca in upstate New York (41% were Apis). Asclepias blossoms in northern Michigan, however, were frequented most by wasps (Vespidae and Sphecidae made up 27% of the nectarivore clientele). Philadelphus hyb. attracted solitary bees (Halictidae -70%) as did Glechoma hederacea (55%). Few bees visited Crataegus phaenopyrum which attracted bee flies (63% were Bombyliidae).

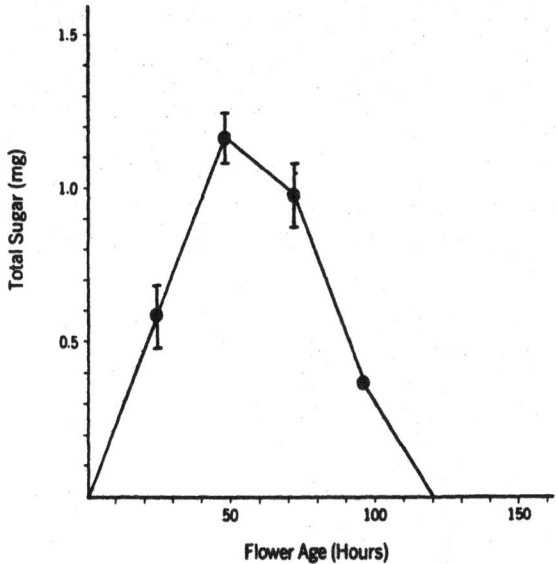

Figure 1.--Age effect on nectar production in bagged blossoms of Asclepias.

DISCUSSION

When nectar is scarce, as it may frequently be in north temperate regions, nectarivores must search more to find food. Many blossoms may be searched with little reward before locating a single blossom with a rich reward. Such "lucky-hits", if sufficiently rewarding, insure continued scrabbling and thereby effective pollinaton by the insect. The reward must be great enough to provide the energy needed to move to the next lucky-hit but not so large as to satisfy the feeder (by filling its crop). At least two conditions could result in lucky-hit blossoms. Such blossoms may merely have been missed by feeders earlier in the day, or the plant may produce copious amounts of nectar in only a few blossoms. The net effect on pollinaton is the same, however in the latter case the host plant conserves energy by producing less total nectar. This lucky-hit phenomenon has been found in hummingbird pollinated flowers in tropical regions by Feinsinger (1978) and limited production is likewise suggested by Heinrich and Raven (1972).

My field data suggest that lucky-hit nectar production may be a true pattern in Asclepias in the north temperate region. Bagged blossoms showed variable 24-hr production with some blossoms outproducing others on the same stem, of the same age, by as much as 20:1. This variation is further enhanced by the flower age dependency of nectar yield. As blossoms on a single stem come into maturity, peak nectar yields (at about 50 hrs after dehiscence in Asclepias) will be distributed likewise. In Asclepias, the pattern of blossoming and nectar presentation is a sequential progression first within the umbel, then from umbel-to-umbel up the stem (Fig. 2.). Attraction to nectar feeders is sustained, by this pattern of sequential peaking in nectar yields, over an extended period (Fig. 2, bottom), yet the feeder must scrabble over the blossoms to locate the lucky-hits. As different stems in a patch (genetic clone) mature, attraction is maintained.

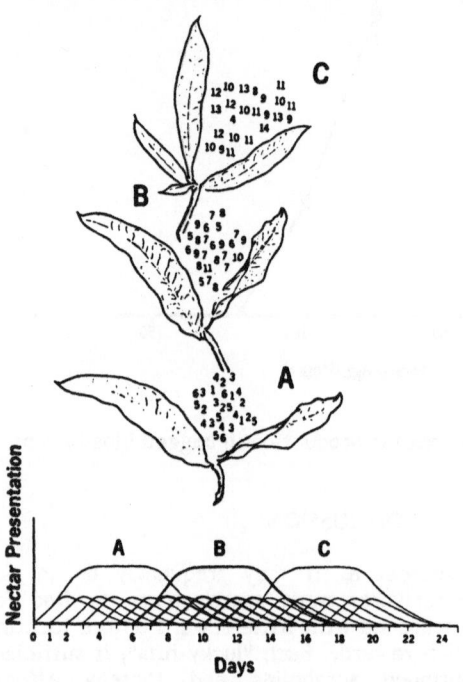

Figure 2.—Pattern of presentation of nectar attractant in a single stem of Asclepias syriaca.

Attractiveness to nectarivores was not correlated with sugar ratios in nectar (Table 1). Baker and Baker (1979) however, have shown general correlations of nectar feeders and sugar ratio groupings. Density of blossoms containing nectar, the volumes and sugar concentrations were positively related to visitation in my study. The importance of non-sugar constituents was demonstrated in Crateagus which attracted few bees even with nectar of high sugar concentration (58 g/100 ml). Faegri and van der Pijl (1979) point out that the pugent aroma is a fly attractant in such fly-pollinated flowers. Sugar or energy concentration and the accessibility of the nectar are probably major factors in visitation. Visitation of all nectarivores was noticeable more correlated with high environmental temperature than any characteristic of the nectar.

ACKNOWLEDGEMENT

I thank Heidi Cannioto for composing and typing this manuscript.

REFERENCES

Baker, H. G., Baker, I., 1979--Sugar ratios in nectars. Phytochem. Bull. 12, 43-45.

Baker, H. G., Baker, I., 1975--Studies of nectar constitution and pollinator-plant coevolution. In L. E. Gilbert and P. H. Raven (eds.), Coevolution of animals and plants, p. 100-140. University of Texas Press, Austin.

Faegri, K., van der Pijl, L., 1979--The principles of pollination ecology. 2nd ed. Pergamon, London.

Fahn, A., 1979--Ultrastructure of nectaries in relation to nectar secretion. Amer. J. Bot. 66, 977-985.

Feinsinger, P., 1978--Ecological interactions between plants and hummingbirds in a successional tropical community. Ecol. Monogr. 48, 269-287.

Heinrich, B., Raven, R. H., 1972--Energetics and pollination ecology. Science 179, 597-602.

Robinson F. A., Oertel, E., 1975--Sources of nectar and pollen, In Dadant and Sons (eds.). The hive and the honey bee, p. 283-302. Dadant and Sons, Inc., Hamilton, Illinois.

Shuel, R. W., 1970--Current research on nectar. Bee World 51, 63-69.

Southwick, A. K., Southwick, E. E., 1982--Aging effect on nectar production in two clones of Asclepias syriaca. Oecologia (submitted).

Southwick, E. E., Loper, G. M., Sadwick, S. E., 1981--Nectar production, composition, energetics and pollinator attractiveness in spring flowers of western New York. Amer. J. Bot., 68, 994-1002.

Waller, G. D., Carpenter, E. W., Ziehl, O. A., 1972--Potassium in onion nectar and its probable effect on attractiveness of onion flowers to honey bees. J. Amer. Soc. Hort. Sci. 97, 535-539.

Information Used in Foraging

Keith D. Waddington, University of Miami

There has been much recent interest in studying from an evolutionary perspective the food-choice behavior of many animals, including insect pollinators (Waddington and Holden 1979). All these studies of "optimal diets" employ mathematical models.

Two basic models have been proposed to predict animal choice patterns. One considers the case of "sequential" encounter of food items (the animal bumps into food items one at a time) (Pulliam 1974) and the second, which was developed for studying honeybee behavior, considers sighting of several flowers simultaneously (Waddington and Holden 1979). In each case much effort has been expended on the mathematics of the models and on testing the predictions of the models, but there are few data to verify the assumptions on which the models are based. If these assumptions are incorrect, predictions of the models will likely be in error. Furthermore, if the assumptions are invalid even a model which yields an accurate prediction under some circumstances will not represent a correct description of the actual processes. My recent efforts (eg., Waddington et al. 1981), and those of some other scientists (eg., Real 1981) have been directed toward filling this gap. The aim is to experimentally verify the behavioral and physiological mechanisms used by animals while foraging before their inclusion in foraging models.

An important assumption in models of optimal diets is that animals can perceive energetic costs and gains while foraging among various food items and explicit assumptions are made as to how the information is processed and what decisions result (Krebs 1978). However, data do not exist which verify the assumptions.

In this paper I attempt to quantify the relative assessments of caloric costs and gains made by honeybees (Apis mellifera L.) by studying the bees' communication round dance. Waddington (unpub. ms.) found that the rate of the round dance, measured as the number of reversals or changes in direction per time, is directly related to rewards gained per flower and inversely related to caloric costs to fly between flowers. I will manipulate the magnitude of costs and gains for bees to forage and then quantify the bees' assessments of these parameters by quantifying the round dance.

MATERIALS AND METHODS

Honeybees (Apis mellifera, L.) were housed in a one-frame, glass-sided observation hive. The bees had free access to the inside of a screen cage measuring 2x4x2m. Colony stores were minimal throughout the experiment.

Tests were carried out on bees, one at a time, on a foraging platform inside the cage. The horizontal platform was made of green Plexiglas^R measuring 0.29 x 3.79m. Two artificial flowers were mounted on the platform. A "flower" consisted of a piece of green Plexiglas measuring 3.75 x 3.75 cm (the foraging surface) laid over another piece of Plexiglas of the same dimensions and mounted on a 5 cm high block of wood. A 2.5 mm diameter hole was drilled through the Plexiglas, centering over a countersunk hole (the 'nectar' well) in the piece of Plexiglas beneath. 'Petals' made of strips of blue tape radiated from the centered nectar well.

Sucrose solution was fed into the nectar well from underneath through polyethylene tubing and was dispensed using a 25 μl syringe associated with a PB-600 Hamilton push-button repeating dispenser.

Bees learned to fly between the 2 flowers to suck 1/2 μl of sugar solution on each visit. Costs and benefits for a bee to forage were manipulated by varying the distance between the 2 flowers and by varying the concentration of the sugar solution, respectively. For bees 13, 14, 16, 19, 20 and 21, two distances, 0.15 and 2.41 m, were used combined with three sugar concentrations, 10%, 20%, and 30% (sucrose, by weight). For bees 31, 32 and 33, three distances, 0.15, 2.41, and 3.56 m, were used and 20% and 30% sugar solutions. Each bee foraged alone on the platform and was tested over several consecutive days. Each day the bee experienced nectar of just one sugar concentration, but the bee experienced one of 2 distances or one of 3 distances (as indicated above) over the separate trials.

A trial consisted of a bee foraging between the flowers until it returned to the observation hive (when presumably its crop was filled). The total time the bee foraged during the trial was determined using a stopwatch and the number of floral visits was counted. Behavior in the hive and real time was recorded on video tape. The bee's dances were later played back and quantified.

Frisch described the round dance as: the "...bee runs in a circle..., suddenly reversing direction and then turning about again to her original course..." In my analysis, a "dance" is considered an uninterrupted segment of running on the comb; the segment is timed and reversals counted so that RATE (reversals/min.) of reversals can be calculated.

A quantitative, 3-dimensional "surface" of a bee's assessment of caloric gains (CALGAIN - calories/flower) and caloric costs (CALCOST - calories expended per flower visit) will be determined by stepwise multiple linear regression analysis, with RATE as the dependent variable.

RESULTS

Results of the stepwise multiple linear regression analyses are presented in Table 1 for the 9 bees.

Table 1. Regression analyses. Y=RATE, X_1=CALGAIN, X_2=CALCOST.

Bee	Regression	F	d.f.	r	Multiple r^2
13	$Y=19.9 + 0.9X_1-33.1X_2$	[@]2.2 n.s.	2,73	.13	.02
		[β]2.0 n.s.		-.12	.04
14	$Y=12.9 + 1.8X_1- 4.7X_2$	53.9**	2,76	.64	.41
		0.7 n.s.		-.09	.42
16	$Y=18.7 + 0.8X_1-25.0X_2$	22.2**	2,133	.38	.15
		1.3 n.s.		-.12	.16
19	$Y=16.3 + 1.5X_1- 4.1X_2$	10.1**	2,69	.35	.12
		0.4 n.s.		-.03	.13
20	$Y=17.8 + 1.7X_1-20.0X_2$	15.5**	2,50	.60	.36
		0.6 n.s.		-.42	.37
21	$Y=13.1 + 1.4X_1- 2.8X_2$	9.6**	2,86	.31	.10
		0.01n.s.		-.02	.10
31	$Y=20.3 + 0.5X_1- 1.1X_2$	2.6 n.s.	2,175	.12	.01
		.02n.s.		-.02	.01
32	$Y=22.9 + 1.0X_1-48.9X_2$	15.9**	2,203	.22	.05
		11.1**		-.17	.10
33	$Y=29.1 - 0.2X_1-24.8X_2$	0.3 n.s.	2,156	-.04	.04
		7.0**		-.21	.04

Statistical significance of H_0: slope = 0. n.s. = not significant, p>0.05; **<p 0.01

[@]F-value for X_1 when entered into the equation
[β]F-value for X_2 when entered into the equation

Rate increases with CALGAIN for all but one bee, and the regression coefficients (slopes), which range between -0.15 and 1.83, are significantly different from zero for all but 3 bees. Even the non-significant coefficients are considered important since they represent a best estimate of the bees' perceptions.

The relationship between RATE and CALCOST is inverse for all bees; however, the coefficients for just 2 bees are significant. The range is between -48.9 and -1.06 with 5 of the coefficients being less than -20.

The ratios of the coefficients ($|X_2/X_1|$) for bees 13-33 are, respectively: 35.2, 2.5, 31.7, 2.7, 12.0, 2.0, 2.2, 48.4, and 165.2.

DISCUSSION

In general, a change in CALGAIN elicits a considerably lesser change in RATE than does the same absolute change in CALCOST. If the dance, specifically RATE, is a reflection of the bees' assessments of gains and costs then a calorie expended is not equivalent to a calorie gained; the results are variable among bees but for each bee the quantity $|X_2/X_1|$ is greater than 1. This suggests that several calories expended are perceptually equivalent to a single calorie gained. This is contrary to usage in most foraging models.

These results do not indicate the nature of decisions based on these perceptions. For example, what will be a bee's pattern of visitation in a mixed patch of plant species that involve different combinations of costs and gains? Will it visit the species that is perceived as "best" more frequently than others, or will it visit this species exclusively? Floral-choice experiments are currently in progress to answer such questions. Once this information is available the suitability of a plant species as forage for a bee relative to other available plant species can be better understood. This has application to understanding pollination biology in natural and agricultural plant communities.

REFERENCES

Krebs, J. 1978. --Optimal foraging: decision rules for predators. In: Krebs and Davies (eds.) Behavioural Ecology: An Evolutionary Approach. Blackwell Scientific Publications, 494p.

Pulliam, H. R. 1974. --On the theory of optimal diets. Amer. Nat., 108, 59-74.

Real, L. 1981. --Uncertainty and pollinator-plant interactions: the foraging behavior of bees and wasps on artificial flowers. Ecology, 62, 20-26.

Waddington, K., Allen, T., Heinrich, B. 1981. --Floral preferences of bumblebees (Bombus edwardsii) in relation to intermittent versus continuous rewards. Anim. Behav., 29, 779-784.

Waddington, K. D., Holden, L. 1979. --Optimal foraging: on flower selection by bees. Amer. Nat., 114, 179-196.

Acknowledgements - This work was supported by NSF grant BNS-8004537.

Random and Systematic Search in Foraging Insects

Rudolf Jander, The University of Kansas

The "theory of optimal foraging", though never concisely defined, presently overdominates the discussion about foraging behavior. Of late, however, the early euphoria of the optimalists has been dampened by a rapidly rising tide of counterexamples, evidence proving non-optimality (e.g. Janetos and Cole 1981). Hence the key question that prompted this contribution: Can this "theory" truly be called scientific; is it a fiction, or can it be deduced from generally accepted evolutionary principles, as have been tried?

For a theory to claim scientific status, first of all, it must be susceptible to refutation. So far, none of the optimalists have explained how his pet theory could be refuted. What is even more disturbing is the fact that citing individual counterexamples is of no avail, they all can be easily explained away as being due to biological constraints. Consequently, the future of this "theory" simply depends on how assiduously the proponents and the opponents accumulate examples or counterexamples; given the complex richness of life, neither group will ever have to worry about disclosing further "evidence" for their cause. The perfect dilemma of an "evolutionary stable controversy"!

Untenable as the claim for scientific status is the evolutionary grounding of the "optimal foraging theory". First, to be consistent, the postulate for optimality must be extended to all adaptations; but whose biological intuition is not revolting at that? More explicitness will disclose the underlying illogic. Let us visualize evolution as an uninterrupted process stretching over billions of years. In addition, let us accept "optimal adaptation" as denoting a state of adaptation that cannot be further improved upon by natural selection. Then, why in the world should all adaptations of millions of species happen to have reached by now this state of optimality? Conversely, if we reject this absurd assumption and predict future evolutionary advances, we cannot but admit that current adaptations are sub-optimal. No matter how we twist it, there is no rational way of meshing "optimal foraging theory" with evolutionary theory.

Nevertheless, the fact cannot be denied that the optimalists successfully advance the scientific understanding of foraging behavior. Somewhere behind the surface must be hidden some undisclosed theoretical truth. Thus we are facing the strange sociopsychological situation, analogous to genetic hitch-hiking: some successful idea pulls along some false idea just as the advantaged gene pulls the disdvantaged one (Hedrick 1980).

Given this tight packaging of theoretical sense and nonsense, one obvious solution is the dissection of the "optimality theory" for the purpose of separating and discarding its false constituents. This prescription is equivalent to dissecting behaviors covered by the theory, as will be illustrated by the following example.

Convoluted local search paths in planar environments, as for instance commonly performed by ants on the ground and by bees at the flower level, are among the most random of all behaviors. Such paths are not strikingly different, yet certainly also not identical, to the paths generated by Brownian motion, the most random of all natural motions. Indeed, no sooner had A. Einstein formulated the mathematical description of Brownian motion during this century's first decade, then zoologists applied these formulas to animal search, though with rather limited success; the catch was some elusive component of non-randomness. It took seven decades for the next theoretical breakthrough to occur, which is based on Mandelbrot's (1977) novel geometry of random shapes. The central procedure in applying this geometry to random search paths is the fractal analysis by means of which the crucial fractal dimension D is evaluated. For more or less random paths D ranges from the maximal value of $D=2$ which also holds for the Brownian motion, to the minimal value of $D=1$, which describes the perfectly straight line. Thus D is a measure for the degree of order in a search path; as D increases randomness increases, and non-randomness concomitantly decreases. Based on my still limited experience I suspect that fractal values of about $D=1.5$ may be typical for locally searching insects. Applying fractal and statistical analysis to the convoluted search paths of <u>Formica pallidifulva</u>, honey bees and other species of insects, a surprisingly simple descriptive and analytical model has emerged. Such convoluted search paths can be closely approximated (modelled) by normotropic random walks (new term): sequences of nonrandom segments of progression, called "steps" in the model, are linked by partially random turns that are members of a wrapped-around normal distribution. Such a distribution results from wrapping a conventional normal distribution around a circle (Mardia 1975). Given this behavioral structure only two parameters have to be specified to describe its particular manifestation, the standard deviation of the wrapped-around normal distribution and the mean "step" length. Both measures can be extracted from recorded convoluted search paths by means of Mandelbrot's fractal analysis.

Ethological reductions of complex behaviors in ways indicated by above dissection of convoluted search, together with evolutionary knowledge, serve as foundations for creating and building the following new "modal tuning theory", which will be briefly and simply explained by deductive argumentation.

Contravening "optimal foraging theory", the argumentation towards the modal tuning theory starts out with the premise of general adaptive sub-optimality. Next, all natural adaptations are conceptually dichotomized into their structural and their measurable aspects. The number of independently varying structures define the complexity of a particular adaptation, the realizations of the independent measures its metricity. The two concepts of complexity and metricity are equivalent to the information-theoretical concepts of logon-content and metron-content (McKay 1969), and are related to the taxonomist's concepts of characters and character-states.

Complexity and metricity serve distinct logical functions within the modal tuning theory by virtue of their distinct evolutionary properties. Note, any individual's adaptive complexity is necessarily limited whereas the environmental structures, to be adapted to, contain virtually unlimited complexity in space and time. Hence the theory's basic proposition that all adaptations, like for instance foraging strategies, when taken as wholes, are bound to be sub-optimal; in other words, there is always the potential for further directional selection once novel heritable structural diversity has been generated. Complexity is known to evolve extremely slowly, metricity, by comparison, fast. Consequently, closely related taxa tend to be much more similar in their complexity than in their metricity. A pertinent illustration is the phylogenetically readily varying neck-length of mammals, contrasting with the ultraconservative underlying seven-vertebrae structure. Also of importance is the fact that virtually no metrical features have been found that did not readily respond to artificial directional selection. From these distinctive evolutionary propertiess of metric (quantitative) features follows the central predictive proposition of the modal tuning theory that under natural conditions metrical features rapidly evolve towards usually unimodal, and sometimes polymodal distributions with the modes then being maintained by stabilizing selection over long periods of time. This very process is called "evolutionary modal tuning" - and it thus serves as the namesake for the new theory. A crude analogy may further clarify the core idea: One easily turns a radio's knobs for setting its metrical states (tuning); yet, enlarging a radio's structural complexity by adding further control circuits is a major costly undertaking. Because the tuning of adpative modes is evidently a speedy evolutionary process, predictions about tuned states are fairly reliable once the evolutionary accessibility of the respective metrical features has been established by prior analysis. Countless predictive rules, many of which are already well established, like Bergmann's rule, subsume under this reasoning; or, to pick an example from foraging behavior, there is the simple rule that preferred prey size increases with the body size of the predator. Modal tuning theory covers vastly more ground and hence is superior in its unifying power to its antecedent "optimal foraging theory." More explicitly, the new theory not only holds across all taxonomic kingdoms but also across morphological, physiological and behavioral adaptations. While being much broader, this new theory is also more specific in its predictions and thus closer to reality than the old one. Finally, whereas the modal tuning theory smoothly squares with evolutionary insight, "optimal foraging theory" is at variance with it.

Citations

Hedrick, P. W., 1980. - Hitchhiking. A comparison of linkage and partial selfing. Genetics 94, 791-808.

Janetos, A. C., and Cole, B. L., 1981. - Imperfectly optimal animals. Behav. Ecol. Sociobiol. 9, 203-209.

Mandelbrot, B. B., 1977. - Fractals. Form, Chance, and Dimension. Freeman, San Francisco, 365 p.

Mardia, K. V., 1971. - Statistics of Directional Data. Academic Press, New Yorm, 357 p.

McKay, D. M., 1969. - Information: Mechanism and Meaning. Cambridge Mass.: MIT Press.

Abstracts

Foraging behavior and sociality in the organization
of a native Rocky Mountain bee community

L. Susan Anderson
University of Arizona

The foraging behavior of native solitary and primitively social
bees was analyzed by identifying individual pollen loads and larval
pollen provisions in trap nests. While in all species individual
bees specialize during single foraging bouts, generalized foraging
may arise over sequential foraging bouts in individuals or simultaneously between individuals at the colony level. Generalized species
have greater flexibility in responding to temporal variation, but
this flexibility is obtained at the expense of less efficient use of
individual resources. Specialized species do not switch resources
and may therefore, have greater foraging efficiency. Social species
can retain both flexibility and efficiency if individual colony members specialize on different resources. However social bees require
a longer period to produce reproductives than do solitary bees, and
may have lowered fecundity if the blooming season is unusually
short. Foraging flexibility also appears to affect the relative
abundance of species in the Colorado bee community. For example,
the proportion of solitary specialists decreases relative to gener-
alists after years of low resource abundance. The relationship between the efficiency and flexibility of foraging in bees with different social behaviors could be important in the evolution of
sociality and the organization of bee communities.

Collection of Cauliflower Pollen and Nectar by Apis cerana indica

Om P. Bhalla and Harmit S. Dhaliwal
Himachal Pradesh Agricultural University, Solan, India

Observations were made on the modus operandi of Apis cerana
indica foragers visiting cauliflower blossoms with a view to estimate the pollinating efficiencies of bees collecting either pollen
on nectar or both. Depending upon how they approached and worked

32

blossoms to obtain these commodities, bees were classified in five groups: (i) <u>Deliberate Pollen Foragers</u>, i.e. bees 'actively' engaged in gathering pollen only; (ii) <u>Incidental Nectar Collectors</u>, i.e. bees chiefly collecting pollen but obtaining nectar also; (ii) <u>Deliberate Nectar Foragers</u>, i.e. bees 'actively' gathering nectar only from within the flowers; (iv) <u>Incidental Pollen Collectors</u>, i.e. bees chiefly collecting nectar but obtaining some pollen also; and (v) <u>Nectar Robbers</u>, i.e. bees collecting nectar only from without the flowers. The <u>modus operandi</u> of pollen collecting bees was found as the most conducive and that of "nectar robbers" as the least likely to cause cross-pollination of cauliflower. Ecological significance of observed differences in the foraging methods of the five <u>A.c. indica</u> forager-groups <u>per se</u> is also discussed in the context of resource utilization and intraspecific coexistance.

Evolution of the Foraging Behavior
of Leaf-cutting Ants (<u>Atta</u> and <u>Acromyrmex</u>)

Harold G. Fowler
Rutgers University

Based on a zoogeographic pattern of species richness and abundance centered in the southern Neotropics, it is probable that the attine genera <u>Atta</u> and <u>Acromyrmex</u> originated in savannas or cataaingas, and not in tropical forests. This inference is supported by a reduced metathoracic leg/mandible ratio and a much simpler fungal substrate preparation procedure in grass-cutting ants, and to the apparent restriction of some taxa to Pleistocene islands of "campo cerrado". Dicot-specialized cutting taxa would thus be derived from grass-cutting taxa, and may explain the depauperate faunas present in northern South America, Central America, and Mexico, and explain the greater contribution of attine genera to the ant fauna of the subtropical southern regions of the Neotropics. If these inferences are borne out by subsequent investigations, the intensive analysis of foraging by Central American leaf-cutting ants may bear little on our understanding of how and why leaf-cutters do what they do, and may shed little light on a more rational basis for their management in agricultural environments.

Patterns of Communication and Recruitment in Stingless Bees

Leslie K. Johnson
University of Iowa

Stingless bees (Apidae: Meliponinae) exhibit considerable diversity in modes of communication about and recruitment to food resources, as revealed by observations of foraging on natural food sources and on grids of food dishes. One set of species exploits small flowers with low food reward. Scout bees of these species excite recruits but communicate only food odor. Recruits fan out from the nest and find other food sources of the same odor, discovering those nearer the nest sooner. Another group of species exploits larger flowers with high food reward. Scouts of these species communicate not only the odor but also the location of food sources, by leading recruits to the discovery site. In these species, recruits searching for additional food sources of the same type spread out, not from the nest, but from the point of initial food discovery. These contrasting foraging patterns are interpreted in light of the fact that high-reward resources are more patchily dispersed than low-reward food resources in the tropical dry forest study site (Guanacaste, Costa Rica). Computer simulations were run to explore further strategies in patchy and diffuse, high- and low-reward resource environments.

The Effect of Optimal Foraging on Intraspecific Resource Partitioning and Competition in the Bumblebee Bombus ternarius

Robert A. Johnson
Museum of Northern Arizona

Proboscis lengths of Bombus ternarius individuals "majoring" on a flower species in one and two flower species (a long and a shqrt corolla species) stands were measured. Proboscis lengths were significantly lower on the short corolla species. Quantification of intraspecific resource partitioning was accomplished using a probability function, calculated as the ratio of individuals of each proboscis length on one flower species to the total number of that length collected on both flower species. The shape of the probability function was fitted to a sigmoidal curve using a probit analysis. Regression of the linearized probabilities was significant. Individuals collected in the single species long corolla stand had shorter probosci than those in the two species stand. These results infer that competition acts more stringently when a variety of flower species are present. Pollinator differences between stands having one and two flower species could influence community structure through pollination success and seed set. The lack of differences found thus far in seed set are attributed to the fact that several species of bumblebees and wasps pollinate these unspecialized flower species and consequently increase pollinator quantity and success.

The Potential of Wild Bees for the Pollination of Fruit Trees

Marianne Klug and Gerhard Bünemann
Institute for Fruit Science and Nursery Management of
Hannover University

In open air observations the bee species participating in pollen transfer in the orchards of the University Experiment Station near Hannover (Northern Germany) were registrated. Andrena albicans and Andrena fulva were the most abundant species, even though they were outnumbered by the honeybee in a relation of more than 1:100. But as individuals the solitary bees seem to be at least as efficient as individual honeybees. They obviously touch the stigma on each

visited flower – in contrast to more than two thirds of the flower-
visiting honeybees. The percentage viability of insect borne
pollen grains "in vitro" was similar for honeybees and solitary
bees. Identification of pollen grains showed that solitary bees
always carry pollen of several varieties on their body hairs. Even
though open air observation showed a tendency of honeybees to prefer certain varieties, pollen from other varieties was also identified on every honeybee examined by scanning electron microscopy.
Attempts are made to establish wild bees in the orchards by providing nesting boxes and a mount of soil for the ground-nesting
species.

Foraging Activity of the Ant Ectatomma ruidum in Plants of Coffee and Cacao.

J. P. Lachaud, J. Valenzuela G., A. López.
Centro de Investigaciones Ecológicas del Sureste.
Tapachula, Chis. México.

Nests of E. ruidum are frequently found at the base of trees of
coffee and cacao in the state of Chiapas in Mexico. Ants are
active during day and night but generally foraging is more important at night where two peaks of activity were found: one during
the first hours after darkness and other a couple of hours before
sun rise. However, in some few nests, diurnal activity can be almost equivalent to nocturnal one. Relatively few preys come to
nest from the ground level. The greatest part of the foraging
activity takes place over the trees, where ants hunt live insects,
specially larvae of Lepidoptera, or collect the death ones that
they find. Most coffee plants where nests are found, have communication with one or several cacaos and an important part of the
activity of these nests takes place over the cacaos, probably because of the presence in these plants of aphids whose secretions
are also collected by the ants.

The Foraging Behaviour of Bumble Bees (Bombus spp.): Flower Complexity and Learning

Terence M. Laverty
University of Toronto

The flower-visiting behaviour of naive and experienced bumble
bees was compared on the following plant species which are listed
in order of the structural complexity of their flowers (Aster spp.,
Taraxacum officinale, Apocynum spp., Vicia cracca, Prunella
vulgaris, Gentiana sp., Penstemon hirsutus, Linaria vulgaris,
Impatiens biflora, Chelone glabra, and Aconitum spp.). On simple
flowers (e.g. T. officinale) naive foragers had little difficulty
in probing the appropriate area of the flowers but their initial
probing rate was significantly slower than that of experienced bees.
On their first visits to more complex flowers, naive foragers made
many "errors" as they learned the location of nectar rewards. In
general, the more complex the flower, the longer naive bees

required until their behaviour approximated that of experienced
bees. The time spent in trial and error learning ranged from 1-15
min. on different plant species with considerable individual variation within any given species. Initial foraging experience with
1 plant species (e.g. V. cracca) facilitated the subsequent
learning of a structurally similar flower (e.g. P. vulgaris), but
did not reduce learning costs on structurally different, more
complex flowers (e.g. Aconitum). Flowers considered to be of high
complexity offered significantly greater nectar rewards than less
complex flowers and, from initial results, greater flower complexity promoted flower constancy in experienced foragers.

Foraging strategies of black termites in Southeast Asia
(Termitidae: Macrotermes, Longipeditermes, Hospitalitermes)

Tadao Matsumoto
The University of Tokyo

In Southeast Asia, three black termites belonging to different
phylogenetic branches of the Termitidae (Macrotermes carbonarius,
Longipeditermes longipes and Hospitalitermes spp.) forage to a distance by the well ordered marchings in the open. The diets are different among the three termites, that of M. carbonarius is new fall-
en litter mainly; L. longipes, decayed litter; Hospitalitermes spp.,
lichen and moss.
  To investigate the foraging strategies of the black termites,
I have observed the foraging behavior and defensive behavior in
tropical rain forests of Malay peninsula and Borneo. Carbon/nitro-
gen ratio and calorific value of the diets were determined for comparing the quality. The foraging strategy is restricted by the
quality, abundance and distribution of food, and influenced by the
defensive system.

Foraging Behaviour and Pollination Efficiency

Josué A. Núñez
Instituto Venezolano de Investigaciones Científicas, IVIC.
Caracas, Venezuela.

Coevolution of the system Plant-Pollinator let expect that
pollination and foraging would be achieved by minimising net rate
of food production by the plant but simultaneously maximising net
rate of food intake by the foraging bee. Although sugar flow at
flowers remains nearly constant during the working day of the bee,
time delay to switch to a new "searching image" by the bee while
foraging, accounts for occasional nectar depletion or accumulation
in flowers. Transient stages of nectar accumulation with maximum profit for the foraging bee followed by a steady-state stage
of depletion with maximum profit for the plant increasing cross-
pollination efficiency is shown with an example (NUÑEZ, J. J. Insect Physiol. 23, 265-275, 1977).

Effects of Intrasecific Competition on the
Foraging Behaviour of Vespula arenaria.

Megan J. Pallett
The University of Toronto

Ten colonies of Vespula arenaria were placed in screen boxes
and transferred to a wooded field station. Observations of foraging behaviour were made on two of the colonies, while the other
eight were used to vary the levels of competitive stress to which
the observed colonies were subjected. The effect of competitive
stress on trip lengths, load types and load sizes was recorded.
The effect of different levels of wasp activity on prey abundance
in the area was also recorded.

Under high levels of competitive stress, trips for prey were
shorter but the prey loads were smaller than when there was no
competition. The total amount of prey coming into the colony did
not change significantly.

Prey abundance in the area varied significantly with levels
of wasp activity. This data indicated that workers from colonies
on a wood-field boundary will preferentially forage in the field
and avoid going into the woods to find prey.

Individual Foraging Specializations of
Desert Seed-harvester Ants

Steven W. Rissing
Arizona State University

Individual foraging specializations of the desert seed-
harvester ants Pogonomyrmex rugosus and Veromessor pergandei were
studied in the field and in laboratory colonies. Individually
marked foragers visiting an experimental patch of mixed grass seeds
specialized to a single grass species even while colony-mates
simultaneously visiting the same patch specialized on other seeds.
Specializations are not related to size of the forager; this variable explains < 4% of the variance in size of seed harvested for
both species of ant. Specializations can hold for several days
but can also be shifted rapidly.

Foraging specializations may enhance individual and colony-wide
foraging efficiency in a manner similar to that of "majors" in
bumblebees and may also result in significant dispersal of seeds
for some desert plants.

The Aggressive Foraging Syndrome in Highly-Social Bees:
Ecological Correlates and Evolutionary Origin

David W. Roubik and Leslie K. Johnson
Smithsonian Tropical Research Institute and University of Iowa

Agressive foraging at flowers is known in two groups of highly-social bees, the sister subgenera Tetragona and Trigona (Trigonini) Apis and Melipona, two other major genera, do not forage agressively. Apis, Trigona and Melipona have species which defend nests with pheromone mediated agression, but this is seldom related to foraging behavior. Trigona s. str. and Tetragona have large mandibular teeth, as do African Meliplebeia and Meliponula. Trigona s. str. has 5 teeth on each mandible, and the other groups have only 2. The former uses them to (1) harvest animal flesh (2) rob flowers (3) damage woody plants to obtain resin (4) attack colony predators, and (5) attack competitors at resources. In Tetragona and Trigona, aggressive foragers have the unique combination of large colonies, medium to medium-large body size, effective recruitment and toothed mandibles. Successful deployment of those traits in combat has produced an aggressive foraging syndrome. Agressive foraging is theoretically possible in African species which have several of the above traits, but the syndrome appears limited to the most recent of stingless bee lineages.

Food Preferences of Apis mellifera

Justin O. Schmidt
Carl Hayden Bee Research Center, Tucson, Arizona

Nurse honey bees, Apis mellifera L., were given a choice of pure species of pollen and a standard pollen mixture in feeding trials. In the tests 15 g of 1-2 day old bees were placed in 9 x 15 x 6 cm screened cages; given water, 30% sucrose solution, and placed in a darkened environmental room maintained at 30° and 70% relative humidity. Feeding preferences of the bees (4 replicates) were determined after 2 and 4 days by weighing the amounts eaten of a pure species and of a standard mixture of 15 species of corbicular pollen moistened with water (the pollen was replaced after day 2).

The results based on tests comparing 23 different species of pure pollen with the test mixture indicate that nurse bees exhibit distinct preferences for some species of pollen over others. These preferences do not appear related to the taxonomic origin of the pollen (i.e. family, order, subclass, etc.), the actual physical size of the pollen grain (within a range of 20- 90 μ), whether or not the pollen is anemophilous or entomophilous, or season of bloom. Preference does appear directly related to protein content and to lack of unfavorable texture, e.g. hard, tough and rubbery, or sticky. These results suggest that protein content and texture are major factors governing honey bee food selection.

Foraging Behaviour and Pollination Ecology of Honey
bees in Relation to Agricultural Cruciferous Crops

R. C. Sihag
Department of Zoology,
Haryana Agricultural University, Hissar- 125004, INDIA

Honey bees and other solitary bees associated with cruciferous
crops have been classified as pollinators or visitors/nectar robbers
depending upon their modus operandi on the flowers. The placement
of floral parts affects the mode of tripping by the bees. On the
basis of abundance of bees and their tripping rate a pollination
efficiency (PE) model is devised. PE is in the order of Andrena
ilerda > Apis florea > Apis dorsata on Brassica campestris var.
toria and B. Campestris var. brown sarson; Andrena leaena > A.
florea > A. dorsata on B. rapa, B. carinata, B.hirta, B.napus and
and Eruca sativa; A.dorsata > A.leaena > A. florea on B.oleracea
var. botrytis; and A. dorsata > A. florea > A.leaena on Raphanus
sativus. Anthesis and foraging activities are controlled by ambient climatic temperature or/and light intensity.

Foraging Behaviour and Pollination Ecology
of Honey bees in Relation to Onion Crop

R. C. Sihag
Department of Zoology
Haryana Agricultural University, Hissar-125004, INDIA

These studies reveal that major population of insects visiting
onion (Allium cepa L.) crop is constituted by three species of bees
viz. Apis dorsata, A. florea and Andrena leaena. However honey
bees are the most efficient pollinators. Due to the structural
resemblance of the inflorescence of onion with unbelliferous crops,
the mode of foraging is alike. The pollination efficiency calculated is in the order of A.dorsata > A. leaena > A.florea. Foraging activity is closely related with ambient temperature and
light intensity. Bees mainly forage for pollen in the morning and
for nectar during the day time. Peak pollen foraging activity
coincides with peak anthesis. The crop is found to be self-fertile
but the bees help augmentation of seed production significantly.

Foraging Behaviour and Pollination Ecology of Honey Bees
in Relation to Agricultural Umbelliferous Crops

R. C. Sihag
Department of Zoology
Haryana Agricultural University, Hissar-125004, INDIA

Honey bees are found to be the chief pollinators of four agricultural umbelliferous crops. Foraging rate of bees on these crops
is determined by their mode of pollen collection, the latter in
turn depends upon the structure of inflorescence. The pollination
efficiency calculated from pollination efficiency model is in the
order of Apis dorsata > Andrena leaena > A.florea on Coriandrum
sativum, Foeniculum vulgare and Cuminum cyminum; and A.florea >
Musca spp. on Daucus carota var. sativa. Foraging activity in bees
is influenced by ambient temperature and light conditions. Increasing temperature in the morning acts as a stimulus for commen-
cement of foraging activity whereas cessation in the evening is
affected by decreasing light intensity. The bees forage for pollen
in the morning whereas they collect nectar during the day time.
Peak pollen foraging activity coincides with anthesis time.

Foraging Distribution of Wood-ants in Relation to Aphids

John Sudd
The University of Hull, England

The branches of young Pinus contorta in N. Yorkshire, England
bore very few aphids (Cinara pini and C. pinea). Most aphids were
solitary apterae or alatae and there was no evidence of reproduction. Native Pinus silverstris of the same age bore dense
colonies of C. pini and smaller ones of C. pinea. The aphids on P.
silvestris were heavily attended by wood-ants, but there was also
an appreciable number of ants on P.contorta branches and beneath
the trees. Some of these may have wandered from wildling trees
which did bear aphids but there were ants on P.contorta which
moved actively and singly and had negligible amounts of carbohydrate in their crops. These might be (a) foragers which have not
yet found aphid groups; (b) predators not searching for honeydew
or (c) collectors of information on the availability or non-avail-
ability of food in that part of the foraging.

Apis mellifera as a bumbling bee: dispelling a myth

Evan A. Sugden
The University of California at Davis

An outline is presented for the critical examination of Apis mellifera as an introduced species. Aspects of consideration include community effects through inadequate/disruptive pollination and competition/interference with legitimate pollinators. Two examples are given. In one study, inundation of a natural community with honey bees produced a significant decline in the density of native flower-visiting species. Resource competition invoked by Apis mellifera is indicated. A second, ongoing study compared pollination efficiency of individual honey bees with that of bumble bees and butterflies. Preliminary evidence indicates Styrax officinalis, a California chaparral shrub, is better adapted for pollination by native species. Honey bees may reduce nectar rewards of the plant to below the level of attractiveness to effective pollinators, thereby depriving the plant of fertilization at otherwise optimal levels.

Foraging Behavior in Lasius neoniger and Monomorium minimum:
The Ecological Consequences of Recruitment

James F. A. Traniello
Boston University

The communication systems of the ant species Lasius neoniger and Monomorium minimum involved in securing prey were studied in the field. Lasius neoniger possess both short and long range recruitment systems used to cooperatively retrieve prey of different sizes. When small prey (10-15 mg) are found, scouts discharge poison gland secretion and hindgut material to attract nearby workers that assist in retrieval. Hindgut material is discharged during trial laying, and is involved in the recruitment of nestmates during trail laying, and is involved in the recruitment of nestmates to large prey. The number of workers recruited is significantly correlated with prey weight. These recruitment systems appear to reduce prey loss due to intra- and interspecific competitors. It was experimentally demonstrated that the foraging success of this species is related to prey size.

Monomorium minimum scouts recruit nestmates to large prey by laying trails composed of Dufour's gland secretion, and the recruitment response is related to prey size. Also, prey are chemically defended from competitors. Large prey induce strong recruitment; the recruitment of a large number of workers increases the number of individuals that chemically defend prey. Because of the relationship between prey size, pheromone deposition, and the effectiveness of chemical defense, workers of M. minimum are able to successfully exploit large prey. The prey size-dependent recruitment/retrieval systems of L. neoniger and M. minimum appear to represent behavioral mechanisms of resource partitioning.

The Social Organization of Foraging in the Neotropical Termite Nasutitermes costalis

James F. A. Traniello
Boston University

Foraging activity in the neotropical termite Nasutitermes costalis is regulated in its initial stages by the soldier caste. Nasutes function as scouts, find food sources, and communicate their location to workers. After discovering food, soldiers recruit additional soldiers by laying chemical trails from the sternal gland. Artificial trails prepared from gland extracts have a recruitment effect restricted to other soldiers; trails prepared from the sternal glands of LW3 workers recruit both workers and soldiers. The recruitment effect of trails originating from workers can be mimicked by increasing the concentration of soldier trail pheromone. The volume of the sternal gland of soldiers is significantly smaller than that of LW3 workers, but is approximately equal to that of SW2 workers. The sternal gland pheromone of these castes is composed of two components: one component is ephemeral in its effect and regulates recruitment; a second component functions as a durable orientation cue. The recruitment effect of a trail prepared from a single LW3 sternal gland lasts about 30-45 minutes; but can serve as an orientation guide for up to 3 hours. The orientation component can be extracted from trails four years in age. Although the differential recruitment of castes appears to be regulated by the concentration dependent effects of a single pheromone present in soldiers and workers, preliminary chemical analysis suggests that some constituents may be caste specific.

Prey Selection by the Ant Formica schaufussi

James F. A. Traniello[*], Rhys V. Bowen[+], and Marty S. Fujita[*]
Boston University[*] and Harvard University[+]

Factors affecting the selection of arthropod prey by the omnivorous ant Formica schaufussi were studied in the field and laboratory. During the spring and summer of 1977-1981, forage samples were collected from workers successfully returning from foraging trips. The size frequency distributions of these selected prey were compared with those of available prey taken in sweep samples. The results of comparisons of selected vs. available prey indicate that prey are not selected in numbers relative to their frequency in the resource distribution. By manipulating the availability of prey, we could show that foragers distinguish between prey of different sizes and are selective when foraging costs are low. A main factor influencing the selection process was ambient temperature, which ranges from 15-40°C during a daily foraging period, and strongly affects metabolic rate, which we measured in respirometry studies. Preliminary studies indicate that O_2 consumption increases approximately twelve fold over 25°C range. Temperature appears to be a principal component of foraging cost. By following marked individuals, we could demonstrate that foraging trip time is influenced by the ambient temperature at the start of the trip. The prior foraging experience of a worker also influences prey selection in combination with temperature.

Dynamics of food gathering in Formica polyctena

Rainer Rosengren
Department of Zoology, University of Helsinki

Diel and seasonal variation in foraging activity was studied in F. polyctena in coniferous forest. The number of foragers was determined by CMR and exits/s, and streams crossing bridges were sampled for crop content and booty (Chauvin-traps). In mid-summer a nest with 3×10^5 foragers harvested 3×10^4 insects and 1 kg honeydew a day, but the number of foragers, mg protein/item and mg protein/mg sugar differed with the season. Running activity and total food gathering were correlated with temperature during the day, but night foraging was much lower than predicted by temperature. The food collectors: total traffic ratio follows a motivational rhythm partly independent of abiotic factors. The highest value of sugar/worker was found at a low temperature, when foragers returned in early morning after a night in the trees. The ants had a functional response to Neodiprion defoliators, but booty selection appeared mainly stochastic and did not fit "individualistic" models of optimal foraging. Body size variance and size-based polyethism were weaker in the mass forager F. polyctena than in F.truncorum, an individual forage living in small colonies lacking trunk trails.

Studies on the foraging behaviour of Apis cerana indica F.

L.R. Verma & V.K. Mattu
Himachal Pradesh University, Simla (India)

Foraging behavior of Indian honeybee, Apis cerana indica F. was studied in Simla hills (31°-06'N, 77°-10'E, 2206m altitude). Data on the foraging behaviour indicated a regular cycle with the maximum activity in summer and autumn as compared to other seasons. Even in winter, when outside air temperature was 6°C, bees collected both pollen and nectar. Percentage of nectar collectors was more than those collecting pollen and pollen + nectar throughout the year. Nectar gatherers showed greater seasonal variations as compared to pollen and water collectors. Average number of foraging trips made by bees per day was greater and weight of pollen pellets collected was heavier in good honey flow seasons (summer & autumn) than in dearth periods (rainy & winter). Duration of each foraging trip was more during the dearth period and yellow coloured pollen pellets were collected predominantly throughout the year. Present data indicated that two good honey crops can be collected in a year in this hilly region and this native bee species resembled Apis mellifera L. in its foraging behaviour.

Foraging Strategy of Honey Bee Colonies
In a Temperate Deciduous Forest

P. Kirk Visscher and Thomas D. Seeley
Cornell University and Yale University

To understand the foraging strategy of honey bee colonies, we measured certain temporal and spatial patterns in the foraging activities of a colony living in a temperate deciduous forest. We monitored the colony's foraging by housing it in an observation hive and reading its recruitment dances to map its food source patches. We found that the colony routinely foraged several kilometres from its nest (median 1.7km, 95% of foraging within 6.0 km), that it frequently (at least daily) adjusted its distribution of foragers on its patches, and that it worked relatively few patches each day (mean of 9.7 patches accounted for 90% of each day's forage). These foraging patterns, together with prior studies on honey bee recruitment communication, indicate that the foraging strategy of a honey bee colony involves surveying the food source patches within a vast area around its nest, pooling the reconnaissance of its many foragers, and using this information to focus its forager force on a few high-quality patches within its foraging area.

Foraging Differences between cross-fostered honeybee workers (Apis mellifera) of European and Africanized races

Mark L. Winston and Susan J. Katz
Simon Fraser University, Burnaby B.C. Canada

Foraging differences between cross-fostered honeybee workers of European and Africanized races in South America are described. Africanized workers began foraging at earlier ages than European workers in colonies of their own races, but cross-fostered workers began foraging at the same age as workers in the colonies in which they were placed. Some differences in the mean time spent foraging per hour and the mean number of flights per hour were also found. The results suggest two major factors determining differences in division of labor between Africanized and European bees: 1) the colony characteristics by which foraging age is determined, and 2) the responses of individual workers to hive environment. A hypothesis to explain these results is presented based on higher levels of foraging stimuli in Africanized colonies as well as a higher stimulus threshold for Africanized workers.

COMPETITION AND POPULATION DYNAMICS IN SOCIAL INSECTS

Organized by Walter Whitford

Introduction: The Wood and the Trees

John H. Sudd, The University of Hull

The main groups of social insects seem at first sight to resemble each other only in their sociality, a feature acquired independently in each group and several times independently within the bees alone. Is it worthwhile discounting their diversity and looking for unified features in their ecology, or would it be better to compare ants with other predators, termites with other litter consumers and bees with - and so on? In fact social insects share important ecological characters, many of which I hope will be brought out in the contributions to this symposium.

POPULATION STRUCTURE

Eusociality with its 3 features of co-operative brood-care, caste differentiation and overlapping generations, has immediate implications in population dynamics. The delay while worker numbers build up, before sexual forms are produced, results in accumulation of biomass, an accumulation which is even more marked if we include in it the energy used or incorporated in nest structures. Darlington (this symposium) describes a population of termite nests with the structure of a mature forest - many old nests, high mortality of young nests and rapid growth of those that survive. Within each mature nest there is a rapid through-put of energy, with perhaps recycling of minerals and a massive production of sexuals. This looks like a K-adapted climax community, what happens in less stable populations?

SPATIAL ARRANGEMENT

The separation of the roles of forager and mother through caste differentiation acts in space as well as in time. Resources are collected peripherally but accumulate centrally where larval growth takes place. Many social insects have two more or less distinct microhabitats, one for foraging and another for nesting. The interaction of these two dimensions of "niche space" is complex since they do not act independently. Two species with a marked overlap in foraging niche may coexist if they have different nest requirements, because of the advantage each holds near its nest.

In the relatively uniform conditions of tropical plantations a
"mosaic" of dominated areas forms in which a few species share
(about 4 species in Papua-New Guinea, Room, 1975) carrying with them
other associated species. Possibly there are slight differences in
habitat preference, but these are often differences in nest
preference (Majer, 1976, Taylor, 1977). In tropical forest where
many niches are available there is much more separation especially
in twig and crevice nesters (Carroll, 1979), and canopy and ground
foragers are more separated (Lévieux, this symposium). The
segregation of nest microhabitat becomes much less in savanna, and
in arid environments nests may be crowded into those sites which
are suitable (though some segregation may still occur - Whitford
et al. 1981) and special behaviour to avoid contests is necessary
(Holldobler, 1979). Spatial partition also occurs in Formica
lugubris in N. England, where it may have a different basis.
The nests of this ant are crowded on the sunny edges of woodland
and their foraging areas within the wood are disjoint, though the
nests are entirely non-hostile and exchange personnel along base
routes. This species overcomes the fading of influence with
distance from the nest by an extreme form of polycalism forming
"super-colonies" (Cherix, 1977).
    The ability to extend over an increased area through aggression
is important in other species too. Leiberburg et al. (1975) ascribe
much of the success of Iridomyrmex humilis in replacing Pheidole
megacephala in Bermuda to aggressive "nest-budding" at a
particular stage of the colony cycle. At other times it appears
that P. megacephala may even hold the advantage in foraging
contests.

DIVERSITY AND COMPETITION

    Finally I should like to turn to the problem of the low
diversity of life style within each major group of social insects.
We have been taught that niche overlap should be reduced by resource
partition, but in ants "there is an inordinately high ecological
overlap" (Lynch et al., 1980). Davidson (1980) has drawn attention
to the part played by indirect competition in moderating interference. Among Bombinae resource partition by tongue length seems
to break down in Finland where too many short tongued species
occur (Raanta & Vepsalainen, 1981) and Koeniger (this symposium)
has studied competition between Asiatic honeybees. The fact is
that many social insects are euryphagous; it is hard to be
oligophagous and perennial. Even detailed multivariate studies
like that of Brian et al. (1976) leave some species only slightly
separated on 3 principal coordinate axes, though we cannot be sure
that these axes relate to those of niche dimension. How far are
differences in foraging microhabitat due to learning? We have
considerable evidence of the importance of learning in the various
degrees of spatial fidelity in ants (Rosengren 1971; Tweed, 1980).
Desert ants show ability to respond to changes in food abundance;
does learning play a part here too?
    No-one would wish to add another persona to the mythology of
the social insects - we already have plenty in the Feminine

Monarchie, the Super-organism and the Factory-fortress! Perhaps the virtue of a comparison of social insects and trees is not just in resemblances in population structure and competition, but in the fact that sometimes one cannot see the wood for trees, and again sometimes vice versa.

References

Brian M.V., Mountford M.D., Abbot T.A., Vincent S., 1976. - Changes in ant species distribution during regeneration of a heath. J. Anim. Ecol., 45, 115-133.

Carroll C.R., 1979. - A comparative study of two ant faunas. Amer. Nat., 113, 551-561.

Cherix D., 1977. - Les grandes colonies de fourmis de bois au Jura. Mitt. Schweiz. Entomol. Gesel., 50, 249-250.

Davidson D.W., 1980. - Some consequences of diffuse competition. Amer. Nat., 116, 92-105.

Holldobler B., 1979. - Territoriality in ants. Proc. Amer. Phil. Soc., 123, 211-218.

Leiberburg I., Kranz P.M., Seipi A., 1975. - Bermudan ants revisited. Ecology, 56, 473-478.

Lynch J.F., Baunsky E.C., Vail S.G., 1980. - Foraging patterns in 3 sympatric forest ant species. Ecol. Entomol., 5, 353-371.

Majer M.D., 1976. - The ant mosaic in Ghana cocoa farms. J. appl. Ecol., 13 , 145-155.

Raanta E., Vepsalainen K., 1981. - Why are there so many species? Oikos, 36, 28-34.

Room M., 1975. - Relative distribution of ant species in cocoa plantations in Papua-New Guinea. J. appl. Ecol., 12, 47-61.

Rosengren R., 1971. - Route fidelity, visual memory and recruitment behaviour in foraging wood ants. Acta Zool.Fenn., 133, 1-106.

Taylor B., 1977. - The ant mosaic on cocoa and other tree crops in W. Nigeria. Ecol. Entomol., 2, 245-255.

Tweed R.L., 1980. - Ph.D. Thesis, University of Hull, 1980.

Whitford W.G., Deprees D.J., Hamilton P., Ettershank G., 1981. - Foraging ecology in seed harvesting ants. Amer. Mdl. Nat., 105, 159-167.

A Comparison of the Ground Dwelling Ant Populations Between a Guinea Savanna and an Evergreen Rain Forest of the Ivory Coast

J. Levieux, Université d'Orléans

From 1963 to 1980, the soil and ground dwelling populations of a guinea savanna (Reserve of Lamto $5°30'N - 6°02'W$) and of an evergreen rain forest (Reserve of Taï $5°50'N - 7°30'W$) were studied by J. Lévieux and T. Diomandé (1981). This paper presents some of the conclusions emerging from the quantitative study of the spatial organization and the role of ants in the food webs.

METHODS

Ant populations were studied by digging 96 quadrats of 16 m^{-2} each randomly dispersed in savanna, 94 in forest (18 x 16 m^{-2} and 76 x 25-^{2}) and two transects (200 m x 0.25 m) in forest. Specimens were determined according to the reclassification of Brown (1975) and Bolton (1980).

RESULTS

In savanna, 117 species belonging to 49 genera were collected (Ponerinae 41%, Myrmicinae 31%, Formicinae 24%, others 5%, mean estimated density: 3500 nests/ha). A multifactorial analysis related to the specific composition of the community shows that two factors mainly interfere with the spatial distribution of the species first, the type of soils, second the passage of the annual bush fire which results eventually in the progression of the drier savannas species. In fact, the progress of the front of fire plays directly a very negligible role on the soil fauna (at -5 cm, the elevation of the temperature is only $3°5$ two hours after the passage of the flames). In the following months, the destruction of the grasses of the burnt savanna modify the microclimatic factors at the soil level, and consequently modify qualitatively and quantitatively the ant spectrum: several species, mainly Ponerinae, disappear in burnt savannas (Amblyopone mutica, Apomyrma stygia, Centromyrmex sellaris etc...); some extend their distribution (Brachyponera senaarensis, Camponotus orthodoxus...), many others exhibit different densities (Acantholepis canescens...).

Leston (1972, 1973) pointed out that in plantations (i.e. with trees uniformly distributed) ant distribution can very often form a

48

mosaic. The nests of ant species of savannas are often interspersed (Lévieux, 1971). The final result of this interspersing depends on the degree of interspecific competition. So, the dominant species, Camponotus acvapimensis, whose nests are spread at random, can exert a certain influence on the spatial distribution of some other ant species. Nevertheless, as the total number of species is high, any free place in the soil can be occupied by a lot of founders which highly diminishes the probability to get a quasi regular distribution of mosaic type.

The estimated total number of workers/ha is higher than 2.10^7. For the commonest species Camponotus acvapimensis the density of workers is about 2.10^6/ha corresponding to a fresh biomass of more than 14 kg/ha or a mean standing crop dry biomass in black clays of $0,52$ g/m^{-2} (energetic equivalent: $1,250$ Kcal/m^{-2}) (Lévieux, 1976)

In forest, 90 species lived in the soil (Ponerinae 43%, Myrmcinae 45%, Formicinae 6%, others 6%). The total density keep comparison with the savanna (3500 nests/ha). Moreover only $3,3 \cdot 10^6$ workers are present, resulting from the presence of many small sized adult colonies of Ponerinae (50 to 350 workers). Expressed in numbers of workers, the Myrmcinae account for 75%, the Ponerinae for 15% and the Formicinae 6%. The eveness of ant community, based on Shannon's (1948) and Pielou's (1975) formulas is better in forest where the environmental factors at the soil level are more equilibrated than in savanna (J = $0,642$ in savanna and $0,832$ in forest).

In these two communities, kept from all human activities, the number of species of ants living in the soil and the density of their nests are similar. In forest, in relation with the lowest densities of individuals, the degree of soil occupation of the superficial ground is slightly lower due to the thinner strata of soil where the ants can dig their nests.

To clarify the function of ants in the food webs, the hunting areas, stratifications, cycles of activity and alimentary diet of 20 species in savanna and 13 in forest have been measured (Lévieux, 1977; Diomandé, unpublished). The ants exhibit several differences whilst exploiting the resources of these two communities.

In savanna two types of feeding strategies are used by the ant when hunting.

First, several Ponerinae (Amblyopone spp. Apomyrma stygia, Centromyrmex sellaris etc...) exhibit a narrow stratification (from -60 cm to -5 cm). This is highly correlated with a specialized diet at the expense of several groups of animals which are active all over the year (Chilopods, termites etc...). The same strategy is used by the species whose stratification is strictly restricted to the ground surface (Megaponera foetens, Leptogenys conradti...). In these cases, the specialized diet of many ants and their low densities keep the workers apart and lessen competition.

In relation with a more omnivorous diet, the other ant species belonging to the subfamilies Myrmcinae, Dolichoderinae, Formicinae, hunt from the soil surface and the vegetation to the top of the trees. Their phases of activitiès are widespread over the 24 hour cycle. Many of them are carnivorous and simultaneously tend the Aleyrodidae. Due to the densities of their nests, containing a high number of workers, competition is often important and a detailed study shows the order of succession of many species when exploiting simultaneously the same resource (Lévieux, 1972).

In evergreen forest, soil fauna can be considered as a block totally separated from the canopy fauna. Stratification of the species is mainly limited to the soil and its surface: none of the soil species typically hunt in the vegetation while some of them climb to 2 meters (Pachycondyla silvestrii, P. gabonensis...). So the same type of strategy is used by the soil ground dwelling ant fauna (Amblyopone, Centromyrmex..) and by the species hunting mainly on the ground floor level (Technomyrmex andrei, Triglyphothrix brevispinosa). On the other hand, ants living in the canopy very rarely hunt on the ground. This differs greatly from the savanna and from the deciduous forest where the soil and ground dwelling ants hunt from - 1 meter to the top of the trees thus exploiting many resources in the vegetation.

Fluctuations are far more important in the horizontal space where mean radius of action varies for the forest fauna from 1 m. (Technomyrmex andrei) to 12 m. (Pachycondyla gabonensis) while in savanna it varies from 1 m. (Strumigenys sp.) to 96 m (Megaponera foetens) (Longhurst and Howse, 1979).

In forest, as a consequence of the regularity of the microclimatic factors at the soil level, workers of many species exhibit during the 24 hour cycle a permanent activity with two peaks (Pachycondyla silvestrii, Psalidomyrmex procerus, Pheidole bucholzi, Technomyrmex andrei), while only two species are strictly nocturnal (Odontomachus assiniensis, Anochetus sp.).

A comparison of the feeding strategies of ants living in these two communities shows that in the soil several similarities can,be drawn (specialized predation, permanence of activities, narrow stratification...). Nevertheless, the density of many Ponerinae and the low numbers of their workers, lower in forest than in savanna, reduced their influence in the community.

In the detail, many differences appear between the members of the two communities. Most often, predation is specialized on one or a few zoological groups. For example, species exclusively preying on Oligochetae are restricted to the forest. Effectively, in relation with the dry season which stops activities of Annelids, these ants are excluded from savanna. In forest ant's predation on termites seems at least to be of the same importance than in savanna (Longhurst et al, 1979). All the omnivorous species eat at least 25% of termites (most often from 50 to 75%). On the contrary, the

Arachnids, present with high densities, seem to be rarely attacked.

To conclude, the originality of the savanna appears to be in the exploitation by the soil fauna of the totality of the present resources even in the vegetation. As far as we know, in forest the interrelations in the food webs seem to be mainly, if not totally, established between the members of the soil fauna. If the same principles of exploitation are used in the two communities, the ants adapt themselves to every particular case.

From the study of the feeding strategies of ants in these two communities it appears that in the savanna, in correlation with a high degree of soil occupation by adult colonies, competition for space or resources remains active between terrestrial ants. In forest due to the lower number of workers and specialized diet of many of them, competition for space and prey seems to play, as far as we know, a less important role.

Table I. -- Feeding habits of highly specialized predators in forest and savanna.

Species of ants	alimentary diet
In the soil	
Amblyopone spp.)	Chilopods (geophilids)
Apomyrma stygia)	
Discothyres oculata	eggs of arthropods
Hypononera gr. coeca	Collembola
Plectroctena subterranea)	eggs of Diplopods
" lygaria)	
Psalidomyrmex procerus	Oligochaete
Centromyrmex sellaris	Termites
Strumigenys spp.	Collembola
At the soil level	
Leptogenys conradti	Isopods (Oniscoidea)
Megaponera foetens	Termites

Table II. -- Examples of alimentary diet of some species of ants hunting mainly on the surface of the soil.

Species of ants	alimentary diet
Pachycondyla silvestrii	Termites 90% Chilopods 10%
Odontomachus assiniensis	Termites 90% Arthropods 10%
Odontomachus troglodytes	Termites 25 to 75% | according to Insects 25 to 75% | the season Isopods, Myriapods, 10% Arachnids
Pheidole bucholzi	Termites 50% Insects 20% Others, 30% (Arachnids, Myria- pods, Isopods).
Technomyrmex andrei	Termites 80% Collembola 8% Insects 12%

References

Bolton, B., 1980. -- The ant tribe Tetramoriini (Hym. Formicidae). The genera Tetramorium Mayr in the Ethiopian zoogeographical region. Bull.Br.Mus.Nat.Hist., Entomology, 40, 193-384.

Brown W.L., JR., 1975. -- Contribution towards a reclassification of Formicidae V. Ponerinae. Search, 5, 1-115.

Diomandé T., 1981. -- Etude du peuplement en fourmis terricoles des forêts ombrophiles climaciques et des zones anthropisées de la Côte d'Ivoire méridionale. Thèse Univ. Abidjan, 288 pp.

Leston D., 1972. -- Ecological consequences of the tropical ant mosaic. Proc. VII Congr. IUSSI, London, 235-242.

Leston, D., 1973. -- The ant mosaic. Tropical tree crops and the limiting of pests and diseases. Pans, 19, 311-341.

Lévieux, J., 1971. -- Mise en évidence de la structure des nids et de l'implantation des zones de chasse de deux espèces de Campono - tus (Hym.Formicidae) à l'aide de radio isotopes Insectes Sociaux, 18 : 29-48.

Lévieux J., 1972. -- Etude du peuplement en fourmis terricoles d'une savane préforestiere de Côte d'Ivoire. Rev.Ecol.Biol.Sol, 10, 379-428.

Lévieux, J., 1972. -- Le rôle des fourmis dans les réseaux trophiques d'une savane preforestiere de Cote d'Ivoire. Ann. Univ. Abidjan, E, 5, 143-240.

Lévieux, J., 1976. -- Densité et biomasses de Camponotus acvapimensis (Hym. Formicidae) dans une savane de Côte d'Ivoire. La terre et la vie, 30, 246-275.

Lévieux J., 1977. -- La nutrition des fourmis tropicales V éléments de synthèse. Les modes d'exploitation de la biocoenose. Ins.Soc. 24, 235-260.

Longhurst, C., Johnson R.A. and Wood T.G., 1979. -- Foraging, recruitment and predation by Decamorium uelense (Santschi) (Formicidae, Myrmicinae) on termites in southern Guinea savanna, Nigeria. Oecologia, 38, 83-91.

Longhurst C. and Howse P.E., 1979. --Foraging, recruitment and emi- gration in Megaponera foetens (Fab.) (Hym. Formicidae) from the Nigerian Guinea savanna. Ins. Soc., 26, 204-215

Piélou E.C., 1975 --Ecological diversity, J.WILEY, New York,165 pp

Shannon C.E., 1948. -- A mathematical theory of communication. Bull. Syst. Techn. J., 27, 379-423.

Population Dynamics in an African Fungus-Growing Termite

Johanna P.E.C. Darlington,
International Centre of Insect
Physiology and Ecology, Nairobi, Kenya

The dominant invertebrate herbivore of the semi-arid grasslands around Kajiado, Kenya, is the termite <u>Macrotermes</u> <u>michaelseni</u> (Sjostedt). As part of a project to quantify the ecological role of this termite its population was studied in detail. The termites build conspicuous, bare earth mounds above compact subterranean nests. Mound populations in the field were monitored for up to seven years. The size and composition of the populations of individual nests were analysed.

METHODS

Nest populations were sampled by fumigating with methyl bromide, then later removing the whole contents of the nest. Termites and fungus comb were separated from soil by flotation in water. The sludge obtained was measured and subsampled, and replicate counts were made of all the castes and instars present. Hence the total population was estimated.

RESULTS

1. Mound distribution, size and survival.
   Mounds occur in desnities of 1.7 to 4.0 per ha in the area studied. A high proportion (80-95%) are of mature size i.e. able to produce winged reproductives. Young mounds grow rapidly, reaching maturity 3-5 years after they first appear above ground, but mortality is relatively high; around 50% per year. After reaching maturity their growth rate declines and they remain more or less the same size from year to year. Few new mounds reach maturity, less than 1% of the existing population per year. Recolonisation of mounds which have died is much higher, probably 5-10% per year, but it is difficult to be precise because moribund nests can sometimes replace their royal pair and revive (Sieber and Darlington, in press). This replacement mechanism may explain the apparent extreme longevity of the nests, of the order of 10-20 years.

Longterm observations in an area of 72 ha which is regularly cleared of nests show that recolonisation, and the appearance and survival of new nests, are at a much higher level where no established termite population is present. Nest density is self-regulating, and if the population is artificially reduced it will regain its initial level in about five years. Density is regulated through the maintenance of defended foraging territories (Darlington, in prep.). Each territory is underlain by a network of passages extending up to 50 m from the nest, and a mature nest may have 6 km of passages associated with it. Where the passage systems of two adjacent nests meet there is evidence of conflict - pits full of the remains of dead termites, which are not found elsewhere in the passage system. Territorial conflict above ground has never been observed.

2. Relationship of nest population to mound size.

Mounds of all sizes were sampled, with populations ranging from less than 100,000 individuals to over $5\frac{1}{4}$ million. The proportions of the four adult castes were fairly constant, with a slightly higher proportion of soldiers in small nests, possibly reflecting their greater vulnerability to predators. The proportions of the three larval instars were also constant, suggesting a steady rate of production of adults. Presoldiers, however, seem to be produced in discrete bursts. The sex ratio in third instar larvae is very close to 1:1. The proportion of larvae in the total population was 35-50% in mature nests, but where less than 40% larvae were present the nest appeared to be in decline, as it was not occupying the whole of its nest chamber. Young, fast-growing nests have 60% or more of larvae in the population.

All the nest parameters measured were highly correlated. Fungus comb weight was linearly related to adult population. Population was linearly related to the queen's weight, a useful relationship because it is very much easier to measure the latter. The relation holds even when there is more than one queen, their weights being summed.

Correlations between population parameters and mound measurements, though not so high, were still good. Over 70% of the variation in populations can be predicted using a combination of linear measurements of the mounds. The reason for this close relationship seems to be that the mound is a cooling system, in which hot air circulates through passages close to the mound surface. The size of the cooling system relates to the amount of heat to be lost, which is a function of the size of the termite population. Thus termite populations can be estimated fairly accurately from a brief survey of mound parameters and density.

3. Seasonality in nest populations.

Mature mounds were sampled at monthly intervals over two years. A single annual brood of reproductives, initiated in about April, was ready to fly in the Short Rains of Nov.-Dec. The number in the brood varied from 3000 to 73000 with a mean of 34,400 (8 observations). The mean dry weight of alates (males and females alike) was about 0.1 g, so the biomass of a large brood of reproductives is substantial and can approach that of the whole sterile population. In spite of this there was no detectable change in the number of larvae or the weight of fungus comb when a growing brood of reproductives was present in the nest.

The seasonal rainfall pattern (two wet and two dry seasons) had no effect on the nest populations in years of moderate rainfall. During the drought of 1976 there was some reduction in larval numbers resulting in a decrease estimated at about 30% in total population of individual nests. Some mature nests died; no young nests survived the drought. Over the same period 85-95% of cattle that remained in the area died of starvation and disease. After the drought the populations of surviving nests recovered within a few months, and the overall population was restored in a few years.

4.   Throughput and productivity in mature nests.
        The sterile population of a mature nest is fairly constant, maintained by steady throughput of juveniles. It is thus not possible to follow through a discrete cohort as with the brood of reproductives. An "absent cohort" was therefore introduced by removing the royal pair, then sampling the nest after an interval. The number missing from each instar after different intervals was calculated, and hence the duration of each stage was deduced. Then from the mean number of each stage in the nest population, the mean throughput was calculated for each stage (Table 1).

The mean throughput of eggs per day is 30% higher than for first instar larvae, suggesting high mortality at eclosion, or that a large number of eggs were infertile. First, second and third instar larvae have almost the same throughput, indicating extremely low mortality during larval development. Assuming a similarly low mortality at the final moult, the daily throughput in the last larval instar will equal the daily production of the corresponding adult caste.

Although the rate of production of the two worker castes is almost the same, the number of minor workers in the nest population is almost twice that of the major workers. The missing workers were either already dead or in the underground passage system outside the nest when it was fumigated. Major workers forage above ground at night and spend the day carrying the forage to the nest and building the fungus combs. A few minor workers and an unknown number of soldiers were also outside the nest. This external population could not be measured, but it could comprise about 400, 000 major workers with a biomass of 1.7 Kg dry weight.

The estimated total annual production of steriles is 26.4 Kg and the mean annual production of alates is 3.4 Kg, giving a total of 29.8 Kg. The mean biomass of steriles in the nest is 4.6 Kg,

Table 1. -- Mean numbers and biomass of sterile castes and instars in a mature nest (mean of 17 nests); estimated durations of juvenile stages; calculated throughput of same; calcuated production of adults.

	Mean number (17 obs)	Biomass in Kg dry wt.	Estimated duration in days	Throughput in numbers per day	Production in numbers per day	Production Kg dry wt. per year
EGGS	771305		27.5	28047		
First instar larvae	300704	.015	15	20047		
Second instar larvae	357655	.06	13 ♀ { 20 ♂	21676		
Female third instar larvae	190889	.09	18.5	10318 } 20576		
Male third instar larvae	188616	.16	18.5	10195		
Large white soldiers	2023	.015	20	101		
Small white soldiers	8300	.01	14	593		
TOTAL LARVAE	1048187	.35				
Major workers (male)	438728	1.87			10195	15.8
Minor workers (female)	792021	1.86			10318	8.9
Major soldiers (female)	17541	.42			101	.9
Minor soldiers (female)	32624	.12			593	.8
TOTAL ADULTS	1280914	4.27			21270	26.4

with a further (very roughly) 1.7 Kg outside, giving a total of 6.3 Kg. (excluding the alates which are not present for long). The ratio of production to biomass is therefore of the order of 4.7.

DISCUSSION

The population of this termite is stable and self-regulating. Individual nests have large populations but low production to biomass ratios c.f. literature. The difference may be real, reflecting adaptation to the grassland ecosystem; but it is also probable the ratio has been overestimated in other studies. If nests are not fumigated before being dug many of the adults will escape, thus reducing the estimated biomass. Production may be overestimated because usually the development time observed in incipient colonies is used to calculate it. This is likely to be too short since incipient colonies produce very small adults. In M. michaelseni, larval development time in incipient colonies is about half that deduced above, though surprisingly the egg stage lasts much longer (Okot-Kotber 1981).

All figures quoted here are provisional, because the work is not yet complete.

REFERENCES

Darlington J.P.E.C., (in prep.). - The underground passages and storage pits used in foraging by a nest of the termite Macrotermes michaelseni (Sjostedt) in Kajiado, Kenya.

Okot-Kotber B.M., 1981. - Polymorphism and the development of the first progeny in incipient colonies of Macrotermes michaelseni (Isoptera, Macrotermitinae) Insect Sci. Applications, 1, 147-150.

Sieber R., Darlington J.P.E.C., 1982. - Replacement of the royal pair in Macrotermes michaelseni. Insect Sci. Applications, in press.

Interactions Among the Four Species of the Genus *Apis*

N. *Koeniger,* J. W. Goethe-Universität

In contrast to the other groups of Apidae (Michener 1974) which consists of 100–300 species each, in the true honeybees, Apis, only four species are generally recognized. Attempts to split this group into more than one genus and many more species have been based on morphological differences found among honeybees originating from different localities (Maa 1953, Sakagami et al. 1980). In the western honeybee (A. mellifera) about 25 distinct forms can be separated biometrically.

These populations interbreed readily, however, and therefore must be classified as subspecies. Morphological differences in allopatric forms of honeybees are likely to occur on a "subspecies level." The discussion whether A. mellifera and the Eastern honeybee, A. cerana, are good species is settled only because strong evidence for incompatibility between the populations is available (Ruttner and Maul 1969).

The Asian honeybees, A. florea, A. dorsata, and A. cerana share a common habitat. Competition and its avoidance can be observed among these species. The importation of A mellifera to Asia has caused a very unstable situation, which has lead to many disastrous effects for the imported species as well as for the indigenous honeybees.

1. Interference during mating

In the four honeybee species the main component of the queen's sex attractant is 9–oxo–trans 2 decenoic acid (Morse et al. 1970). Honeybee queens or their extracts attract drones of other Apis species (Butler 1967). Observations on the three sympatric Apis species in Sri Lanka show that this kind of interference is avoided entirely there. The daily periods of drone flight are well separated: A. florea 12.00–14.00, A. cerana 16.15–17.15 and A. dorsata 18.00–18.45. It is not known whether different daily mating periods are generally present in sympatric honeybees (Koeniger and Wijayagunasekera 1976). In bumblebees the odor trails of the drones and the mating places are locally separated (Haas 1946).

In A. mellifera drone congregations occur at the same place every year (Ruttner and Ruttner 1965). Imported A. cerana drones visited a mellifera congregation place in Germany at the same time. They were caught using a mellifera queen as bait with a frequency similar to mellifera drones (Ruttner 1973). Thus mating of A. cerana failed in the habitat of A. mellifera. Only after isolation of the cerana colonies from A. mellifera was natural mating achieved (Ruttner et al. 1972).

In Pakistan and India the reverse problem occurred with imported A. mellifera and led to the extinction of the imported bees. These observations demonstrate that interspecific competition seems to be an important factor influencing the coexistence of honeybee species.

2. Competition for nesting sites.
In many Apoidea the number of available nesting sites is a factor which limits the size of the population. In the three Asian honeybees competiton for nesting sites has no significance. A. florea and A. dorsata nest in the open - a very advanced mode of nesting among bees (Koeniger 1976). A. florea builds its single comb around a small twig whereas A. dorsata needs a more or less horizontal support, a large branch or a rock ceiling under which it constructs the comb. So there are clear differences in their needs. A. cerana nests in hollows.

3. Competition for food.
Intraspecific competition for food is known to many beekeepers. Honey yields are reduced if many colonies are brought to one location. A study in interspecific competition is available from Sri Lanka (Keoniger and Vorwohl 1979). In the Anuradhapura district the three Asian honeybees and the stingless bee, Trigona iridipennis, were found in abundance. Pollen spectra of honey samples showed that many flowers were visited by all the bee species. Further correlations were found between the frequency of some types of pollen found in honey samples and the body size of the bee species from which the sample originated. The number of different pollens was highest in Trigona and smallest in A. dorsata honey. Experiments at an artificial feeding dish resulted in direct interspecific reactions: Trigona attacked mainly A. florea and A. cerana; A. florea reacted to A. cerana; and A. cerana induced the flight of A. dorsata. In the further event of these experiments one species became dominant and excluded the other from the feeding dish. Smaller bees were more successful than the larger species. The sequence was Trigona, A. florea, A. cerana and A. dorsata. This study resulted in a still rather hypothetical picture. Small bees exploit many plant species within their limited flight range. When disturbed during foraging they react aggressively. Larger honeybees retreat. For them it seems to be more profitable to look for another flower within their large flight range than to defend. They also forage on fewer, and maybe richer,   plant species.

Competition for food during the summer drought is prevented by an annual migration. In Sri Lanka Apis dorsata leaves the Anuradhapura district in June, July and migrates to the mountain regions. The dorsata swarms return only after the rains in December at the beginning of a good nectar flow (Koeniger and Koeniger 1980). A. florea and A. cerana are present in the area for the whole year.

4. Interspecific robbing.

In times of scarce food sources honeybees search for forage wherever they are able to find it. During these periods A. mellifera scout bees are regularly found trying to enter nearby colonies. The guard bees of strong colonies normally succeed in warding off all bees which do not belong to their own colony. In case of weaker colonies sometimes scout bees are able to enter and to take honey from another colony. This successful scout recruits large numbers of foragers in its own colony and directs them to the weaker colony. These recruited bees fight their way into the colony. Most of the defending bees and the queen are killed and the colony's honey stores will be taken. This robbing behavior is an important factor in the intraspecific selection of Apis mellifera. Interspecific robbing among the sympatric honeybee species has not yet been reported in the literature. In Pakistan and Sri Lanka we occasionally saw bees of different species trying to enter a comb. Normally these trials were not successful and in some cases when it was, it did not result in a behavior comparable to intraspecific robbing in A. mellifera or A. cerana.

Robbing between allopatric honeybees occurs frequently and causes many problems. In Germany we kept A. cerana for many years. Regularly at the end of the nectar flow in July and August A. mellifera scout bees discovered the A. cerana colonies. At the hive entrance the cerana guard bees stopped and attacked the A. mellifera scout. In this situation some interspecific "misunderstanding" occurred: the attacked A. mellifera released feeding behavior in the cerana guard bees. The mellifera scout soon returned to its colony with a full honey crop and directed a wave of numerous new mellifera bees to the cerana colony which overcame the guard bees and killed the cerana colony.

On the average A. cerana has a smaller body size than European A. mellifera. We closed the entrance of the cerana colonies with a metallic screen that prevented normal size mellifera bees from passing. However, there was an overlap in size between the two species and soon a group of smaller mellifera specialized in robbing A. cerana. Ultimately, the only way we succeeded in protecting the cerana colonies was to isolate them from A. mellifera.

In Pakistan imported A. mellifera colonies were robbed by A. cerana and A. dorsata soon after their arrival. At the same time robbing mellifera foragers were observed at A. cerana colonies nearby. After this experience the A. mellifera colonies were always kept far from A. cerana.

In India many attempts were made to introduce A. mellifera. Only in the early 1960's was mellifera successfully established in Himachal Pradesh province (Sharma et al. 1980). Similar experiences were reported as we had in Germany: "This species (A. mellifera) robs cerana colonies very badly and many times the latter perish."

5. Influence of predators on the interspecific balance.

Predators normally prey on a large variety of species. But in some cases they prefer one species to another. In Sri Lanka we dissected 15 bee-eaters (Merops orientalis), which interfered with our work. The chitin remains of 114 A. dorsata, 23 A. cerana and 3 A. florea were found. These birds seem to prefer A. dorsata in a place where all three honeybee species were nearly equally common. The honey buzzard (Pernis apivorus) is specialized to prey on nests of A. dorsata. The nests of A. florea and A. cerana are not accessible for this large bird. No information is available about the many important mammals, reptiles and arthropods preying on honeybees but it can be assumed that they have an effect on the honeybees. In the case of imported A. mellifera, hornets (Vespa spp.), ants, and birds are known to be disastrous for A. mellifera.

6. Exchange of parasites.

There are some parasites that are found in all honeybees. The wax-moth (Galleria mellonella) is found in the Asiatic species as well as in A. mellifera. It is not known whether this moth changes readily from one species to another or whether there are differences and adaptations to the host species. Such adaptations are known for ecto- as well as for endoparasites. In A. mellifera a parasitic sporozoa, Nosema apis, can be found in nearly every colony. The spores are constant in size and have a volume of $30\mu^3$. In A. florea Nosema spores occur with a volume of half the size. This florea Nosema was experimentally transferred to A. mellifera. The size of the spore remained constant (Mautz 1975). In A. cerana an ectoparasitic mite, Varroa jacobsoni, is found. Another species of the same family, Euvarroa sinhai, is restricted only to A. florea. The two species have similar life cycles. They multiply only on drone larva and the adult mites survive for longer periods (when no drone brood is available) on adult bees (Koeniger et al. 1981). Another mite, Tropilaelaps clarae, lives naturally only with A. dorsata (Delfinado 1963).

A survey of the parasites of honeybees in Sri Lanka indicates that there are efficient natural barriers which prevent the exchange among the indigenous honey bee species (Koeniger et al. 1982). This situation changes drastically when exotic bees are imported. Varroa jacobsoni was transferred from A. cerana to A. mellifera. It is disastrous in the colonies of its new host. Varroa reproduces on worker pupae of A. mellifera which leads to an unlimited multiplication. Recently Varroa has spread to all continents except Australia. It is today one of the worst problems of beekeeping (A. mellifera) throughout the world.

In Asia <u>Tropilaepaps</u> <u>clarae</u> switched over from its native host
<u>A</u>. <u>dorsata</u> to imported <u>A</u>. <u>mellifera</u> and is causing severe damage.
There are no reports on parasites of <u>A</u>. <u>mellifera</u> which have changed
over to the Asiatic species. In cannot be assumed that the transfer
of parasites in a "one-way road." One has to realize that most of
the Asiatic honeybees are not under the care of scientifically
oriented beekeepers and it may take a long time until reports of
damage from parasite transfers are published.

REFERENCES

BUTLER C.G., CALAM D.H., CALLOW R.K., 1967. - Attraction of <u>Apis</u>
<u>mellifera</u> Drones by the Odours of the Queens of two other
Species of Honeybees. <u>Nature</u> 213, 423-424.

DELFINADO M.D., 1963. - Mites of the Honeybee in South-East Asia.
<u>J. Apic. Res</u>. 2, 113-114.

HAAS D., 1946. - Neue Beobachtungen zum Problem der Flugbahnen bei
Hummelmännchen. <u>Zeitschrift f. Naturforschung</u> 1, 596-600.

KOENIGER N., KOENIGER G., 1980. - Observations and Experiments on
Migration and Dance Communication of <u>A</u>. <u>dorsata</u> in Sri Lanka.
<u>J. Apic. Res</u>. 19, 21-34.

KOENIGER N., 1976. - Neue Aspekte der Pylogenie innerhalb der
Gattung Apis. <u>Apidologie</u> 7, 357-366.

KOENIGER N., KOENIGER G., WIJAYAGUNESEKERA H.N.P., 1981. -
Beobachtungen über die Anpassung von <u>Varroa jacobsoni</u> an ihren
natürlichen Wirt <u>A</u>. <u>cerana</u> in Sri Lanka. <u>Apidologie</u> 12,
37-40.

KOENIGER N., VORWOHL G., 1979. - Competition for Food among the four
Species of Alpini in Sri Lanka. <u>J</u>. <u>Apic</u>. <u>Res</u>. 18, 95-109.

KOENIGER G., BAKER M.D., KOENIGER N., 1982. - Survey on Bee Mites
of Sri Lanka. In Prep.

KOENIGER N., WIJAYAGUNESEKERA H.N.P., 1976. - Time of Drone Flight
in the three Asiatic Honeybee Species. <u>J</u>. <u>Apic</u>. <u>Res</u>. 15,
67-71.

MAA T.C., 1953. - An Inquiry into the Systematics of the Tribus
Apidini or Honeybees. <u>Treubia</u> 21, 525-640.

MAUTZ D., 1975. - Untersuchung der in <u>A</u>. <u>florea</u> gefundenen neuen
<u>Nosema</u>. <u>Der Imkerfreund</u> 30, 76-77.

MICHENER C.D., 1974. - <u>The Social Behavior of the Bees</u>. Belknap
Press, Cambridge.

MORSE R.A., SHEARER D.A., BOCH R., BENTON A.W., 1957. - Observations
on Alarmsubstances in the Genus <u>Apis</u>. J. <u>Apic</u>. <u>Res</u>. 6, 113-
118.

RUTTNER F., RUTTNER H., 1965. - Untersuchungen über die
Flugaktivität und das Paarungsverhalten der Drohnen. 2.
Beobachtungen an Drohnensammelplätzen. <u>Zeitschrift f. Bienenf</u>. 8,
1-8.

RUTTNER, F., 1973. - Drohnen von <u>Apis cerana</u> Fabr. auf einem
Drohnensammelplatz. <u>Apidologie</u> 4, 41-44.

RUTTNER F., MAUL V., 1969. - The Cause of the Hybridization barrier
between <u>Apis mell</u>. L. and <u>A</u>. <u>cerana</u> Fabr. <u>XXII</u>. <u>Int</u>. <u>Beekeep</u>.
Congr., 561.

RUTTNER F., WOYKE J., KOENIGER N., 1972. - Reproduction in Apis
    cerana. J. Apic. Res. 11, 141-146.
SAKAGAMI S.F., MATSUMURA T., ITO K., 1980. - Apis laboriosa in
    Himalya, the little known world largest Honeybee (Hymenoptera,
    Apidae). Insecta Matsumurana 19, 47-77.
SHARMA O.P., MISHRA R.C., DOGRA G.S., 1980. - Management of Apis
    mellifera L. in Himachal Pradesh. II. Int. Conf. Apiculture in
    Tropical Climates, New Delhi.

Abstracts

Manipulation of Ants in Maya-type Raised Fields
in Mexico

Awinash P. Bhatkar

Department of Entomology (CSAT), H. Cardenas, Tab.,
86500-Mexico

Three ha area divided into 6m wide, 60–120m long alternating
raised platforms and canals, based on the historical Maya agriculture, was monitored for ant colonization and distribution in
corn and bean cultivars. Red and black forms of Solenopsis geminata
occurred in mutually exclusive areas, predominated the hypogeic,
epigeic and arboreal foraging strata, yet adapted by intercolonial
food exchange on confrontation to the same foraging time. The
species of Tapinoma, Monomorium and Pheidole foraged at 36–42°C and
at the spots not inhabited by S. geminata. At these temp S.
geminata foraged hypogeically. The hypogeic predation of soil
insects by Diplorhoptrum, Monomorium and Pheidole spp. at the depths
of 6–10 cm superceded that of other species. These species inhabited the corn cobs during seed formation and controlled the seed
insects effectively. Raising or lowering the water table could be
used to augment ants to the arboreal strata to implement an effective pest control.

Competition and Community Organisation in an Australian Ant Fauna

P.J.M. Greenslade
CSIRO, Division of Soils, Glen Osmond, South Australia

A diverse ant community was studied in open Acacia woodland
on the southern edge of the Australian arid zone in the Gawler
Ranges, South Australia. Ants were sampled by means of seed and
sardine baits and pitfall traps. This work is part of a continuing
project on community structure amongst Australian ants. It confirmed and amplified previous findings and raised new points,
demonstrating inter alia: a recurring faunal composition with
many species but few genera: effects of habitat on diversity; an
inverse relationship between this diversity and mean niche breadth;
dominance by species of the dolichoderine genus Iridomyrmex; the
importance of competition, especially diffuse competition which con-
strains the abundance of broadly adapted species; the nature of competitive relationships between the opportunist ponerine, Rhytidoponera metallica, and members of the genus Pheidole and between
Pheidole and Monomorium species.

Population Ecology of <u>Reticulitermes</u> spp. in Southern Mississippi

Ralph Howard, Susan Jones, Joe Mauldin, and Raymond Beal
Forest Service-USDA, Gulfport, Mississippi

Species distribution, abundance, and colony size were studied for subterranean termites in southern Mississippi. An exhaustive census of mature termite colonies in 24 one-hectare plots was made. The average abundance of <u>Reticulitermes flavipes</u> (Kollar) was 4.42 $^+_-$0.43 ($\bar{X}$ $^+_-$ SE) colonies/hectare. <u>Reticulitermes virginicus</u> (Banks) colonies occurred at a density of 2.38 ± 0.29/hectare. Spatial relationships of colonies were studied by nearest neighbor analysis. The mean distance between <u>R. flavipes</u> colonies was 22.48 $^+_-$ 3.09 m, between <u>R. virginicus</u> colonies 26.19 $^+_-$ 6.52 m, and between colonies irrespective of species, 16.80 $^+_-$ 2.01 m. Six colonies of <u>R. flavipes</u> were selected by a ranked set sampling method, and exhaustive efforts to remove all termites yielded an estimate of 244,445 $^+_-$ 53,156 termites/mature colony.

Adaptive strategies in colony foundations
of two Termitidae.

Guy Josens
Université Libre de Bruxelles, Belgium

The foundation strategies of two species of Termitidae are compared with the usual one within the same family.

In <u>Gnathamitermes perplexus</u>, the founding pairs distinguish themselves by a digging activity which lasts a long time. Digging may even continue after the first eggs have been laid. This may be considered as an adaptive strategy for a desert inhabiting species; it allows the founding pairs to settle deep into the soil, at a level where they will remain at a suitable humidity.

In the humid tropics, the flights of <u>Pseudacanthotermes militaris</u> occur by day and at the end of the rainy season, on the contrary of the other Termitidae. Then the laying of eggs starts after 15 days or even later, egg hatching lasts more than 6 weeks and the development of the first brood is also slow. This peculiar slowness involves that the alates carry away a large amount of energy; but it also allows the young societies to have their first worker ready at the beginning of the following rainy season.

Population Dynamics of Harvester Ant
Workers (Pogonomyrmex)

Sanford D. Porter and Walter R. Tschinkel
Florida State University

Population dynamics of Pogonomyrmex workers were studied in
Florida and Idaho. Pulses of workers were spray-marked with in-
visible fluorescent ink and released into field or lab colonies.
Capture samples were then taken at periodic intervals to determine
the proportion of marked workers remaining and the duties which
these workers performed. Aging workers followed a general progression from interior to exterior work. Dry weight increased as
callows matured into dark workers and then decreased as dark workers
progressed to foraging. Mortality rates of exterior workers were
much higher than interior workers. In the case of Pogonomyrmex
badius, mortality rates were also a function of worker size.

Factors Influencing the Annual Abundance of Yellowjackets

Robert E. Wagner
University of California, Riverside

Populations of a three species complex of yellowjackets, Vespula
pennsylvanica, V. atropilosa and V. sulphurea, have been monitored
with heptyl butyrate baited traps for the past seven years in a
heptyl butyrate baited traps for the past seven years in a southern
California canyon. The number of queens of each species trapped
during nest initiation during April through July of any given year
was not indicative of subsequent colony success as indicated by
workers trapped later in that year. Timing of peak queen emergence
appeared to be correlated with accumulated heat intake during the
hibernation period. Early queen activity was observed after relatively warm fall-spring temperatures regardless of rainfall during
the hibernation period and was followed by early appearance of
workers. High temperatures during the period of early nest development in 1981 appeared to be very deleterious for annual worker catch
was less than 20% of normal. In 1977, after the coolest hibernation
period in the years studied, queen emergence was delayed until late
May. Worker populations trapped were 25% higher than normal in that
year. During years of lowest worker populations, all three species
were observed scavenging pet food which may indicate that prey availability was also a limiting factor in population growth.

THE ROLES OF SOCIAL INSECTS IN ECOSYSTEMS
Organized by W. L. Nutting

Introduction

W. L. Nutting, The University of Arizona

The roles of the termites, ants, and wasps are considered here; the bees, perhaps of more obvious interest, are taken up in the symposium on Foraging Behavior and Pollination. Among these social insects, the ants are completely social and unquestionably dominant geographically, numerically, and ecologically. Their numerical superiority is probably a resultant of the development of a wingless worker caste which is wholly female, a generally flexible and catholic diet, and a relatively simple soil-nesting habit.

The termites, so divergent from the Hymenoptera in origin, are fully comparable in the complexity of their caste system (both sexes) and the organization of their societies. Their nests, subterranean, surficial or arboreal, afford considerable environmental control and they feed on cellulosic materials in all its different forms. Indeed, these two groups of dominant social insects have partitioned large areas of tropical and temperate ecosystems between them.

The minority of wasps which are unsocial have largely substituted temporal for morphological polyethism, remained predatory with some addiction to sweets, and have abandoned ancestral burrows for a variety of elaborate subterranean or above-ground constructions.

Termites must lay much of their success to the choice of cellulose as their main source of energy, for it is probably the most abundant of the constantly cycled organic materials. They are thus "micro-ruminants", both primary consumers and decomposers, sharing this resource with a variety of soil invertebrates, fungi, and micro-organisms. It is well known that termites plague man by damaging crops, denuding pastures, and destroying timber in forests and wood in buildings. However, we are increasingly aware of their many functions in natural ecosystems such as recycling dead plant material, turning and enriching the soil, and producing seasonal swarms of winged reproductives that form a food source of high caloric value for a host of predators including man.

The ants, not so single-minded in their choice of food and nesting sites, affect ecosystems in more diverse and, generally, more subtle ways. They are certainly significant turners of the soil and decomposers of litter. However, they are probably most important as predators of other insects and thus of value to man in urban, agricultural, and forest ecosystems, barring the fungus-growers which are serious pests in those situations.

Except for a very few species, the wasps have received little ecological attention. Considering their smaller numbers and colony size, they are probably most conspicuous as predators. In agricultural and urban ecosystems, they serve to reduce the ravages of many important crop pests but they may be serious pests themselves when they prey on honeybees or attack man and interfere with his activities.

These three great groups of social insects may thus have far-reaching effects on ecosystems through modification of habitat, their contributions to energy flow and cycling of nutrients, and their often strong interactions with other organisms. The following contributions will serve as a measure of our progress toward understanding the complex, sometimes complementary, roles of the termites, wasps, and ants on selected ecosystems: Pastoral, rain-forest, and urban/agricultural, foci of the most recent and intensive ecological studies.

Ecological Role of Termites in a Tropical Rain Forest

Takuya Abe, University of the Ryukyus

Information on the role of termites in the tropical rain forest ecosystem is very scarce in comparison with that in the savannah ecosystem. We studied the termites in a tropical rain forest, Pasoh Forest Reserve, of West Malaysia within the framework of IBP. Most of the results were published by Matsumoto (1976), Abe (1978), Abe (1979), Matsumoto & Abe (1979) and Abe (1980). In this paper their results are compared with those obtained in other areas and the ecological role of termites in the tropical rain forest is discussed from some new aspects.

STUDY SITE

The study was made in Pasoh Forest Reserve located at about 140 km southeast of Kuala Lumpur in Malaysia. The vegetation is described as lowland Dipterocarps forest. The forest received an annual precipitation of about 2,000mm. The air temperature on the forest floor ranged between 27° and $23^{\circ}C$. Relative humidity was almost always 100% from evening to morning, suddenly falling to 70-80% in the afternoon for a short period (Aoki et al, 1975).

RESULTS

1. Distribution, feeding habits and abundance of termites in Pasoh Forest Reserve

Fifty-seven species of termites were found in·Pasoh Forest and 52 species in the 1 ha plot. They occupied all strata of the forest from the height of more than 30m on and in the standing trees down to the depth of 30cm in the soil.

Food contents of termites were various in Pasoh Forest: lichen, dead leaves, living wood, dead wood and humus. The termites in Pasoh Forest ate all kinds of materials described by Wood (1978). They have relations with almost all stages of plant litter decomposition in the forest. In spite of the richness in the number of species and biomass, most termites did not attack the living trees.

71

Table 1. Density (n/m^2) and biomass ($g.w.w./m^2$, in parenthesis) of termites with reference to their feeding habits in the tropical rain forests and savannah

Ecosystem	Tropical rain forest			Savannah
Locality	Gunung Mulu*		Pasoh**	Mokwa***
Latitude	4°N		3°N	9°N
Altitude (m)	220	130	130	205
Precipitation (mm)	5100		2000	1175
Authors	Collins (1980)		Abe (1979)	Woods & Sands (1978)
Feeding habits				
Lichen feeders	+	+	50(0.14)	0
Soil feeders	375	1579	1505(2.52)	163(0.66)
Fungus growers	9	13	960(6.12)	2193(6.39)
Others	465	11	970(0.64)	1652(3.54)
Total	880	1603	3485(9.41)	4008(10.59)

* Sarawak; **West Malaysia: ***Nigeria

Table 2. Density (n/m^2) and biomass ($g.w.w./m^2$, in parenthesis) of soil macrofauna in the tropical rain forests and subtropical rain forest in Iriomote-jima, Japan

Locality	Pasoh	Gunung Mulu	Irimote*
Author	Kondoh et al (1980)	Collins (1979)	Abe (1981)
Soil macrofauna			
Termite	3485(9.41)	1125(1.78)	144(0.06)
Earthworm	25(0.18)	26(0.65)	66(9.45)
Ant	1624(0.82)	457(0.47)	172(0.20)
Others	253(2.07)	280(1.36)	196(4.34)
Total	5387(12.48)	1888(4.26)	578(14.05)

* Small island consisting of subtropical rain forest dominated by Castanopsis sieboldii located at 24°N.

Density and biomass of the termites are shown in Table 1 with reference to their feeding habits together with those in the other tropical rain forest (Gunung Mulu forest in Sarawak) and savannah. Though both of the forests at Pasoh and Gunung Mulu have features characteristic of tropical rain forest, having giant emergent trees reaching a height of 50-55m, they have different annual precipitation (Gunung Mulu:5100mm). As shown in Table 1, soil-feeding termites (mainly consisting of Termitinae) increase and fungus-growing termites (all consisting of Macrotermitinae) decrease with increase in annual precipitation. Richness in the number and biomass of soil feeders seems to be one of the most important characteristics of the termites in the tropical rain forest.

Table 3. Food consumption by a fungus-growing termite, Macrotermes
carbonarius, in Pasoh Forest

Items		References
Number of mound	20/ha	Abe & Matsumoto(1979)
Population	88,500/mound	Matsumoto(1976)
Termite biomass	1.95g. w.w./m^2	Abe & Matsumoto(1979)
Natural food	Leaf litter and small wood	Abe(1978)
Leaf litter consumption	2.71g.d.w./m^2/week	Matsumoto & Abe(1979)
Daily consumption	199mg d.w./g.w.w. termite	Present paper
Leaf litter supply	6.3t/ha/year	Ogawa(1978)
Percentage of leaf litter consumed by M.c.	22.4%	Matsumoto & Abe(1979)

Abundance of termites is compared with those of other soil
macrofauna in Table 2. It is evident that termites occupy an
overwhelmingly dominant position among soil macrofauna in Pasoh
Forest. On the other hand earthworms, which are usually abundant
in temperate and subtropical regions, are scarce in Pasoh and Gunung
Mulu forests.

2. Food consumption by termites in Pasoh Forest Reserve
     To estimate the consumption rate of leaf litter by termites,
newly fallen leaves were marked and distributed on the forest floor.
The loss of leaf area due to termites was determined photometrically.
The leaf litter consumption by a dominant fungus grower, Macrotermes
carbonarius, is shown in Table 3. Daily consumption by the species
was very high (199mg d.w.per g.w.w.termite). Wood & Sands (1978)
indicated that fungus growers (Macrotermitinae) have higher weight-
specific consumption rates than other species, undoubtedly due to the
fact that their consumption is required to maintain their symbiotic
fungi in addition to themselves. An amount equivalent to 22% of
daily leaf litter fall was transported by M. carbonarius to its
mound. The role of fungi cultivated by this species on their fungus
combs was discussed in relation to the nutrition of termites and
decomposition of leaf litter with special reference to their
nitrogen metabolism by Matsumoto (1976).
     The bole and branch samples of fallen trees were placed on the
forest floor and their weight loss was measured with reference to
the attack of termites. More than 20 species of termites attacked
the wood and dominant fungus-growing wood consumers such as
Macrotermes malaccensis and Odontotermes spp. brought a large
amount of soil into wood. Most of the breakdown of wood seemed to
be attributed to the action of termites or both termites and
microbes.

DISCUSSION

Two questions are interesting, though they are very difficult to answer. "Why don't termites attack the living trees in Pasoh Forest? "Why are termites so prosperous in the tropical rain forest?"

1) Why don't termites attack the living trees in Pasoh Forest?

This question is interesting, because the termites attack the living trees outside Pasoh Forest. In the cultivated area of Malaysia, the termites attack living plants such as tea, rubber and oil palm (Harris, 1966). These plants are exotic to Malaysia. It is evident that the period of interaction between the termites and the exotic plants are extremely short in comparison with that between the termites and trees in the virgin forest such as Pasoh Forest. It is not so unreasonable to imagine that the natural selection favored the termites which did not attack living trees in the tropical rain forest, for the termites attacking the living trees might destroy the forest and perish because of their ability. The fact that soil-feeding termites are most numerous in the tropical rain forest among various habitats of termites is interesting from this viewpoint.

2) Why are termites prosperous in the tropical rain forest?

Nielsen (1963) investigated the range of carbohydrases possessed by soil animals and concluded that, of the animals studied, probably only snail and slug can decompose significant quantities of plant structural polysaccharides, suggesting that the microorganisms such as protozoa, bacteria and fungi are of primary importance in the process of litter decomposition.

In Pasoh Forest about 65% of termites are fungus-growers in biomass and most of them carry much soil into wood. They are excellent go-betweens which combine organic matter with microorganisms in three different ways. At first they bring much soil, which contains microorganisms, into wood (organic matter) on the forest floor. Secondly they take litter and bring it to the bacteria in the digestive tube of termites. Thirdly they put their feces containing organic matter on the fungus combs where fungi decompose organic matter. As these termites take the older part of fungus combs, the combs are, as it were, another digestive apparatus outside the termite body.

Only a little information is available on the factors limiting the activity of microorganisms in the tropical rain forest where climatic factors such as temperature and humidity are adequate (Jenny et al, 1949). If the speed of combination of microorganisms with organic matter is the factor limiting the decomposition rate in the tropical rain forest, it is reasonable that termites are the most prosperous litter consumers in the tropical rain forest, because they seem to be the most excellent go-betweens which combine microorganisms with organic matter.

REFERENCES

Abe T., 1978.-Studies on the distribution and ecological role of termites in a lowland rain forest of West Malaysia (1) Faunal composition, size, colouration and nest of termites in Pasoh Forest Reserve. Kontyu, 46, 273-290.

Abe T., 1979.-Studies on the distribution and ecological role of termites in a lowland rain forest of West Malaysia (2) Food and feeding habits of termites in Pasoh Forest Reserve. Jap. J. Ecol, 29, 121-135.

Abe T., 1980.-Studies on the distribution and ecological role of termites in a lowland rain forest of West Malaysia (4) The role of termites in the process of wood decomposition in Pasoh Forest Reserve. Rev. Ecol. Biol. Sol., 17, 23-40.

Abe T., 1981.-Human impact on the soil macrofauna in Kume-jima and Iriomote-jima, the Ryukyu Islands, Japan. Man's impact on the Island Ecosystems in the Ryukyu Islands, II, 173-192.

Abe T., and Matsumoto T., 1979.-Studies on the distribution and ecological role of termites in a lowland rain forest of West Malaysia (3) Distribution and abundance of termites in Pasoh Forest Reserve. Jap. J. Ecol., 29, 337-351.

Aoki M., Yabuki K. and Koyama H., 1979.-Micrometerology and assessment of primary production of a tropical rain forest in West Malaysia. J. Agric. Meteor., 31, 115-124.

Collins N.M., 1979.-A comparison of the soil macrofauna of three lowland forest types in Sarawak. Sarawak Museum Journal, 27, 267-282.

Collins N.M., 1980.-The distribution of soil macrofauna on the West ridge of Gunung(Mount)Mulu, Sarawak. Oecologia, 44, 263-275.

Harris W.V., 1966.-Termites; their recognition and control. Longmans Green and Co., London, 186p.

Matsumoto T., 1976.-The role of termites in an equatorial rain forest ecosystem. I. Population density, biomass, carbon, nitrogen and calolic content and respiration rate. Oecologia, 22, 153-173.

Matsumoto T., and Abe T., 1979.-The role of termites in an equatorial rain forest ecosystem. II. The role of termites in the decomposition of leaf litter on the forest floor. Oecologia, 38, 261-274.

Kondoh M., Watanabe H., Chiba S., Abe T., Shiba M. and Saito S., 1980.-Studies on the productivity of soil animals in Pasoh Forest Reserve, West Malaysia V. Seasonal change in the density and biomass of soil macrofauna: Oligochaeta, Hirudinea and Arthropoda. Mem. Shiraume Gakuen Coll., 16, 1-26.

Ogawa H., 1978.-Litter production and carbon cycling in Pasoh Forest. Malay. Nat. J., 30, 367-273.

Wood T.G., 1978.-Food and feeding habits of termites. "Production ecology of ants and termites" (ed. Brian M.V.), 55-80. Cambridge University Press.

Wood T.G. and Sands W.A., 1978.-The role of termites in ecosystems. "Production ecology of ants and termites" (ed. Brian M.V.), 245-292. Cambridge University Press.

A Population Dynamics Hypothesis for *Vespula* in England, U.K.

Michael E. Archer, The College of Ripon and York St. John, York

Bree (1849) casually noted that an abundant flight of spring queens was followed by a summer scarcity of wasps and vice-versa. Malaise trap catches of queens and workers of Vespula vulgaris and V. germanica between 1972 and 1980 supported Bree's observations.

Beirne (1944) examined casual observations of summer wasp abundance in biological journals from 1864 to 1931. During these 68 years there were ten years of summer wasp abundance and ten years of summer scarcity. Abundant and scarce years were associated as year-pairs more than would be expected by chance (X^2=3.8, p=0.05).

Fox-Wilson (1946) recorded the number of colonies destroyed in 40.5 hectares at Wisley in Surrey from 1921 to 1949. Colony numbers varied from 2 to 84 per year. Beirne (1944) and Fox-Wilson (1946) suggested that summer abundance was related to spring rainfall although another factor, probably disease, could override the spring rainfall effect. Recent correlation analysis of Fox-Wilson's data shows no statistically significant relationship between spring rainfall or summer weather and colony abundance.

Fox-Wilson's data do, however, show a yearly oscillation above and below the mean value more than would be expected by chance (X^2 = 8.5, p $<$0.05). Moran plots (ΔN on N, Williamson 1972) give a linear correlation coefficient (r = -0.84, p$<$0.001) and regression slope (-1.41). These results have been interpreted as indicating a cyclic pattern in the data, the nest counts approaching an equilibrium through damped oscillations. Further the mechanism controlling the cyclic pattern acts in an over-compensating manner. The auto-correlation function gives a negative coefficient at 1-year lag, followed by a lesser positive coefficient, indicating a strong 2-year cycle with a damped waveform. The waveform is similar to that expected from an endogenous regulatory mechanism (Pielou 1974). Further positive peaks at 7 and 14 years lag which are near equality indicate a superposed undamped waveform function suggestive of an additional exogenous controlling agent (Pielou 1974).

Fox-Wilson's data mainly refer to three species:- V. vulgaris, V.germanica, V.rufa; counts for each species are not available. Queens and workers of V. vulgaris and V. germanica have been collected by malaise and suction traps at six sites in England from 1972 to 1980. Moran plots of ΔN on N all yield regression coefficients between -1 and -2 indicating a 2-year cyclic pattern in the data, the adult counts approaching an equilibrium through damped oscillations and as such supports Fox-Wilson's data. Despite the short period of time covered by these data most of the correlation coefficients have reached the statistically significant levels of either 5% or 1%.

An examination of the life-table for V. vulgaris (Archer, unpublished) suggests that the endogenous mechanism operates in the spring during nest initiation and the perturbating exogenous factor later in the year. The large mean queen production per colony of 962 queens demonstrates the capacity for the species to show an over-compensation reaction.

Between 1967 and 1977 Archer (1981) collected nest data for V. vulgaris from the York area, England. Summer wasp abundance did not correlate with spring rainfall or summer weather. Nest site characteristics, via a possible microclimatic effect, also could not be correlated with nest size. However the presence of successful colonies producing many queens was associated with summer weather above the median and unsuccessful colonies with summer weather below the median more than would be expected by chance ($X^2 = 8.5$, $p < 0.005$) (= summer weather effect).

Percentage use of large cells to rear sexuals (mainly or entirely queens) was higher in large colonies and in colonies that reached their demise earlier in the autumn. It is considered that large colonies and earlier maturing colonies were better able to compensate for deteriorating autumn weather conditions (= autumn weather effect).

Other possible exogenous influences have been considered but can probably be dismissed as unimportant.

Queen activity as measured by the number of cells she builds correlates with the eventual mature size of the nest ($r = 0.69$, $p < 0.001$) indicating that queen behaviour could be the source of the endogenous mechanism. Potter (1964) found that the major factor influencing worker foraging was the presence of the queen. Archer (1981) reviewed the observational and experimental evidence that shows that: the size and success of a mature colony can be linked to queen influence and that possibly a pheromone or pheromones mediate the influence.

Queen influence as measured by colony success can be related to the 2-year cycle by the increasing and decreasing year concept. An increasing year is one in which summer wasp abundance in the current year is greater than the previous year; vice-versa for a decreasing year. Then decreasing years have more unsuccessful colonies and increasing more successful colonies than would be expected by chance ($X^2 = 22.4$, $p < 0.001$).·

It is suggested that intra-specific usurpation between spring queens for early nests is density related. Decreasing years would show higher frequencies of queen usurpation leading to either colony failure to develop or abnormal development of colonies which are unsuccessful. Increasing years would show lower frequencies of queen usurpation leading to normal and successful colony development.

Archer (1980) reviewed the evidence for intra-specific queen usurpation and showed that usurpation is a frequently occurring behaviour.

Summary (Fig. 1)

Populations of V. vulgaris and probably V. germanica in England show a strongly pronounced 2-year cycle of summer wasp abundance which damps to equilibrium driven by an endogenous mechanism which depends upon density-related intra-specific usurpation between spring queens for early nests. The equilibrium can be perturbed by an exogenous mechanism controlled by summer and autumn weather.

Acknowledgements

D. Bunn, J. Douglas, A.J. Halstead, G. Nixon, D. Owen, L.R. Taylor, M. Williamson.

References

Archer, M.E., 1980 Population dynamics, in Edwards, R.- Social Wasps. The Rentokil Library, East Grinstead.

Archer, M.E., 1981 Successful and unsuccessful development of colonies of Vespula vulgaris (Linn.)(Hymenoptera:Vespidae). Ecol. Entom., 6,1-10.

Beirne, B.P., 1944 The causes of the occasional abundance or scarcity of wasps (Vespula spp.). Entom. mon. Mag., 80, 121-124.

Bree, W.T., 1849 Seasonal abundance or scarcity of the Common Wasp. Zool., 7,2614-2615.

Fox-Wilson, G., 1946 Factors affecting populations of social wasps, Vespula species, in England (Hymenoptera). Proc. R.ent. Soc. Lond. (A),21, 17-27.

Pielou, E.C., 1974 Population and Community Ecology.Gordon and Breach, New York.

Potter, N.B., 1964 A study of the biology of the Common Wasp, Vespula vulgaris L., with special reference to foraging. Ph.D. Thesis, Bristol Univ.

Williamson, M., 1972 The Analysis of Biological Populations. E. Arnold, London.

Fig. 1 - A population dynamics model for _Vespula vulgaris_
in England.

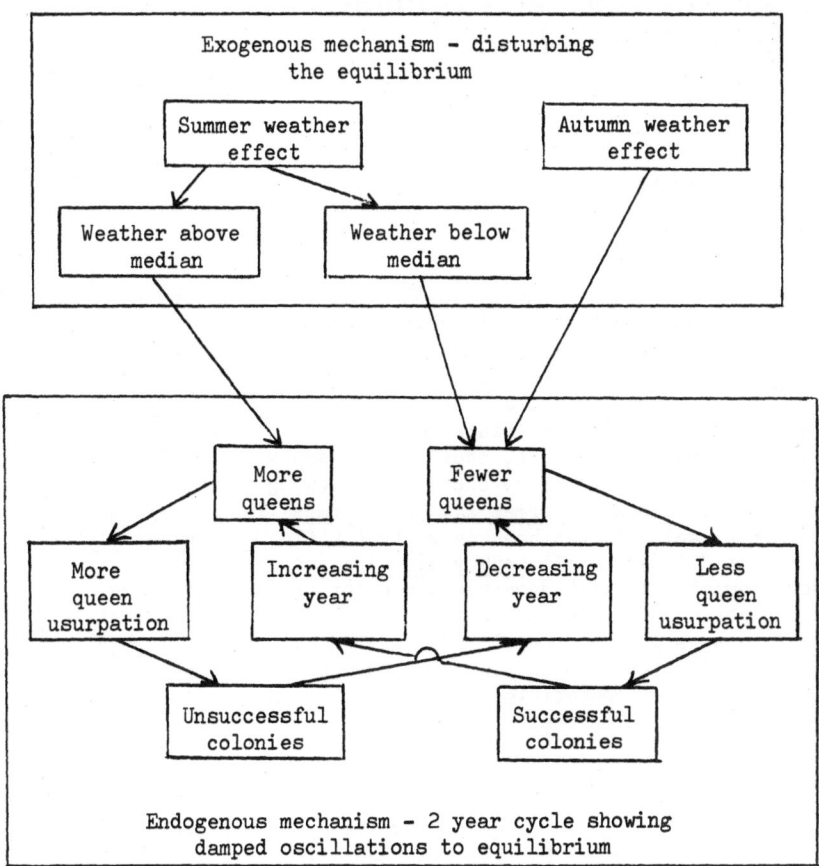

The Interaction and Impact of Domestic Stock and Termites in a Kenyan Rangeland

N. M. Collins,
The International Centre of Insect
Physiology and Ecology, Nairobi, Kenya

Kajiado District in Kenya is 90% semi-arid rangeland with rain-fall between 400 and 800 mm a ^{1} falling mainly during two rainy seasons, November/December and April/May (Norton-Griffiths 1977). The vegetation is grassland and wooded grassland with Themeda, Cynodon and Pennisetum as dominant grasses, and Acacia tortilis (red soils) or A. drepanolobium (black cotton soils) the dominant trees. The land is grazed by cattle, sheep, goats and wild ungulates. The termite fauna is dominated by the fungus-growing genera Macrotermes, Odontotermes, Synacanthotermes and Microtermes (Isoptera:Termitidae: Macrotermitinae). It has been shown that while Macrotermes michael-seni (Sjöstedt) (as M. aff. subhyalinus) is basically a litter-feeder, competition with domestic stock for standing grass may result from drought or over-grazing (Lepage 1979, 1981). The present study seeks to assess whether the closely-related Macrotermes subhyalinus (Rambur) is similar in behaviour, and to describe the mechanism of the development of competition, if any. Ensuing results are prelimi-nary and field work is continuing.

METHODS

The study site is on reddish brown sandy clay loams near Bissel, 32km S of Kajiado town. Macrotermes subhyalinus nests dominate M. michaelseni, which are more common in higher altitude/rainfall areas (Van Der Werff 1981). Only M. subhyalinus is present in the study area, at a density of 4.1 nests ha^{-1} in an over-dispersed distribu-tion. The site consists of 4 60 x 40m plots, 2 fenced and 2 open to normal grazing patterns. The 4 plots did not differ significantly in terms of grass or litter biomass at the beginning of the experi-ments in August 1980. Results are calculated from the combined data of each pair of plots, all differences being accredited to the treatments. Productivity measurements cover the period October 1980 to September 1981, which approximates to the November to October climatic year.

Grass and litter biomass and production were estimated at 10 stratified random locations in each plot, from which 1m^2 collections

were made monthly. Foraging patterns of M. subhyalinus were followed on 24 10x1.5m strips in stratified random positions in the open
plots only. For 7-14 days each month the number of utilised foraging holes was estimated by marking holes with white powder (as used
for wall repairs) and checking for evidence of termite activity the
following morning. This method could not be used in the fenced
plots since trampling would have been unavoidable. However, a
single comparison of the total number of foraging holes inside and
outside the fences was done in December 1981. Five $20m^2$ strips were
checked in each of the four plots. Total grass consumption was
extrapolated from estimates of the amount of litter carried into one
foraging hole per day. These were achieved by encircling foraging
holes with 20cm diameter tin arenas and replacing any grass or litter
inside with 5g of dry grass and litter. If the hole was opened over-
night the grass was removed, re-weighed after drying, and the weight
loss calculated with allowance for losses from unutilised control
arenas.

RESULTS

1. Impact of grazing stock
        Grazing in the open plots was mainly in the dry season,
while other areas of the ranch were used in the rainy season. Table
1 is a summary of the results of productivity and grazing measurements. Full data and calculations cannot be given, but the main
relevant conclusions may be summarised as follows:
    a. Total NAPP is similar in open and fenced plots
    b. The overall annual increment of grass standing crop is the
        result of higher rainfall (569mm) than in the previous
        corresponding year (439mm), and reduced grazing pressure.
    c. Litter production is actually and proportionately higher
        in open plots, presumably as a result of trampling.
    d. Despite lower stocking levels than in the previous year,
        the open plots were overgrazed. Maximum offtake according
        to local stocking procedures should be 876 kg $ha^{-1}a^{-1}$
        (0.2 stock units (S.U.) ha^{-1} @ consumption rate of 12 kg
        $S.U.^{-1}d^{-1}$). The 600 ha ranch actually carried 183 S.U. in
        October 1981, with a calculated consumption of 1336 kg
        $ha^{-1}a^{-1}$. Actual consumption was measured as 1964 kg ha^{-1}
        a^{-1}, indicating overuse of the ranch as a whole, and even
        greater overuse of the study area due to poor rotation.
        This conclusion is supported by grass composition data,
        which indicate an over-abundance of the unpalatable pere-
        nnial tussock-grass Pennisetum stramenium, (ca. 50% of both
        cover and biomass).

2. Impact of Macrotermes subhyalinus
        M. subhyalinus usually feeds only on grass litter and dung.
During this study the termites were observed feeding on standing
grass only during two periods, January and April/May. These were
both times of high grass production, low litter production and
biomass, and low foraging intensity (ca. 20kg $month^{-1}$, see below).

Table 1: Net above-ground primary production (NAPP) of grass, with estimates of losses to litter and stock. Biomass figures in kg. ha^{-1}a^{-1} dry weight.

Treatment	Open		Fenced	
	Biomass	% NAPP	Biomass	% NAPP
Minimum standing crop	1550	–	2200	–
Maximum standing crop	6110	–	6800	–
Total NAPP	7478	100	7446	100
Litter production	1762	24	1334	18
Increment over year	2490	33	3450	46
Resorption, loss to insects	1262*	17	2662	36
Losses to stock and game	1964*	26	0	0

*Loss to stock is the sum of monthly biomass decrements in open plots, minus estimated losses to resorption etc. calculated from fenced plot data. Grazing was only in the dry season in this area.

The only standing grasses selected by the termites were tussocks of Pennisetum stramenium, a species unpalatable to cattle.

The level of foraging varied considerably throughout the year, with a main peak during the long dry season (June – October) and a minor peak in the short dry season (February–March, Figure 1).

Figure 1: Density of foraging holes used/day/month, with estimates of litter consumed (histogram). Monthly rainfall is shown (broken line).

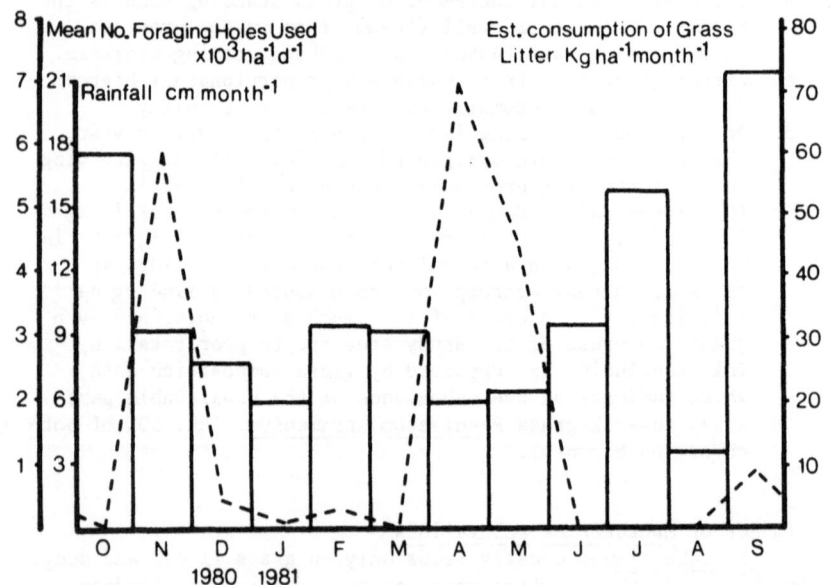

In seven consumption trials (n=14-60 foraging holes per trial) mean consumption varied between 0.21g and 0.57g, with a mean of means equal to 0.34g d^{-1} hole^{-1}. Analysis and collection of consumption data is continuing, but interim monthly consumption figures may be calculated from 0.34g x 30 x no. foraging holes opened d^{-1}. Monthly consumption estimates vary between 12 and 73kg ha^{-1}, with an annual total of 413kg ha^{-1}a^{-1} (Figure 1). This represents 23.4% of litter production in grazed plots, and 31.0% of litter production in fenced plots.

The single estimate of total foraging hole densities in the open and fenced plots in December 1981 (17 months after the fences were put up) gave significantly different values of 32,600 and 9,350 holes ha^{-1} respectively (F = 32.84 @ 1 & 17 d.f., P =<0.001).

Figure 2: The relationship between density of utilised foraging holes and grass litter biomass. The data represent 17 monthly measures of foraging hole density (7-14 days month^{-1}), aligned at the single estimate of litter biomass in open plots for that month, ±95%c.l. = y ± 1.96s(y)

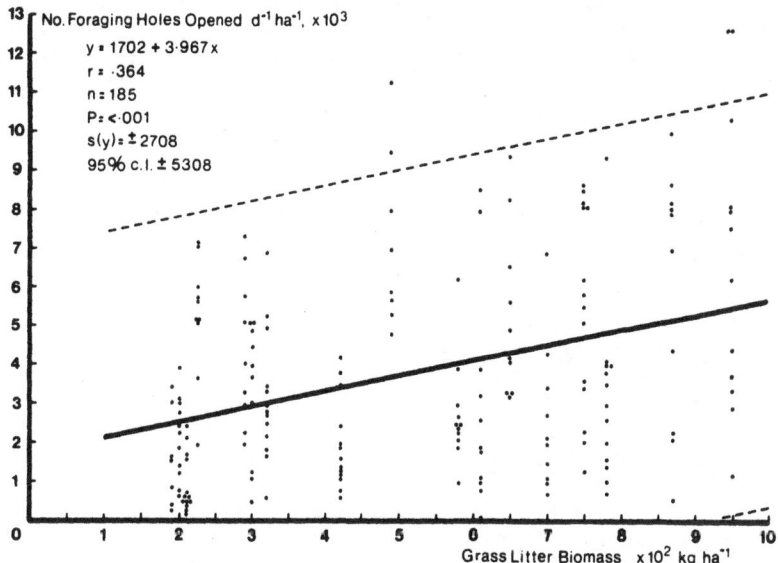

DISCUSSION

For most of the year M. subhyalinus and cattle feed in different sub-systems (decomposer and herbivore respectively). The seasonal pattern of termite foraging levels takes advantage of periods of high litter biomass. This primary dependence on litter is illustrated in Fig. 2, which shows the correlation (P = <0.001) between

foraging levels and litter biomass over time. The wide 95%c.1.
result from patchy foraging, small-scale weather patterns and colony
condition (e.g. ± alates). The observed higher foraging hole
density in grazed plots indicates that the termites take advantage
of high litter levels spatially as well as temporally. However, the
termites are adaptable, and if litter levels fall below about 2-300
kg ha^{-1}, they may climb and cut standing grasses. The interaction
between cattle and termites during the study period is limited to
the fact that grazing causes high litter levels and increased termite foraging. More complex interactions may be envisaged if over-
grazing persists to the level of reducing NAPP, or a drough occurs.

M. subhyalinus colony size, and therefore consumption rate, is
finite (see J.P.E.C. Darlington, this volume). Therefore, as litter
biomass levels rise with increasing grazing intensity, the required
foraging area of colonies will be reduced. In time, gaps in the
network of nests will appear, permitting ingress of new colonies
and an overall increase in population density. Should grazing
reach a level at which offtake is so high that litter production is
curtailed, M. subhyalinus would resort to standing grass and compete
directly with cattle. Drought would have a similar effect. If termite populations were already artificially high through limited
overgrazing in previous years, their impact on standing grass at
this critical time might be substantial. Conversely, if grazing is
kept within recommended levels and reduced during periods of low
rainfall, the termites will feed almost exclusively on litter, and
competition will be trivial.

References

Lepage M., 1979. --La recolte en strate herbacée de Macrotermes aff.
    subhyalinus (Isoptera:Macrotermitinae) dans un écosystème semi-
    aride (Kajiado-Kenya). C.R. UIEIS sct.
    Francaise - Lausanne.
Lepage M., 1981.-- L'impact des populations recoltants de
    Macrotermes michaelseni (Sjöstedt) (Isoptera:Macrotermitinae)
    dans un écosystème semi-arid (Kajiado-Kenya). 1. L'activite
    de recolte et son determinisme. Insectes soc. 28, 297-308.
Norton-Griffiths M., 1977.--Aspects of climate in Kajiado District.
    UNDP/FAO Kenya Wildlife Project (Ken/71/526) Working Document
    13, Nairobi:FAO.
van Der Werff P.A., 1981.-- Two mound types of Macrotermes near
    Kajiado (Kenya): Intraspecific variation or interspecific
    divergence? In Systematics Association Vol. 19 Biosystematics
    of Social Insects, ed. P.E. Howse, Academic Press, London &
    New York.

This work was supported by the United Nations Environment
Programme.

Comparative Early Growth and Foraging of Two Naturally Established Vespine Wasp Colonies

Albert Greene, University of Maryland

This paper examines the early development of two vespine wasp colonies, one of Vespula germanica (F.) and one of V. vidua (Sauss.). These species are at opposite ends of the spectrum of vespine biology. The former is a member of the subgenus Paravespula, characterized by large, long-lived colonies whose omnivorous foraging habits include extensive scavenging on human food and garbage. Indeed, V. germanica is undoubtedly the world's most pestiferous social wasp, due to its colonization of numerous countries in both hemispheres, its pronounced synanthropy in some regions, and the ability of its colonies under certain conditions to become polygynous, persist for a second season, and at times grow far larger than has been recorded for any other social wasp or bee species (Thomas, 1960; Edwards, 1980; MacDonald et al., 1980). In contrast, V. vidua belongs to the subgenus Vespula, comprised of mostly innocuous species with relatively small, short-lived colonies whose foraging is restricted mainly to live arthropod prey. V. vidua has a limited distribution in eastern North America and its biology is virtually unknown (Akre et al., 1980).

MATERIALS AND METHODS

Two identical observation nest boxes similar to those described by Akre et al. (1976), and containing envelope remnants from vespine colonies which had previously occupied them, were placed at adjacent, east-facing windows inside the apiculture building on the University of Maryland campus. Remaining open throughout the winter, the nest box entrance holes were 2.7 m apart and 1.7 m from the ground. A V. germanica queen initiated her nest in one of the boxes on 19 or 20 April 1980; a V. vidua queen initiated her nest in the other box on 5 May 1980. These fortuitous circumstances thus allowed a unique comparative study of naturally established colonies in the same season and location.

Both nests were observed 1-3 hr daily. Other than occasional envelope aperture enlargement with a flattened wire, the colonies were not disturbed until 4 weeks after the first adult worker eclosion in each nest. At that time, the nest box was sealed and removed from its stand at night, refrigerated, the colony and its

nest analyzed, then transplanted into an adjustable-depth box,
returned to its stand and reopened the same night.

A full account of the demographic results of this study, as
well as details of nest box design and other techniques for the
maintenance of naturally established vespine colonies, will be
described elsewhere.

RESULTS

1. Colony growth

The comparative early growth of the two colonies can be
summarized as an interesting paradox: despite equivalent rates of
eclosion from first generation cells, total production of the V.
germanica colony nevertheless far exceeded that of the V. vidua
colony at the time of analysis.

The V. vidua queen nest period, from initiation to first adult
worker eclosion, lasted 21 days, at which point the foundress had
constructed over 70 cells. The V. germanica queen nest period
lasted 23 or 24 days, with the foundress constructing only 54 cells.
This early lead in production by V. vidua primarily resulted from a
higher rate of cell construction and oviposition in the latter part
of the queen nest period. Throughout the 4 weeks following the end
of this period, worker emergence from first generation cells in both
nests paralleled each other to a remarkable extent, and in fact
totaled an identical 82 individuals for each colony at the time of
analysis. However, the V. germanica nest consisted of 1167 cells
when analyzed, whereas the V. vidua nest had only 531 cells. In
addition, 99 living V. germanica workers were present compared to
only 69 living V. vidua workers, apparently reflecting a difference
in mortality rate as well as a difference in second generation
production (i.e. cell reuse).

Despite the ergonomic superiority of the V. vidua foundress,
the V. germanica colony obviously began operating with much greater
efficiency than did the V. vidua colony soon after eclosion of the
first workers. The following section discusses one of the probable
reasons for this difference.

2. Foraging

Observations of solid food carried by incoming workers of both
nests were made during May and June. Allowing natural colony growth
with a minimum of interference was this study's highest priority,
thereby precluding any systematic sampling of forager loads.
However, since much of the food brought into a vespine nest is either
severed arthropod body parts or has been malaxated beyond
recognition, on-the-spot inspection provides only a limited
perspective of colony diet. Additional identifications were
therefore made of inedible body parts dropped to the glass nest box
floor, which were retrieved with a hooked wire without disturbing the
wasps. Despite the shortcomings of this mainly qualitative survey,
major differences between the two colonies were readily apparent.

Two factors allowed a more precise evaluation of the V. vidua
colony's diet: 1) total influx of prey was much less than in the
V. germanica nest, enabling a more complete tabulation; 2) like

several other species of Vespula s. str., V. vidua has very poor
nest sanitation, thus leaving an extensive record of accumulated
prey remains beneath the combs. During the 5 weeks following first
worker emergence, by far the majority of these remains, as well as
observed loads carried by incoming foragers, consisted of various
membracids. Six species could be identified from retrieved pronota,
listed in decreasing order of abundance: Ophiderma flava Goding, O.
flavicephala Goding, O. salamandra Fairmaire, Smilia camelus (F.),
Telamona monticola (F.), and T. ampelopsidis (Harris). Several
other distinctive pronota, probably representing additional species,
could not be retrieved. In the latter part of June, the elytra of a
cercopid, Aphrophora parallela (Say), also began routinely appearing
on the nest box glass. The following tabulation of prey individuals
represented by remains collected daily from 5-30 June, provides an
approximate measure of the V. vidua colony's restricted prey
selection: Membracidae-58, Cercopidae-11, Diptera-3, Coleoptera-2,
Lepidoptera-1, Orthoptera-1.

Extensive analysis of prey remains was not possible with the V.
germanica colony as this species has excellent nest sanitation, with
most solid and liquid waste quickly removed by outgoing workers.
Recording forager loads was also difficult, due to the volume of
incoming food and general worker activity level, both of which were
much greater than in the V. vidua nest. Although these handicaps
prevented any firm conclusions on prey selection, almost all
identifiable prey were Diptera, with only one membracid observed
during the 7 weeks following first worker emergence. However, a
clear majority (roughly estimated at 75%) of incoming food during
the second half of this period obviously did not consist of prey,
but rather of two other types of collected material.

The first observed non-prey item in the nest was a honey bee
(Apis mellifera L.) thorax on 30 May, 17 days after first worker
emergence. At this point, 26 workers had eclosed, although the
actual nest population was definitely less than this number (but
probably at least 20) due to natural mortality. By the second week
in June, the easily recognized honey bee parts, mainly thoraces and
gasters, had apparently become the colony's primary source of
protein. Although there are a few records of successful yellowjacket
attacks on honey bees (De Jong, 1978), the latter are not normal prey
for these wasps, probably due to the bees' relatively large size in
combination with their activity, thick integument, and capacity for
self defense. However, scavenging of dead bees from the ground in
front of hives by Paravespula workers is well known to beekeepers
(De Jong, 1978, 1979). It is almost certain that the V. germanica
foragers were obtaining their bees from hives near the apiculture
building. These were only ca. 35 m from their nest entrance, all
had abundant bee carcasses in front of them, and in any case, were
the only honey bee colonies within the wasps' flight range.
Nevertheless, presumably due to the small number of foragers
involved, several hours of observation around the 15 hives in early
June failed to detect any yellowjacket activity there (although
their scavenging became conspicuous by the end of the month).

This difficulty in obtaining direct field observations of early
season foraging behavior hindered identification of the other

important non-prey material brought into the young V. germanica
nest. Starting on 4 June and continuing throughout the month,
numerous workers (often at the approximate rate of one every 2
minutes) began returning with large, amorphous lumps of a pale to
bright red substance which was fed directly to the larvae. The
question of whether this was animal or plant matter was apparently
resolved on 13 June, when one of the incoming pieces had a
distinctive stalk attached to it. The only fruit in the study area
with stems and flesh matching the unknown material is that of
several crab apple trees (Malus floribunda Sied.), the nearest of
which was ca. 70 m from the V. germanica nest. Small (7-13 mm
diam.) but abundant, the mature fruit has a very hard, smooth skin
which does not bruise easily during the period when the flesh is
red. As no feeding scars from birds or other insects were apparent,
it is likely the wasps made the initial wound themselves. To
confirm the circumstantial evidence of the mystery substance's plant
origin, an assay (Updegraff, 1969) was performed on one of the lumps
(retrieved from a netted forager) which revealed a 12% cellulose
content.

DISCUSSION

One of the more surprising findings of this study was the
almost complete lack of overlap in foraging spectra of the two
adjacent colonies during the first few weeks of their growth. The
V. vidua specialization in cryptic Homoptera was intriguing in view
of the prevailing conception that "wasps and hornets are great
opportunists - they will collect almost any soft-bodied insect,
spider or other small animal they can find" (Edwards, 1980).
Although prey selection undoubtedly diversifies as any vespine
colony increases in size, relatively narrow preferences may be
typical for young colonies of some species. A similar study of
foraging was conducted in 1981 with a colony of Vespa crabro L., a
non-scavenging hornet which, like V. vidua, has extremely poor nest
sanitation. The first workers eclosed on 4 June, and for the next
6 weeks, the abundant prey remains consisted almost entirely of
honey bees, wasps (yellowjackets, eumenids, and chrysidids), and
aposematic mimics such as syrphid flies and lepturine cerambycid
beetles. The colony's foraging spectrum only expanded substantially
in mid-July (Greene, unpublished data).

No less unexpected was the young V. germanica colony's heavy
exploitation of food sources other than live prey. Although the
belief that vespine brood consume only protein has been previously
disproved (Edwards, 1980), the quantities of carbohydrate fed to the
V. germanica larvae indicate a degree of nutritional flexibility
that is unusual for aculeate wasps. The most significant aspect of
the colony's foraging, however, was its reliance on scavenging and
frugivory so early in the season. It is widely accepted that this
behavior only becomes prevalent in late summer, principally because
of the "increasing nutritional demands of the colonies associated
with queen rearing, and compounded by a decline in natural prey as
the season progresses" (MacDonald et al., 1980; also cf. Thomas,
1960; Edwards, 1980). However, as suggested by this study, broad-

spectrum foraging by young colonies is not necessarily detected even in limited areas by field observation alone. While it is true that proximity to honey bee hives and crab apple trees is not a universal characteristic of young Paravespula colonies, it is probable that many are not far from some source of non-prey food, particularly those near human habitations. (In this regard, a V. maculifrons (Buysson) queen has been observed feeding on cooked meat loaf in late May - A.E. Giraldi, pers. comm.) If this study has indeed involved typical colonies, then it provides some perspective on the almost immediate, striking disparity in their growth rates despite an equivalent initial rate of adult worker recruitment. Since hunting for prey is one of the most arduous, time-consuming worker tasks, a longer adult life span and the ability to maintain greater numbers of brood might both be a consequence of the more efficient foraging behavior of V. germanica.

ACKNOWLEDGEMENTS. - I thank R.H. Racusen for the cellulose assay of the unknown V. germanica forager load, and T.J. Cooke for the determination of its probable source; T.K. Wood and R.F. Denno for determinations of the membracid and cercopid prey of V. vidua; and D. De Jong for technical assistance with the two colonies.

References

Akre R.D., Garnett W.B., MacDonald J.F., Greene A., Landolt P., 1976. - Behavior and colony development of Vespula pensylvanica and V. atropilosa (Hymenoptera: Vespidae). J. Kansas Ent. Soc., 49, 63-84.

Akre R.D., Greene A., MacDonald J.F., Landolt P.J., Davis H.G., 1980. - The Yellowjackets of America North of Mexico. USDA Agric. Handbook 552, 102 p.

De Jong D., 1978. - Insects: Hymenoptera (Ants, Wasps, and Bees)., In: R.A. Morse (Ed.). Honey Bee Pests, Predators, and Diseases. Cornell Univ. Press, 430 p.

De Jong D., 1979. - Social wasps, enemies of honey bees. Amer. Bee J., 119, 505-507,529.

Edwards R., 1980. - Social Wasps. Their Biology and Control. Rentokil Ltd, 398 p.

MacDonald J.F., Akre R.D., Keyel R.E., 1980. - The German yellowjacket (Vespula germanica) problem in the United States (Hymenoptera: Vespidae). Bull. Ent. Soc. Amer., 26, 436-442.

Thomas C.R., 1960. - The European Wasp (Vespula germanica Fab.) in New Zealand. N.Z. Dept. Sci. Indust. Res., Inf. Ser. 27, 74 p.

Updegraff D.M., 1969. - Semimicro determination of cellulose in biological materials. Analyt. Biochem., 32, 420-424.

Ant Manipulation in Agro- and Forest-Ecosystems

Jonathan D. Majer,
Western Australian Institute of Technology

Ants, because they are frequently ecologically dominant, tenders of Homoptera, avid predators, vectors of pathogens, cutters of leaves, or simply domestic nuisances, are often a central consideration in pest control schemes. As far back as the 12th century attempts were made to use the tree nesting ant, **Oecophylla smaragdina**, to limit citrus pests in southern China (Way, 1954). The potential use of ants in biological or integrated pest control schemes has been reviewed by Leston (1973) and Room (1973) for the tropics and by Adlung (1966) for wood ants (**Formica** spp.) in temperate forests. For the record, these reviews have not included the more recent studies of Finnegan (1975 and earlier papers) on the role of **Formica** spp. in Canadian forests, Kim and Murakami (1980 and earlier papers) on **Formica yessensis** in Korean pine forests and of Laine and Niemela (1980) on **Formica aquilonia** in. Finnish mountain birch woodland.

The role of ants in pest control has not always centred around promoting species in order to limit pests. For instance Taylor and Adedoyin (1978 and earlier papers) investigated the role of Nigerian ants involved in cocoa blackpod disease (**Phytophthora palmivora**) spread while Haines and Haines (1979 and earlier papers) looked at the use of baits to control the crazy ant, **Anoplolepis longipes**, in the Seychelles where, for various reasons, it is considered a pest.

The complexity of the interactions which exist in ant communities are often not fully understood. This has necessitated an empirical approach to ecosystem manipulation rather than one with a sound theoretical basis. This paper will review both the mechanisms which influence and maintain the spatial pattern of ant communities and some attempts which have been made to manipulate them. It will then list some of the practical considerations which need to be considered in future ant manipulations. The review will largely confine itself to the tropics; leaf-cutter ants will not be considered.

THE MOSAIC OF DOMINANT ANTS

Surveys of Ghanaian cocoa have revealed almost 300 species of ants. About 14 of these sometimes numerically predominate over other ant species in the area (Leston, 1973). Majer (1972) divided the tropical ant fauna into four status groups; dominants – ants which predominate numerically in an area, normally to th exclusion of all other dominants; co-dominants – dominants which, for various reasons, are able to co-exist; sub-dominants – species which may reach dominant status if a dominant ant is removed; and non-dominants – species with generally low colony size which occur within or between the territories of dominant ants. Under certain circumstances the dominants exhibit a dominance hierarchy based on their ability to replace each other (Greenslade, 1971). This hierarchy exhibits flexibility under changing conditions. Evidence suggests that this ant status system is common to other parts of the tropics (Room, 1975; Leston, 1978; Taylor and Adedoyin, 1978).

It follows from the definition of dominance that such ants are distributed in a three-dimensional mosaic. This has been verified in West Africa by various surveys (Leston, 1973; Majer, 1972, 1976a, 1976b; Taylor, 1977), in tropical Asia (Greenslade, 1971; Room, 1975) and now in the New World neotropics (Leston, 1978). Wilson's (1958) work in New Guinea suggests that there is a geographic patchiness in distribution superimposed over this local distribution pattern.

In order to understand how ants might be manipulated some general points on the maintenance of the mosaic need to be described. Majer's (1976b) findings are drawn upon here although they are not at variance with the findings from elsewhere in the tropics.

The heterogeneity of the environment is one factor contributing to the patchy ant distribution. For instance, different species of ants occur under particular cocoa canopy density regimes. Ants such as **Tetramorium aculeatum**, which are associated with dense cocoa, exhibit niche flexibility by moving into thinner canopy when adjacent dominants are removed. This suggests that the interspecific ant mosaic is maintained by a combination of competition and habitat requirements. Interspecific competition may be for food, for nesting and foraging sites or it may take the form of aggressive competition between adjacent colonies or mature colonies and founding queens.

Aggressive behaviour is particularly intense between interspecific blocks of the mosaic. **T. aculeatum** is able to reduce competition with certain **Crematogaster** spp. within its territory by spacing out its foraging time while **Crematogaster castanea** may co-exist with **Oecophylla longinoda** by adopting a similar colony odour.

New colony establishment is rare in mature cocoa since changes

in the ant mosaic are usually compensated for by lateral spread of
colonies. The role of dispersing queens is probably more important
in developing cocoa farms. O. **longinoda** may aid species segregation
by having queens which select their habitats when dispersing.

Climate and weather influence the structure of the mosaic by
directly influencing the features of the habitat, the availability
of certain types of food or by physically weakening colonies.

It remains to be mentioned that as each dominant has a suite of
non-dominants associated with it (Room, 1975), these are also
distributed in a mosaic fashion.

EXPERIENCES WITH ANT MANIPULATION

Practical attempts at ant manipulation have included the
reduction of pest species (Haines and Haines, 1979), the reduction
of inefficient predators in order to encourage more effective
species (Stapley, 1971), and the direct encouragement of beneficial
species. All approaches can be considered as part of the same
problem since the encouragement of a new species will lead to
displacement of residents and species removal will lead to lateral
spread of adjacent dominants; one approach is simply the reverse of
the other. Some ant manipulation approaches are now reviewed under
separate categories.

1. Introduction of ants from other regions
The introduction of ants into new geographic areas is generally
considered to be too risky (Room, 1973) although Finnegan (1975)
successfully introduced the Italian wood ant, **Formica lugubris** into
Canadian forests. Establishment was generally successful although
competition with native ants and predation by birds interfered with
the programme's success. These problems were ameliorated by
chemically treating native ants in the area and by excluding
predators from nests with wire mesh. Other examples in this
category have involved accidental introductions which turned out to
be beneficial. **Wasmannia auropunctata**, accidentally introduced into
Cameroun from the neotropics, has been used to limit cocoa pests.
Farmers encourage its spread by distributing artificial nests formed
in bundles of raffia leaves (Bruneau de Mire, 1969); it is then able
to outcompete some of the smaller native ants in the mosaic. The
cosmopolitan tramp species, **A. longipes**, has been artificially
introduced in some New Guinea cocoa plantations and proved to be of
benefit (Room, 1973).

2. Introduction of ants from within the region
This category describes the transfer of ants from one mosaic
block to another area which is often held by other dominants. In
Indonesia, nests of **Dolichoderus bituberculatus** are artificially
established in bamboo or bundles of leaves and then hung in cocoa
trees to protect them against mirid and other pest attack (Meer
Mohr, 1927). Maintenance of the nests depends on their being placed

in shady conditions and the provision of a suitable honeydew supply;
without this the ant is outcompeted by adjacent dominants. The
transfer of **O. smaragdina** nests on to citrus is only temporarily
successful due to competition from resident species and the absence
of queens in transferred nests. Brown's (1959) work in coconuts
indicated that even if queens were included, nest transfer only met
with temporary success.

3. Selective chemical treatment of ants
      Here the mosaic blocks of unfavoured species are sprayed or
poisoned in order to encourage adjacent beneficial residents.
Phillips' (1956) trials in Solomon Island coconuts showed that
Pheidole megacephala and **Iridomyrmex cordatus** were reduced by
organochloride application to the lower tree while the beneficial **O.
smaragdina** and **A. longipes** were less affected. The beneficial
species were further encouraged by specifically spraying the areas
nested by the other two species.   Similar observations were made in
Tanzanian coconuts by Vanderplank (1960) where a repeated DDT
application resulted in a drastic reduction of the predominantly
ground nesting **A. longipes** and the spread of tree nesting **O.
longinoda** and **Pheidole punctata.** Many trees remained without
dominants following spraying indicating an inadequate capacity of
ants to rapidly fill lacunae.

      The examples from coconut generally involve the spread of
arboreal nesters, which are untouched by the pesticide, and the
decline of soil or tree base nesters. A similar but less desirable
trend was observed in Ghanaian cocoa when blanket spraying caused
the spread of tall shade tree nesting **Crematogaster** sp. into the
lower cocoa territory originally held by **O. longinoda** and **T.
aculeutum.** Selective spraying of all non-**O. longinoda** trees led to
a localised replacement of **Crematogaster** sp. by more beneficial ants
(Majer, 1978).

4. Selective mechanical removal of ants
      Majer (1976b) ran a field trial in which nests of four dominant
species were selectively mechanically removed from separate,
replicated, 40 x 40 m plots. Ant elimination was invariably
followed by the spread of adjacent residents over distances as great
as 50 m. Species varied in their capacity to spread into new
lacunae and, in one case, large gaps remained without dominants for
at least 16 months. This was attributed to the unsuitability of the
habitat to adjacent residents. Incidence of incipient females
increased in proportion to the area of dominant ant removal although
very few colonies were established in the 16 months following ant
elimination. Therefore the greatest changes in ant distribution
were acounted for by lateral spread of existing colonies.

5. Alteration of ant segregation
      The early Solomon Islands coconut work suggested that **O.
smaragdina** was encouraged by a dense understorey since it was then
able to forage independently of ground living antagonists (Phillips,

1956). Removal of understorey or artificial introduction of
vertical segregation by providing bridges between trees did not
produce the predicted outcomes so other factors may be involved.
Majer (1972) observed a natural experiment in cocoa in which a shade
tree supporting a **Crematogaster depressa** colony fell into O.
longinoda territory on cocoa. A combat immediately resulted and **C.
depressa** was eliminated.

6. Habitat alteration
     The most extensively studied example in this category is in
Solomon Island coconuts following the last war. During the war
dense unergrowth built up which was subsequently removed. Post-war
vegetation changes were probably responsible for the series of
replacements; **P. megacephala** ≯ **O. smaragdina** > **A. longipes** > **O.
smaragdina** and other species. Greenslade (1971) summarised the
reason forthis succession in terms of the complex changes in
vegetation, its floristic composition and structural complexity, the
climatic regime, and net production. The work on Western Australian
bauxite mine rehabilitation (Majer et al., 1981) provides an
interesting comparison. Here the diversity and species composition
of the colonising ant fauna was strongly influenced by plant species
richness and diversity, vegetation cover in particular strata, the
thickness and patchiness of litter and the availability of dead
wood. Rehabilitated areas acquired quite different ant species
depending upon the type of rehabilitation which was performed.

7. Crop interplanting
     A specialised example of habitat alteration is the
interplanting of crops. Way (1954) noted that where coconut was
interplanted with clove or citrus the queen of **Oecophylla** was
invariably situated within the latter species. This was possibly
associated with the favoured Homoptera population on these plants
and suggested a new option for encouraging **Oecophylla**. Stapley
(1971) extended this observation in the Solomon Islands by
interplanting with sowersop. It was then possible to establish O.
smaragdina on the sowersop provided that sufficient Homoptera of
the appropriate species were present. Leston (1973) and Majer
(1974) have suggested using scale insect-rich coconuts in cocoa
plantations in order to encourage the establishment of **O. longinoda**.

PRACTICAL CONSIDERATIONS

     If an ant is a pest then it is obvious which species needs to
be manipulated. Where a beneficial species needs to be selected for
promotion the criteria given by Finnegan (1971) and Room (1973) are
useful; they will not be repeated here.

     This review of ant mosaic dynamics and past manipulation
attempts has revealed a number of practical considerations which,
for brevity, are here summarised in note form. In what follows a
distinction is made between species which are physically
'introduced' and residents which are 'encouraged'. The term

'promoted' collectively refers to both methods.

* Where a pest complex is involved, will the promotion of a
  beneficial ant bring desirable changes in all members of the
  complex?
* Will the promotion of a seemingly beneficial dominant encourage
  undesirable Homoptera?
* Will promotion of a beneficial ant be accompanied by that of a
  less desirable co-dominant?
* Is a particular beneficial species amenable to artificial
  introduction?
* Is the ant which is to be introduced limited by its geographical
  range?
* Is the mosaic as densely packed in other parts of the world as it
  is in Ghanaian cocoa or, are undominated (unprotected) lacunae
  more common elsewhere?
* How permanent is the effect of a pesticide which is used to
  remove undesirable species?
* Is time of spraying non-beneficial ants compatible with
  alate flight times of beneficial species?
* What options other than pesticides are available to remove
  undesirable species?
* Is it possible to capitalise on climatically induced changes in
  ant abundance to promote or reduce particular species?
* How rapidly can adjacent dominants or introduced species fill
  artificially created lacunae?
* Does a promoted ant have sufficient niche flexibility to colonise
  the new area?
* When a lacuna is created in order to encourage a resident will
  the desired adjacent species occupy the area?
* Do the ants which colonise lacunae hold the territory on a first
  come, first to dominate basis?
* Will the promotion of a species fail due to its position in the
  dominance hierarchy in relation to that of the adjacent
  dominants?
* Does the size of an ants colony limit its capacity for range
  extension?
* Can an area of crop or forest support large monospecific expanses
  of one dominant or is the small-block mosaic necessary for
  adequate food provision?
* How does intrinsic growth and decline of ant colonies fit in
  with the desired ant distribution and abundance pattern?
* A mosaic of beneficial ants may represent a dysclimax - how can
  it be maintained at this stage?
* Will incipient females preselect areas known to be suitable nest
  sites and hence reverse attempts to change ant distribution?
* Is it necessary to supplement food in areas where a new species
  has been promoted, especially if they have recently been sprayed?
* Is the climate suitable for maintaining a promoted species
  throughout the year?
* What other aftercare needs to be performed when the mosaic has
  been manipulated?

* If the crop habitat is modified to encourage an ant, is the habitat compatible with the farming technique?
* If the habitat alteration approach is used to encourage species, how predictable is the change in dominants present?
* Is the promotion of particular ants more likely to be successful if new plantations are cultivated in order to encourage initial colonisation of beneficial species?
* When interplanting is employed, how economically compatible is the inter-crop?

These considerations will be discussed during the presented paper and then guidelines for manipulating ants will be proposed.

REFERENCES

ADLUNG, K.G., 1966. - A critical evaluation of the European research on the use of red wood ants (Formica rufa Group) for the protection of forests against harmful insects. Z. angew. Ent., 57, 167-189.

BROWN, E.S., 1959. - Immature nutfall in coconuts in the Solomon Islands. I. - Distribution of nutfall in relation to that of Amblypelta and of certain species of ants. Bull. ent. Res., 50, 97-133.

BRUNEAU DE MIRE, P., 1969. - Une fourmi utilisée au Cameroun dans la lutte contr des mirides du cacaoyer Wasmannia auropunctata Roger, Cafe, Cacao, Thé, 13, 209-212.

FINNEGAN, R.J., 1971. - An appraisal of indigenous ants as limiting agents of forest pests in Quebec. Can. Ent., 103, 1489-1493.

FINNEGAN, R.J., 1975. - Introduction of a predacious red wood ant, Formica lugubris (Hymenoptera: Formicidae), from Italy to Eastern Canada. Can. Ent., 107, 1271-1274.

GREENSLADE, P.J.M., 1971. - Interspecific competition and frequency changes among ants in Solomon Island coconut plantations. J. appl. Ecol., 8, 323-352.

HAINES, I.H., HAINES, J.B., 1979. - Toxic bait for the control of Anoplolepis longipes (Jerdon)(Hymenoptera: Formicidae) in the Seychelles. III. Selection of toxicants. Bull. ent. Res., 69, 203-211.

KIM, C.H., MURAKAMI, Y., 1980. - Ecological studies on Formica yessensis Forel, with special reference to its effectiveness as a biological control agent of the pine caterpillar moth in Korea. J. Fac. Agr. Kyushu Univ., 25, 119-133.

LAINE, K.J., NIEMELA, P., 1980. - The influence of ants on the survival of mountain birches during an Oporinia autumnata (Lep., Geometridae) outbreak. Oecologia (Berl.), 47, 39-42.

LESTON, D., 1973. - The ant mosaic - tropical tree crops and the limiting of pests and diseases. PANS, 19, 311-341.

LESTON, D., 1978. - A neotropical ant mosaic. Ann. ent. Soc. Am., 71, 649-653.

MAJER, J.D., 1972. - The ant mosaic in Ghana cocoa farms. Bull. ent. Res., 62, 151-160.

MAJER, J.D., 1974. - The use of ants in an integrated control scheme for cocoa. Proc. Fourth Conf. of West African Cocoa Entomologists, pp. 181-190.

MAJER, J.D., 1976a. - The ant mosaic in Ghana cocoa farms: Further structural considerations. J. appl. Ecol., 13, 145-155.

MAJER, J.D., 1976b. - The maintenance of the ant mosaic in Ghana cocoa farms. J. appl. Ecol., 13, 123-144.

MAJER, J.D., 1978. - The influence of blanket and selective spraying on ant distribution in a West African cocoa farm. Rev. Theobroma (Brasil), 8, 87-93.

MAJER, J.D., DAY, J.E., KABAY, E.D., PERRIMAN, W.S., 1981. - Recolonisation by ants in bauxite mines rehabilitated by a number of difference methods. J. appl. Ecol., in press.

MEER MOHR, J.C., 1927. - Au sujet du role de certaines fourmis dans les plantations coloniales. Bull. agric. Congo Belge, 31, 97-106.

PHILLIPS, J.S., 1956. - Immature nutfall of coconuts in the British Solomon Islands Protectorate. Bull. ent. Res., 47, 575-595.

ROOM, P.M., 1973. - Control by ants of pest situations in tropical tree crops: a strategy for research and development. Papua New Guin. agric. J., 24, 98-103.

ROOM, P.M., 1975. - Relative distributions of ant species in cocoa plantaions in Papua New Guinea. J.appl.Ecol., 12, 47-61.

STAPLEY, J.H., 1971. - Field studies on the ant complex in relation to premature nutfall of coconuts in the Solomon Islands. Proc. Conf. on Cocoa and Coconuts in Malaysia, pp. 345-354.

TAYLOR, B., 1977. - The ant mosaic on cocoa and other tree crops in Western Nigeria. Ecol. Ent., 2, 245-255.

TAYLOR, B., ADEDOYIN, S.F., 1978. - The abundance and interspecific relatons of common ant species (Hymenoptera: Formicidae) on cocoa farms in western Nigeria. Bull. ent. Res., 68, 105-121.

VANDERPLANK, F.L., 1960. - The bionomics and ecology of the red tree ant, Oecophylla sp., and its relationship to the coconut bug Pseudotheraptus wayi Brown (Coreidae). J. Anim. Ecol., 29, 15-33.

WAY, M.J., 1954. - Studies on the life history and ecology of the ant Oecophylla longinoda Latreille. Bull. ent. Res., 45, 93-112.

WILSON, E.O., 1958. - Patchy distribution of ant species in New Guinea rain forests. Psyche, 65, 26-38.

Behavioral Ecology of the Army Ant
Neivamyrmex nigrescens in
a Desert-Grassland Habitat

Howard Topoff, Hunter College of C.U.N.Y.

Colonies of the nearctic army ant Neivamyrmex nigrescens (subfamily Ecitoninae) exhibit a behavioral cycle composed of two
distinct, regularly alternating phases. In the nomadic phase,
colonies change nesting sites frequently and forage intensively for
food (Rettenmeyer, 1963; Schneirla, 1958, 1961, 1963). During the
statary phase, by contrast, the colony may not forage for several
days and typically remains in the same subterranean bivouac. According to Schneirla's theory of brood stimulation, the colony's shift
from one behavioral phase to another is regulated by the developmental condition of the synchronized brood. Thus the nomadic phase
is triggered by stimulation of eclosing pupae (Topoff et al., 1980a),
and maintained by comparable stimulation from developing larvae. The
statary phase results from the decrease in brood stimulation as the
larvae pupate, and colony activity remains low until the pupae
eclose.

NOMADIC BEHAVIOR IN A DESERT-GRASSLAND HABITAT

It is well known that Schneirla developed brood-stimulation
theory as a result of his studies in Central America on the epigaeic
species Eciton hamatum and E. burchelli. In extending his results
to nearctic populations of N. nigrescens in southeastern Arizona,
Schneirla's base of operations was the Southwestern Research
Station, 8 km west of Portal. At an altitude of 1,600 m, the
Station is in an oak-juniper habitat, dominated by Arizona oak,
Emory oak, alligator juniper, and Chihuahua pine. In this habitat,
the modal length of the nomadic phase is 18 days (range = 18-21
days), and the modal length of the statary phase is 19 days (range =
18-21 days) (Schneirla, 1958).
During the past several years, we studied N. nigrescens in a
desert-grassland habitat at an elevation of 1,250 m. This community
receives approximately 287 mm of rain annually, and has mixed vege-
tational cover (Lowe, 1964). For most of the year, shrubs such as
mesquite, yucca, snakeweed, and mormon tea dominate the landscape.
Following the heavy rains in July, however, grasses and a wide
variety of annual flowering plants come to predominate in this
relatively ungrazed portion of valley bottomland. The modal length

of the nomadic phase in this desert-grassland habitat is only 13
days (range = 11-14 days), and that of the statary phase is 16 days
(range = 15-19 days) (Mirenda and Topoff, 1980). As shown in Fig.
1, these differences in phase length correlate with the fact that
larvae from desert colonies grow faster than those from colonies at
higher elevations. This difference in growth rate, in turn, might
be due to temperature differences in the two habitats. At 1,600 m,
the maximum and minimum temperatures average 29° C and 13° C during
the summer months. In the desert-grassland study site, maximum and
minimum temperatures for the same time period average 33° C and 18°
C, respectively. The correlation between increased larval growth
rates and shortened nomadic phase length is clearly consistent with
Schneirla's theory of brood stimulation.

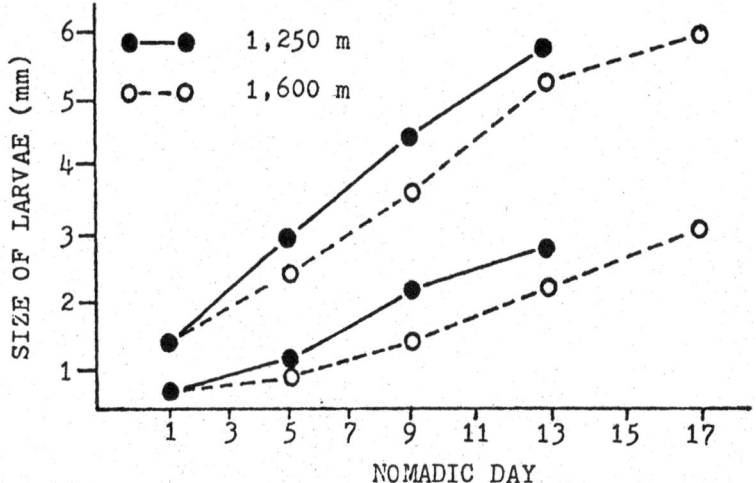

Fig. 1. Comparison of larval growth during the nomadic phase of
N. nigrescens in two habitats: desert-grassland (1,250 m);
oak-juniper (1,600 m). Upper and lower curves show growth
of largest and smallest larvae, respectively. Figure
modified from Mirenda and Topoff (1980).

THE IMPACT OF ARMY ANTS ON THE DESERT-GRASSLAND ECOSYSTEM

1. The use of mass chemical recruitment for predation.

Studies of communication in ants have recently focused on
patterns of recruitment that contribute to ecologically adaptive
strategies of foraging (Hölldobler, 1978). In the mass recruitment
system utilized by army ants, a chemical trail serves both to arouse
and orient nestmates. Additional arousal stems from tactile interactions among recruiters and between recruiters and foragers
(Chadab and Rettenmeyer, 1975; Topoff et al., 1980b). This derived
pattern of recruitment behavior is utilized by N. nigrescens to prey
upon adults and (especially) brood of numerous ant species (Table 1).
Some of these prey species, such as Pheidole desertorum and

Novomessor cockerelli, respond with "protean" behavior (Humphries and Driver, 1970), in which adults pick up their brood and abandon their nest. The success of army ant predation depends, therefore, upon the ants' ability to have a critical striking force available for recruitment to all potential prey sites. N. nigrescens accomplishes this, first by utilizing mass chemical recruitment, a communicatory process clearly "designed" to enable a few individuals to initiate a chain reaction that can quickly arouse thousands of nestmates to trail following.

Equally significant is the fact that mass recruitment can occur before booty is located. The first surge of army ants out of the bivouac occurs at dusk. At this time the principal source of arousal is the presence of new foraging ground. For N. nigrescens, this phase of recruitment may play an important role in generating the multi-branched trail system that ensures the availability of a large striking force within easy reach of any booty site that is eventually located. When a suitable prey nest is found, the mass recruitment to this food source clearly takes precedence over the recruitment to new foraging grounds. In the field, this results in ants in all outlying columns "draining" into the column that leads to the site being raided.

2. Prey selection and foraging strategy.

The most abundant prey items taken by N. nigrescens are immature stages of other ant species. There is also considerable evidence that N. nigrescens selects certain species of ants preferentially. Colonies of all species of Pheidole, for example, account for over 50% of the nests raided by army ants (Table 1). Although Pheidole is indeed the most abundant genus in the desert (Davidson, 1977), it is still preyed upon by Neivamyrmex more than twice as frequently as would be expected if it were raided in proportion to its density (Mirenda et al., 1980).

The foraging strategy of N. nigrescens must be analyzed, not only by considerations of prey abundance, but by an understanding of the relationship between raiding and emigrations. Although these behaviors can be uncoupled (such as during bivouac disturbance), a colony typically emigrates into an area of successful raiding. During the nomadic phase, workers in exploratory columns are responsive to both prey and nest sites, and will readily recruit to either (Topoff and Mirenda, 1980b). When two or more raid columns are active, the colony usually emigrates over the most successful route. Nevertheless, there are instances when the emigrate-to-food strategy is not evident. This was manifest when colonies sent out only one or two raid columns that made a narrow sweep of an area before emigrating (Mirenda et al., in press). In these instances the emigrations occurred so early in the activity period that the ants were hardly sampling the availability of food before the onset of the emigration. In addition, the movements of colonies during the nomadic phase are often irregular in direction and distance. Thus some colonies stayed within a few meters of one another for several nomadic days; other colonies would emigrate into an area that was just vacated by a different colony. Finally, a series of emigrations by a colony would occasionally bring it right back to its original area. Overall, it seems that colonies do not always change their

Table 1.   Prey of two colonies of N. nigrescens during summer,
1977 (Modified after Mirenda et al., 1980).

Prey species	Colony 1		Colony 2	
	Nests raided	% of total	Nests raided	% of total
Pheidole desertorum	18	31.0	26	18.4
Pheidole sciophila	5	8.6	16	11.3
Pheidole rugulosa	0	0.0	20	14.2
Pheidole hyatti	5	8.6	13	9.2
Pheidole sitarches	0	0.0	2	1.4
Unidentified Pheidole spp.	2	3.4	11	7.8
Leptothorax sp.	4	6.9	1	0.7
Conomyrma insana	4	6.9	2	1.4
Paratrechina melanderi	2	3.4	0	0.0
Aphaenogaster huachucana	1	1.7	1	0.7
Pogonomyrmex californicus	1	1.7	1	0.7
Myrmecocystus mimicus	0	0.0	1	0.7
Camponotus festinatus	1	1.7	0	0.0
Unidentified ants	6	10.3	18	12.8
Gnathamitermes sp.	9	15.5	24	17.9

foraging area as completely as optimal foraging theory would predict.
Factors which may account for the discrepancy include high prey
density on our study sites, small colony size, or a shortage of nest
sites on the pavement-like substrate of the desert.

REFERENCES

Chadab R., Rettenmeyer C., 1975. -- Mass recruitment by army ants.
    Science, 188, 1124-1125.
Davidson D., 1977. -- Species diversity and community organization
    in desert seed-eating ants. Ecology, 58, 711-725.
Hölldobler B., 1978. -- Ethological aspects of chemical communication
    in ants. Adv. Study. Behav., 8, 75-115.
Humphries D., Driver P., 1970. -- Protean defence by prey animals.
    Oecol., 5, 285-302.
Lowe C., 1964. -- The Vertebrates of Arizona. Univ. Ariz. Press,
    Tucson (Ariz.).
Mirenda J., Topoff H., 1980. -- Nomadic behavior of army ants in a
    desert-grassland habitat. Behav. Ecol. Sociobiol., 7, 129-135.
Mirenda J., Eakins D., Gravelle K., Topoff H., 1980. -- Predatory
    behavior and prey selection by army ants in a desert-grassland
    habitat. Behav. Ecol. Sociobiol., 7, 119-127.
Mirenda J., Eakins D., Topoff H. (in press). -- Relationship between
    raiding and emigrations in the army ant Neivamyrmex nigrescens.
    Insectes Sociaux.
Rettenmeyer C., 1963. -- Behavioral studies of army ants. Univ.
    Kansas Sci. Bull., 44, 281-465.
Schneirla T., 1958. -- The behavior and biology of nearctic army
    ants. Insectes Sociaux, 5, 215-255.
Schneirla T., 1961. -- Behavior and biology of nearctic doryline
    ants. Z. Tierpsychology, 18, 1-32.

Schneirla T., 1963. -- Springtime resurgence of cyclic function in army ants. Anim. Behav., 11, 583-595.

Topoff H. Mirenda J., Droual R., Herrick S., 1980a. -- Onset of the nomadic phase in the army ant Neivamyrmex nigrescens. Distinguishing between callow and larval excitation by brood substitution. Insectes Sociaux, 27, 175-179.

Topoff H., Mirenda J., Droual R., Herrick S., 1980b. -- Behavioural ecology of mass recruitment in the army ant Neivamyrmex nigrescens. Anim. Behav., 28, 779-789.

Abstracts

Nest Ecology in Polycalic Colonies
of Formica polyctena Foerst.

Robert Ceusters
Katholieke Universiteit Leuven

Four polycalic colonies of the polygynous ant species Formica polyctena Foerst. were studied during periods of 15 to 38 years. Individual nests in the colonies number 20 up to 130. New nests are founded by fission or removal of the whole nest to a suitable new site. Colony density ranges from 1.5 to 6.3 nests per hectare according to the characteristics of the habitat. Nearest neighbour statistics reveal a contagious distribution pattern of nests during rapid colony growth but after a stabilization period the occupation of the habitat evolves to regularity. The exponential survivorship curves of the nests show exponents of -0.15 to -0.30. Potential longevity of nests on the same site adds up to more than 20 years; an individual nest age of 28 years was observed. Annual turnover of nests ranges from 20% to 30%. At this moment in all four colonies average annual mortality exceeds natality; some reasons may be found in climatic fluctuations and changing forest exploitation. Although the number of nests in the colonies decreases the total overground nestvolume may be stable. Size and form of the overground nest are correlated with the vegetative cover and the associated degree of insolation. Fish eye photography reveals that nests with sparse insolation grow high and large (up to 1.5 m and 5 m^3) while nests with a high degree of insolation remain low.

Altitudinal Distribution of Japanese Serviformica

Masaki Kondoh
Shiraume Gakuen College

The Japanese Formica (Serviformica) ants have been recognized as five species, Formica gagatoides, Formica lemani, Formica japonica, Formica sp. (Hayashi-kuroyama-ari), and Formica transkaucasica, which are only black in colour. The first three distribute under typical altitudinal segregation in Japanese mountain

region, which is corresponding with horizontal distribution from
Kyushu to Hokkaido. *Formica gagatoides* distributes above 2700m,
associates with *Vaccinio-Pinetum pumilae* association and makes the
nest in open gravel ground at the edge of the vegetation.
F. lemani distributes 1450-3000m, associates with *Cirsio-Campan-
ulatum hondoensis*, *Carexo-Stellarietum nipponicae* associations
and wayside of *Abietum veitchii* association etc., and makes the
nest in open ground at the edge of the vegetations and its rhizo-
spare. *Formica japonica* distributes from seashore to 1500m, associates with many types of grass lands especially sparse vegetation.
In Mt. Fuji, good segregation has been observed between *F. lemani*
and *F. japonica*. Economic differences of these species are
analysed. *Formica sp.* have nearly the same distribution as *F.
japonica*, but the social behaviour and nesting behaviour are
different, which is found in light wood.

The last *F. transkaucasica* spread their distribution from lowland
to mountainous field especially wet grass land.

Termites among Soil Animals
in Savanna and Forest Ecosystems of West Africa (Ivory Coast)

Michel G. Lepage
Laboratoire de zoologie, Ecole Normale Supérieure, Paris, France

Annelida and Macroarthropoda were studied in a joint program
in 12 forest and savanna soil sub-ecosystems of the Guinea and
Sudan vegetation zones in Ivory Coast. Vertical and seasonal populations figures and biomass of the feeding categories were obtained by comparing two sampling methods: cylindral cores (86 cm^2)
and subsequent flotation and hand-sorting from 1/2 m^2 pits. Termites, one of the main component of the fauna (together with Earthworms), showed important vertical populations movements and food
reserves fluctuations between dry and rainy seasons. Termite biomass within the upper soil layer (0-20 cm) was correlated to the
available food through the biotopes. Opposite feeding strategies
were illustrated in the two groups, fungus-growing and humivorous
species, as the first overcame the second in changing environments.
Meaningful comparison between Termites and Earthworms, as the dry
season expanded from south to north of the gradient, showed large
Termites soil-feeders to increase their numbers while the large
deep-dwelling Earthworms disappeared, expressing differences in
ecosystems dynamics.

Energy Flow in Harvester Ant Populations

William P. MacKay and Emma E. MacKay
Colegio de Graduados, Escuela Superior de Agricultura
Ciudad Juarez, Mexico

This investigation compares energy flow in pupulations of three
species of Pogonomyrmex harvester ants. All three species (P.
rugosus Emery, P. subnitidus Emery, and P. montanus MacKay) demonstrate similar seasonal patterns, with higher foraging activity

and respiratory costs in mid summer and little or no activity
during the winter. A higher percentage of production is invested
in workers than in reproductives. Each species partitions similar
proportions of energy between the production of males and females,
although greater numbers of males are produced. Respiration
accounts for most of the energy assimilated by the nests. The investigation suggests that production may not be correlated with
food input. Worker care provided to the brood may be one of the
most important determinants of brood production. As a consequence,
food input may not be a direct limiting resource in harvester ants.

Ants - useful bio-indicators of the minesite rehabilitation, land-use and land conservation status

J. D. Majer
Western Australian Institute of Technology

In terms of their numerical abundance, size and species richness ants are a prominent taxonomic group in many terrestrial ecosystems. This, and the fact that ants occupy higher trophic levels
and often specialized niches, suggests that they may be good bio-
indicators of various environmental parameters.

This paper develops the rationale for using ants as bio-
indicators and reviews examples of their use. Biological surveys
and minesite rehabilitation studies, which have included ants, are
described and the effectiveness with which ants may be used to
describe certain facets of the environment are investigated.

The influence of _Lasius_ species on some ecosystems in Denmark

Mogens Gissel Nielsen
Institute of Zoology and Zoophysiology
University of Aarhus, Denmark

Lasius _alienus_ (Först) is often the most important animal in
dry sandy heath areas in Denmark. Densities of more than 5000
worker ants per m^2 are not unusual, and a very high proportion of
the energy flow goes through this species. The diffuse gallery system of the nest improves the soil aeration, and the digging activity
of the ants is the main factor of the turnover of soil.

In many biotopes _Lasius_ _flavus_ (F.) has a visible influence
on the ecosystem caused by the high number of the characteristic
nests, which can cover more than 18 percentage of the locality. In
such biotopes the density of _L._ _flavus_ can exceed more than 10000
worker ants per m^2. Since the environment of the nests is quite
different from that of the surroundings, the nest dome is often covered with specific plant communities. Further, a specific micro-
and macrofauna is associated with the nests.

It is unusual that an extreme high density of one animal
species, like _L._ _flavus_, gives rise to increasing diversity of
plants and animals.

Some Tropical Termites and Mineral Cycling

M. Vikram Reddy
North Eastern Hill University, India

Termites are one of the most important groups of soil animals which consume a major quantity of plant litter in tropical ecosystems and in turn play a significant role in mineral cycling. Different termite originated structures collected from various habitats were analyses for mineral elements in relation to the underlying soils and their dwelling-in wooden substrates in order to ascertain the nutrient status of such materials. Analysis of one of such structures i.e., the carton of the wood-destroying Coptotermes kishori, consisting of frass and fragments of comminuted wood and organic excreta mixed with mineral soils revealed that is was slightly acidic in nature with low % of organic carbon possessing less P_2O_5 content and significantly high K_2O content. The % of carbon and P_2O_5 contents were less whereas the pH and K_2O contents were high in carton materials compared to the underlying soils. Nest and associated structures of Microtermes obesi, Odontotermes distans and another termite were also analysed in relation to their wooden habitats and the underlying soils. The mineral elements of these structures were correlated.

Contributions of Social Insects to Desert Ecosystem Processes

Walter G. Whitford
New Mexico State University

Ants and termites are the most conspicuous and potentially the most important arthropods in desert ecosystems. Termites play a major role in the cycling of nutrients especially nitrogen and interract with ants (predators) in this process. Termites account for a major part of the turnover of dead grasses and wood. Gallery carton left by termites and their subsoil structures greatly affect percolation and run-off. Harvester ants affect seed reserves especially of preferred seed species and compete with rodents and birds for resources. Excavation activity by ants also affects geomorphic processes and soil formation in arid ecosystems. Termites provide an important source of protein to the many species of ants that prey on termites. Some termite species account for major turnover in root biomass of grasses and ephemeral plants and termites contribute significantly to the rapid decomposition of perennial plant leaves and grasses as well as being extremely important in dung decomposition.

ECONOMICALLY IMPORTANT SOCIAL INSECTS
Organized by S. Bradleigh Vinson

Introduction

S. Bradleigh Vinson,
Texas A & M University

The following series of papers were brought together to emphasize the economic importance of social insects. Because a large number of workers can concentrate on a small area, damage is often locally heavy and can be in conflict with the interests of man. Many social Hymenoptera are capable of stinging and are protective of their colonies, causing direct attacks on men that interfere with their colonies or food gathering activities. These stings are sometimes just a nuisance but at other times they are life threatening. To determine the economic importance of social insects is complicated by the fact that they are usually beneficial as well as harmful. As Akre points out, the yellow jackets are one of the most beneficial of the social insects although this aspect of their biology is nearly undocumented. While Lofgren and Adams, in their discussion of the imported fire ant, did not discuss the beneficial aspects of this insect, the imported fire ant is known to reduce tick populations and control the sugarcane borer and certain cotton pests (Negm and Hensley 1969; Sterling, 1978; Harris and Burns, 1972). Gillaspy suggests that the beneficial aspects of Polistes wasps outweigh their negative impact. Even the Africanized bee, which has generated many scare stories in the popular press, is not a serious problem if properly managed. As pointed out by Gonçalves, the Africanized bee is a much better producer of honey and other bee products than the bee it is replacing.

The termites, discussed by Mauldin and by Sen Sarma, are an important and necessary part of the ecosystem in that they help return minerals to the soil and serve as an important food source for many species of wild life. The negative impact of termites is largely due to their destruction of cellulose products used by man. However, as both Mauldin and Sen Sarma point out, it is difficult to put a dollar value on the economic damage they inflict. Not only does the cost of termite control and the repair of termite damage have to be considered, but the value of irreplaceable artifacts must be taken into account.

Stings may require medical attention and this expensive is difficult to document. People who are very sensitive to insect venom may become ill or die.

The threat of being stung repeatedly may result in restricting the activities of these people, as noted by Akre, Lofgren and Adams, Gillaspy, and Gonçalves. Thus the economic impact of stinging insects is very hard to determine.

Many social insects damage crops (Akre, Cherrett, Lofgren and Adams, Mauldin and Sen-Sarma). Such damage can be extensive as documented by Cherret in his discussion on the leaf cutter ants. While agricultural damage is easier to assess Lofgren and Adams point out that there has been little attempt to do so until recently. Pharaoh's ant discussion here by Edwards, poses a different problem in that they are found world wide in human habitats and are suspected of being carriers of human pathogenic bacteria.

A social insect colony represents a super organism with the queen at the center, often protected from external perturbation. Thus, the control of social insects is much more difficult than the control of many other pests.

The biological control of social insects has not been very successful. This again appears to be due to protection afforded the queen by the presence of workers. While pathogens of social insects are known, they often result only in the weakening of a colony by killing a few workers which are soon replaced. There are of course, exceptions as seen with the domesticated honey bee. The persistent contact insecticides, widely available after World War II were effective, killing off workers over an extended period of time. However, such persistence lead to environmental problems and the persistent toxicants have generally been removed from the market place. Thus, the method of choice for social insect control has been baits since they are carried by workers into the colony. However, even the use of baits is not a simple matter. If the toxicant is too effective the workers may be killed before the control agent is carried into a colony and distributed to other workers and the queen. Since it is important to kill the queen, the control agent must either be slow acting, not effective on workers, or the workers must be induced to carry the toxicant into a colony without becoming seriously contaminated themselves. Also, the baits must be attractive to the social insects in question. Some of these problems are pointed out by Edwards in his discussion of Pharaoh's ant.

References

Harris, W. G. and E. C. Burns, 1972. - Predation on the lone star tick by the Imported fire ant. Environ. Entomol. 1:362-365.
Negm, A. A. and S. D. Hensley, 1969. - Evaluation of certain biological control agents of the sugarcane borer in Louisiana. J. Econ. Entomol. 62: 1008-1013.
Sterling, W. L, 1978. - Fortuitous biological suppression of the boll weevil by the imported fire ant. Environ. Entomol. 7: 564-565.

Economics and Control of Yellowjackets (*Vespula, Dolichovespula*)

Roger D. Akre,
Washington State University

Yellowjackets are primarily north temperate species occurring in Asia, Europe, northern Africa, and North America. Over the northern areas of their range they are undoubtedly the dominant social wasps in numbers of colonies and individuals (Akre 1982).

Most pestiferous species belong to the Vespula vulgaris (L.) species group which typically have large colonies and a long colony cycle (Akre et al. 1981). The pest status and economic impact of yellowjackets have been covered by Akre and Davis (1978), Akre et al. (1981), Davis (1978), and Edwards (1980). This paper will concentrate mostly on economics of yellowjackets of the United States, particularly the Pacific Northwest, and on previously unpublished data.

ECONOMICS

1. Beneficial

Yellowjackets are probably one of the most beneficial social insects, second only to the ants, as natural biological control agents of forest and crop pests (Akre 1982). However, this aspect of their biology is nearly completely undocumented, and only recently has the Forest Service recognized their possible beneficial role as predators (Roush and Akre 1978a, b). Another unappreciated facet of their ecological role is as scavengers of both invertebrate (Seastadt et al. 1981) and vertebrate carrion. Their role as scavengers is probably just as important in urban and suburban areas as it is in rural areas or in the forest.

2. Agricultural Crops

While yellowjackets can be beneficial predators in fruit orchards, they can also be pestiferous by feeding on the fruit and by disrupting harvesting operations by stinging the laborers (Akre et al. 1981). This is especially true in peach orchards, although it also occurs to a lesser extent in pear, apple, and plums. Vespula pensylvanica (Saussure) is primarily responsible for these problems in the PNW, and some problems, especially during pruning operations, are caused by colonies of Dolichovespula arenaria (F.) and D.

maculata (L.) nesting in the trees. Yellowjackets are also a problem in grape vineyards and can be responsible for nearly total devastation of the crop.

A continuing problem to homeowners and commercial field workers are yellowjackets feeding on berries such as strawberries and raspberries. Stings are common, and some people refuse to pick these berries during abundant yellowjacket years. This problem is greatly magnified at commercial plants where flats await processing. For example, during September 1978 workers of Vespula germanica (F.) formed writhing balls on strawberries awaiting processing at Smuckers Jam Company in Ohio. A similar problem occurs at fruit processing plants such as the Del Monte Plant in Toppenish, Washington where sweet juices are available.

Although yellowjackets have been present in Hawaii since 1919, it is only since 1978 that V. pensylvanica has become a serious problem. The population explosion of V. pensylvanica has caused the Departments of Health, Agriculture, and Land and Natural Resources to combine efforts with the Hawaii Sugar Planters Association to combat the problem. One especially important problem is that workers in sugarcane fields are being stung while undertaking routine operations.

3. Forest Products

Yellowjackets are also a stinging problem for loggers, sawmill operators, and Forest Service personnel. Indeed, during a 5 year period, insect stings accounted for over 5% of all Forest Service medical treatment and lost time accidents. This is especially true during drought years when forest fires are common and problems of forest fire fighters precipitated writing a special manual on procedures and control (Putnam 1977). The USDA, Office of Safety and Health Management, tabulated information on lost time accidents due to insects and ticks in the 12 agencies under their direction for 1974 through the first two quarters of 1979. Results showed 92% (1,960 of 2,134) of accidents occurred in the Forest Service. Most were probably due to stinging Hymenoptera, particularly yellowjackets.

Yellowjackets are attracted to Christmas tree farms probably because of honeydew produced by aphids. Also, species of Dolichovespula and Vespula nest in or under the young trees. Information collected in Washington state showed Christmas tree farming has the highest accident rate of any type of farming (Fanning 1980). Most accidents are due to shearing knife accidents, while an average 16% (1978-1980) are directly attributed to yellowjackets (Antonelli 1980, Fanning et al. 1981). However, many of the knife injuries also result when personnel are harassed by yellowjackets. A dramatic drop in accidents during the third quarter of 1980 as compared to previous years was attributed to low yellowjacket populations.

4. Recreational

With the current emphasis placed on recreational activities, more people are experiencing stinging episodes from yellowjackets. Financial loss and human misery is considerable (Akre et al. 1981). This was clearly illustrated during 1979, a year of great yellowjacket abundance in the PNW, when parks and recreation areas were nearly

deserted (Akre and Reed 1979). During these years of high yellowjacket populations, mosquito abatement district personnel turn their
attention to yellowjacket control, as do employees of the Forest and
Park Services. Recently, the Park Service has been releasing con-
tracts for control of yellowjackets in their recreation areas. All
these problems and attempts at control must result in considerable
financial expense.

5. Urban
     Yellowjackets are a principal concern to homeowners in urban,
suburban, and rural areas. Indeed, from 1976-1979 calls for information about wasps exceed 1,000 per year, and were ranked number one
of the twenty most frequent calls to the Minnesota Information Service for 3 of the 4 years (Ascerno 1981).
     Information on problems caused by yellowjacket nests in or near
human structures is unavailable. However, the magnitude of the
problem is probably reflected by calls to Pest Control Operators
(PCOs) for help in killing colonies. Records of a company in the
Seattle Tacoma area showed 153 colonies were killed during 1980.
Large cities usually have 30 or more pest control companies, and
assuming that they all control a similar number of colonies, home
owner expenses incurred in controlling yellowjackets can be estimated.
Nearly all PCOs charge a minimum $60 for controlling a colony.
Assuming an average of 60 colonies per year, each company makes
$3,600-4,000 per year on yellowjackets. Thus in large cities, at
least in the PNW, homeowners are paying over $100,000 per year for
yellowjacket control.
     Vespula germanica (Fab.) is of special concern to homeowners
and business since nearly all colonies in the United States are
located inside structures (MacDonald et al. 1980). These colonies
sometimes chew through ceilings or walls in their continuing efforts
to expand the nest. This, of course, releases many workers into the
main rooms of the house. V.·germanica is rapidly expanding its
distribution westward across Canada and the United States.

6. Medical
     Yellowjackets are responsible for a number of human and animal
medical problems ranging from fear to allergic responses, and in
extreme cases, death. They have been declared the number one public
health problem in the Southeast, and number 2 in the PNW, second only
to biting flies (Akre 1978). It is probably not as well known that
yellowjackets can cause terror in animals. For example, 15 horses
in El Cajon Valley, California, developed abscesses from V. pensylvanica stings. The yellowjackets then foraged for flesh from this
area, and drove the horses crazy (sic) (Hawthorne 1974).
     The extent of human misery and economic loss from medical
problems resulting from stings is unknown. However, the latest
estimate is that one million Americans per year suffer potentially
serious reactions to an insect sting, ca. 50 die (Valentine 1981),
perhaps more (Akre et al. 1981). Wicher et al. (1980) stated that
Vespula are responsible for most stings. However, many stings are
also due to Dolichovespula arenaria and D. maculata, with most other
stinging problems probably due to V. maculifrons, V. pensylvanica,

V. squamosa, and V. vulgaris.

An increasing amount of research is being directed toward
investigating components and medical effects of vespine and other
insect venoms (Edery et al. 1978, Schmidt 1982). Additionally, many
severely allergic people are undergoing desensitization. Most
doctors are using whole body extracts for this purpose even though
pure venom is superior. The market for venoms and extracts is highly
competitive, but currently Vespa labs is probably supplying most
venom preparations to Hollister Stier and Pharmacia. Volume of sales
is considered an industrial secret, but this is unquestionably a
huge market, probably responsible for several millions of dollars
from yellowjackets alone. Hollister Stier is also financing research
into economic methods of collecting pure venom from yellowjackets
and paper wasps. Recently a new market was opened as Sigma Chemical
Company of St. Louis, Missouri offered yellowjacket venoms for
research purposes at $25-40 per mg.

CONTROL

Control of yellowjacket colonies includes nest destruction, the
use of poison baits and chemical lures, traps, and the management of
garbage (Akre et al. 1981). Biological control is not considered
feasible. No method is without drawbacks, nor are current control
methods useful for all species. However, control methods that should
be investigated include pathogens (Akre and Reed 1981), growth reg-
ulators (Roush and Akre 1978b), juvenile hormone analogs, more
effective baits and toxicants, chemicals that might interfere with
the queen pheromone so colony cohesion is lost, and chemicals that
interfere with nestmate recognition.

SUMMARY

Damage and economic losses due to yellowjackets in the United
States probably run into millions of dollars each year. However,
this estimate is arrived at only by making many assumptions based on
piecemeal evidence from limited areas. Sorely needed are data on
crop losses, losses and problems in recreational and urban areas,
and disclosure of the magnitude of medical problems.

In the future, perhaps a computer could be used to store reports
on crop losses, life threatening stings, and colonies controlled by
PCOs in urban areas. These data, coupled with full disclosure by NIH
(and the commercial companies under its guidance) on venom uses,
would reveal that yellowjackets cause much greater damage and
economic losses than previously realized.

LITERATURE CITED

Akre, R. D. 1978. Yellowjacket problems. Proc. 18th Ann. Conf.
NW Mosq. and Vector Cont. Assoc. Boise, Idaho, p. 91.
Akre, R. D. 1982. The social wasps. In Hermann, H. R., ed.,
Social Insects. Vol. 4, Chapt. 1. Academic: New York (in press).
Akre, R. D. and Davis, H. G. 1978. Biology and pest-status of
venomous wasps. Ann. Rev. Ent. 12:19-42.

Akre, R. D., Green, A., MacDonald, J. F., Landolt, P. J., and Davis, H. G. 1981. The yellowjackets of America north of Mexico. USDA Agric. Handbook 552.

Akre, R. D. and Reed, H. C. 1981. Population cycles of yellowjackets (Hymenoptera: Vespinae) in the Pacific Northwest. Environ. Ent. 10:267-274.

Antonelli, A. L. 1980. Recognition and management of Christmas tree pests. Coop. Ext. College Agric. Wash. State Univ. EB 0735 17 p.

Ascerno, M. 1981. Diagnostic Clinics: More than a public service. Bull. Ent. Soc. Amer. 27:97-101.

Davis, H. B. 1978. Yellowjacket wasps in urban environments. In Frankie, G. W. and Koehler, C. S., (Eds.) 1976. Perspectives in Urban Entomology. XI Int. Congr. Ent. Academic: New York. pp. 163-185.

Edwards, R. 1980. Social wasps Their biology and control. Rentokil: Sussex, England. 398 p.

Edery, H., Ishay, J., Gitter, S., and Joshua, H. 1978. Venoms of Vespidae. pp. 691-771 in Bettini, S. (Ed.), Arthropod Venoms. Springer-Verlag: Verling. 977 p.

Fanning, P. K. 1980. Christmas tree farm accidents and what to do about them. Coop. Ext. Coll. Agric. Wash. State Univ. EM 4512 9 p.

Fanning, P. K., Symons, W. M., and Buhaly, J. 1981. Christmas tree industry safety update. No. 301-2. Coop. Ext. Wash. State Univ. 2 p.

Hawthorne, R. M. 1974. Cooperative Economic Insect Report for California. Week ending 18 Oct. 4 p.

MacDonald, J. F., Akre, R. D., and Keyel, R. 1980. The German yellowjacket (Vespula germanica) problem in the United States. Bull. Ent. Soc. Amer. 26:436-442.

Putnam, S. E., Jr. 1977. Controlling stinging and biting insects at campsites. Project Record. ED & T 2689. Control of stinging insects in Forest Service Camps. USDA, Forest Service Equip. Dev. Center, Missoula, MT. 22 p.

Roush, C. F. and Akre, R. D. 1978a. Nesting biologies and seasonal occurrence of yellowjackets in northeastern Oregon forests (Hymenoptera: Vespidae). Melanderia 30:57-94.

Roush, C. F. and Akre, R. D. 1978b. Impact of chemicals for control of the Douglas-fir tussock moth upon populations of ants and yellowjackets (Hymenoptera: Formicidae, Vespidae) Melanderia 30:95-110.

Seastadt, T. R., Mameli, L., and Gridley, K. 1981. Arthropod use of invertebrate carrion. Amer. Midl. Nat. 105:124-129.

Schmidt, J. O. 1982. Biochemistry of Insect venoms. Ann. Rev. Entomol. 27:1-61.

Valentine, M. D. 1981. Insect stings - what to do about them. U.S. News & World Report 90:71.

Wicher, K., Reisman, R. E., Wypych, J., Elliott, W., Steger, R., Mathews, R. S., and Arbesman, C. E. 1980. Comparison of the venom immunogenicity of various species of yellowjackets (genus Vespula) J. Allergy Clin. Immun. 66(3):244-249.

The Economic Importance
of Leaf-Cutting Ants

J. M. Cherrett,
University College of North Wales

The economically important leaf-cutting ants are restricted to
the genera <u>Atta</u> and <u>Acromyrmex</u>, are fungus-growing ants of the Tribe
Attini (Hymenoptera, <u>Formicidae</u>) and are found only in the New World.
Members of the genus <u>Atta</u> build large nests over 100m^2 in surface
area, and containing millions of individuals, whilst <u>Acromyrmex</u> spp.
build smaller nests, only a few m^2 containing tens of thousands of
ants.

TYPES OF DAMAGE CAUSED

The cutting activity of the workers defoliates living plants
and with large <u>Atta</u> colonies, the effect can be spectacular, large
citrus trees being stripped in 2 or 3 nights. As the workers take
only sap directly from the leaves, and feed the remains to their
fungus, part of which forms the principal diet of the developing
brood, they are enabled to circumvent the defence mechanisms of many
plant species, and so are highly polyphagous (Cherrett 1980),
attacking an exceptionally wide range of useful plants.
1.   Agricultural and horticultural crops
In a questionnaire survey of 27 countries, Cherrett and Peregrine
(1976) reported that 47 agricultural crops suffered leaf-cutting ant
damage, and a survey of the literature on ant damage and control has
shown that the 6 crops most frequently mentioned are the same, with
slight differences in order. Citrus heads both lists, with the other
woody plant crops cocoa and coffee included. It seems likely that the
small amount of tillage required for these semi-permanent crops once
established permits a slow build-up of leaf-cutting ant populations.
Cotton and manioc both appear in the top 6, with maize the only
monocotyledonous crop. All are especially vunerable to defoliation
when young or flushing, and in cocoa, the loss of young flowers may
be important for fruit set. Garden flowers, especially roses are cut.
2.   Pastures
Cherrett and Peregrine (1976) reported 13 species of range plants
attacked by the ants, whilst the literature survey placed pastures as
the third most frequently cited crop (after citrus and cocoa). There
are 4 types of damage: 1. The ants cut and remove grass which might
otherwise be eaten by domestic animals. The number of nests required
to consume as much grass as one cow being 8 <u>Atta capiguara</u> Gonçalves

(Amante 1967) or a median figure of 15 A. vollenweideri For.
(Jonkman 1977). 2. By defoliating and killing desirable grass species,
less desirable weedy, often broad leaved species establish, lowering
pasture productivity. 3. The spoil heaps on the surface of the nests
are free of vegetation, and as the mean surface area of old A.
vollenweideri nest sites is 87m^2, large nest populations render
appreciable areas of the pasture unproductive (Jonkman 1977). 4. Dead
nests of both A. capiguara and A. vollenweideri may collapse, and the
centre may fill with water (Jonkman 1977). An appreciable number of
cattle die in accidents involving collapsed nests.
3.   Forestry
      Cherrett and Peregrine (1976) recorded pines and teak as being
attacked, but Eucalyptus, rubber and Gmelina are also reported from
the literature. Defoliation reduces wood yield and can kill large,
but more especially young trees. Although a wide variety of tree
species are attacked in tropical rain forest, the impact of leaf-
cutting ants is not obvious so long as forestry is based on extracting
timber from natural stands of trees. Plantations of exotic species
can suffer badly and may prove impossible to establish without ant
control.
4.   Stored products
      Leaf-cutting ants will pick up and carry back to their nests a
variety of dried food stuffs such as cereals, flour, dried beans and
breakfast foods, and it is this behaviour which allows poison baits
based on such materials to be used so successfully. Consequently
when stored products are accessible to the ants they are likely to be
taken, and there are even records of rice being removed from a ship
via its mooring ropes. Although with modern storage methods, leaf-
cutting ants are only a minor irritation, they must have been a
powerful deterrent to any attempts by the hunter gatherers living in
tropical rain forests to store food.
5.   Subsidence
      Jonkman (1977) excavated an A. vollenweideri nest which had a
surface area of 36m^2, a depth of 5m and which contained over 3000
chambers of total volume 5m^3. We have seen that in pastures when
such a nest dies and collapses, cattle may fall in, but when similar
sized nests of other species collapse under roads or under the
foundations of buildings, considerable structural damage can result.
Bondar (1927) claimed that in Bahia, 300-500 repairs to buildings
were required annually, and that the reinforced foundations necessary
to avoid subsidence could add 20% to the costs of construction.

SPECIES RESPONSIBLE

      There are 14 species of Atta and 23 of Acromyrmex, and
representatives can be found from the Southern United States to
Northern Argentina (Weber 1972). Each species will attack a wide
range of vegetation, the principal division being into those which
primarily cut monocots or dicots. In the genus Atta. A. bisphaerica
For., A. capiguara and A. vollenweideri cut monocots; A. goiana
Gonçalves and A. laevigata (F. Smith) cut both dicots and monocots,
whilst the remainder cut dicots. In the genus Acromyrmex, of the 19
species for which information is available, the subgenus Moellerius

seems to have specialised in cutting monocots (A. heyeri (For.), A. landolti (For.) and A. striatus (Roger)), with the last named also taking some dicots. A. lobicornis (Emery) also cuts both groups of plants, but all the rest take dicots.

The 3 species of Atta most frequently cited as pests in the literature are A. sexdens (L.), A. cephalotes and A. laevigata. They are the 3 most widely distributed species, and A.sexdens is often quoted as a species encouraged by agriculture. The 3 species of Acromyrmex most frequently cited are A. octospinosus (Reich), A. subterraneus (For.) and A. landolti, the first and last of these again being the most widely distributed species in the genus.

THE EXTENT OF LOSSES

St. Hilaire, a French naturalist travelling during the period 1816-22 wrote "Ou o Brasil mata a saúva ou a saúva mata o Brasil" (Either Brazil kills the saúva (Atta spp.) or the saúva will kill Brazil) (Mariconi 1970), and the literature abounds in general statements saying that leaf-cutting ants are the most important economic problem to growing crops in South America.

Closer analysis however reveals considerable confusion about the way in which leaf-cutting ant losses should be calculated, and this can be illustrated by reference to the citrus industry in Trinidad. Lewis and Norton (1973) calculated an average loss of $27,600 U.S. per annum for the 3 years 1968-71 which represented the cost of replacing the young trees which the ants killed. Cherrett and Sims (1968) by contrast quoted $169,700 U.S. based on a larger estimate of the citrus acreage, but on loss estimates within the range recorded by Lewis and Norton. This latter figure however also included the costs of the control measures being taken against leaf-cutting ants, and so was an estimate of the savings to be expected if leaf-cutting ants were eliminated. However as both studies were carried out in commercial citrus orchards operating routine insecticidal ant control, the impact of the ants was not obvious and appeared to be confined to the death of young trees. Cherrett and Jutsum (in press) however were able to study nearby citrus orchards 4 to 7 years after they had been abandoned as a result of changes in the economics of citrus growing. Without control, the population of Atta cephalotes increased greatly, and over 65% of the mature trees were killed, probably by repeated defoliation. It seems clear that the potential loss without any control could be the total value of the industry in Trinidad, estimated at $3,200,000 U.S. in 1967. We therefore have 3 loss estimates of 1, 5 and 100% respectively of the crop value.

Leaf-cutting ants are especially damaging to oranges and grapefruit, and in 1979, the countries where they are reported as pests produced for export some 377,000 metric tons of fruit, valued at $99.5 million U.S., whilst total production including home consumption was 40 times this. By multiplying up from the Trinidad loss estimates, and by adding in all the other types of damage, it would not be difficult to arrive at the much-quoted figure of $1000 million U.S., originally suggested by Townsend (1923) for total damage in tropical America, and used ever since without any corrections for inflation.

The usefulness of such global figures is questionable except to stress the importance of the leaf-cutting ant problem, and to draw comparisons with other better known pests such as locusts which are thought to be responsible for comparable losses world wide.

Two general points are however worth noting:

1.   The difference between potential and actual loss to citrus in Trinidad is very wide, and is due to the success of modern methods of ant control. This was not always so, and early accounts of leaf-cutting ant depredations (Mariconi 1970) have a ring of desperation about them. The nomadism of some Amerindian tribes is said to have been a response to upsurges in Atta populations once forest is cleared and Bates (1891) wrote "In some districts it is so abundant that agriculture is almost impossible, and everywhere complaints are heard of the terrible pest". Legislation was invoked, and as early as 1785 the authorities of Salvador, Bahia, passed legislation requiring farmers to destroy ant nests on their property, or face a fine and 30 days in prison (Mariconi 1970), whilst in 1815 quarantine laws were set up to prevent the possible introduction of Atta into the island of Puerto Rico. There are numerous accounts of local ant eradication schemes, and even of a competition with a substantial prize for the best control technique. Without toxic chemicals, a large Atta nest containing perhaps 3-4 million workers must have seemed an almost insuperable problem, and Belt (1874) conveys this sense when he writes "Again and again have I been told in Nicaragua, when inquiring why no fruit trees were grown at particular places, "It is no use planting them; the ants eat them up." ...they are one of the greatest scourges of tropical America, and it has been too readily supposed that their attacks cannot be warded off". Today leaf-cutting ant nests can be destroyed relatively easily so that only the poor who cannot afford insecticides still appreciate the full extent of the damage they can do.

2.   The potential loss to leaf-cutting ants continues to increase as more land is brought under agriculture and managed forestry, and as both become more intensive. This is especially true in the Amazon Basin, prime Atta country. As Jonkman (1977) has emphasisied, the impact of leaf-cutting ants in parts of South America may currently be exaggerated because the pastoral potential of the land is not fully utilised. Reducing the amount of grass taken by the ants would not necessarily increase stocking rates. However the increasing demand for food is causing the replacement of extensive grazing by intensive management of sown, fertilised pastures, and higher yields of all crops are increasingly expected. In addition, extraction of timber from natural forests is being replaced by the management of exotic trees such as pines, Gmelina and Eucalyptus.

Despite modern control technology, leaf-cutting ants still cause considerable damage; the control industry is a substantial one (in 1974, 3000 tons of one bait, selling for $4.5 million U.S. was being manufactured per annum in São Paulo, at a time when at least 11 other baits were available in the area), and for the reasons outlined, the need to maintain control is more important than ever.

References

Amante E., 1967. -- A formiga saúva Atta capiguara, praga das pastagens. Biològico, 33, 113-120.

Bates H.W., 1891. -- The naturalist on the river Amazon. Clodd, London (1st edition 1863).

Belt T., 1874. -- The naturalist in Nicaragua. Bumpus, London.

Bondar G., 1927. -- A formiga saúva na Bahia. Correio agric., 5, 99-104.

Cherrett J.M., 1980. -- Possible reasons for the mutualism between leaf-cutting ants (Hymenoptera: Formicidae) and their fungus. Biologie-Ecologie méditerranéenne, 7, 113-122.

Cherrett J.M., Jutsum A.R., In press. -- The effects of some ant species, especially Atta cephalotes (L.), Acromyrmex octospinosus (Reich) and Azteca sp. (Hym. Form.) on citrus growing in Trinidad. I.U.S.S.I., Cocoyoc, Mexico.

Cherrett J.M., Peregrine D.J., 1976. -- A review of the status of leaf-cutting ants and their control. Ann. appl. Biol., 84, 124-128.

Cherrett J.M., Sims, B.G., 1968. -- Some costings for leaf-cutting ant damage in Trinidad. J. agric. Soc. Trin., 68, 313-324.

Jonkman J.C.M., 1977. -- Biology and ecology of Atta vollenweideri, Forel 1893 and its impact in Paraguayan pastures. Thesis. Universiteitsbibliotheck, Leiden.

Lewis T., Norton G.A., 1973. -- Aerial baiting to control leaf-cutting ants (Formicidae, Attini) in Trinidad. III. Economic implications. Bull. ent. Res., 63, 289-303.

Mariconi F.A.M., 1970. -- As saúvas. São Paulo: Editôra Agronômica 'Ceres'.

Townsend C.H.T., 1923. -- Um inseto de um bilhão de dollares e sua eliminação. A formiga saúva. Almanaque Agricola Brasileiro, São Paulo, 12, 253-254.

Weber N.A., 1972. -- Gardening ants the Attines. Mem. Am. phil. Soc., 92, 1-146.

Control of *Monomorium pharaonis* (L) with Methoprene Baits: Implications for the Control of Other Pest Species

John P. Edwards, Ministry of
Agriculture, Fisheries and Food,
Slough Laboratory, Berks., U.K.

Pharaoh's ant, Monomorium pharaonis (L) is a tropical species
first reported in England in 1828 (Donisthorpe, 1927). In temperate
regions the species is found only indoors in permanently heated
buildings e.g. hospitals, kitchens and increasingly in domestic
appartment blocks. In such situations worker ants are frequently
found foraging at drains, toilets and other insanitary areas and
have also been recorded in sterile supplies and even feeding under
wound dressings on post-operative patients (Beatson, 1973,
Cartwright and Clifford, 1973). Furthermore, worker ants have been
shown to be capable of carrying a variety of pathogenic bacteria
(Beatson, 1972) and infestations are therefore a considerable
potential hazard to public health. Infested premises usually
contain several nests (polydomic colonies) between which there is
free in⁺erchange of ants. The nests usually inaccessibly situated
in wall cavities, foundations and ducting systems, contain several
queens and large numbers of workers which leave the nest to forage
along well-defined trails. Such trails often extend several metres
from the nest. Colony dispersal is by sociotomy (budding) whereby
groups of worker ants carry brood stages to a new nest site.
Although queens sometimes accompany these emigrating groups, they
are not essential for the survival of the new nest since workers
can rear new sexual stages from the existing brood. Often, these
emigrating groups will form a temporary nest whilst searching for a
more permanent abode. Such temporary nests are highly mobile and
are probably the main method by which infestation is spread both
within an infested building and, with the inadvertent transporta-
tion of nests in equipment and personal belongings, further afield.

During the last few years, we have developed a technique for
eradicating infestations of this species in hospitals and similar
premises. The technique involves the use of attractive baits
containing the insect juvenile hormone analogue methoprene (Edwards
and Clarke, 1978). This compound, when taken back to the nest by
foraging workers, prevents the normal development of brood stages
and sterilizes queens (Edwards, 1975). During the course of our
studies we have attempted to extend this technique for use against
some other house-infesting ant species (Edwards et al. 1981). From
these experiments we have been able to identify some of the aspects

of behaviour and life-styles which may render ant species susceptible control with juvenile hormone analogues.

MATERIALS AND METHODS

Several ant species have been investigated in both the laboratory and the field. These include Monomorium pharaonis, Pheidole megacephala, P. siniatica, Paratrechina longicornis, P. vividula and Iridomyrmex humilis. In laboratory studies of species other than M. pharaonis we have mainly investigated the attractiveness of various food materials in an attempt to produce an acceptable bait formulation. In the field studies we have incorporated methoprenne (isopropyl-11-methoxy,-3,7,11-trimethyl-2, 4-dodecadienoate) at 0.5% w/w into an appropriate bait and monitored the effects of the·treatment on the population.

RESULTS

To date we have completed a total of 8 field trials using methoprene baits against infestations of M. pharaonis and in all cases we have achieved eradication of the infestations (Edwards and Clarke, 1978; Edwards et al., 1981). In addition we have successfully used the same technique to eradicate infestations of Pheidole megacephala and P. siniatica (Edwards et al., 1981). However, for reasons which will be discussed below, the same technique appears to be ineffective against Paratrechina spp and Iridomyrmex humilis.

DISCUSSION

The death of colonies of M. pharaonis exposed to hormone-based baits is a result of the morphogenetic (development-disrupting) effects of the hormone on the brood stages together with the action of the compound on the ovaries of queens which become atrophied and stop producing oocytes (Edwards, 1975). The resulting sterility of queens may be due entirely to the physiological effects of the hormone on the ovaries but may be compounded by the disruption of social interactions necessary for egg-laying as a result of the absence of brood stages killed by the morphogenetic action of the hormone. Whatever the exact mechanism, it is clear that both effects are important in achieving the eventual eradication of the nest. Little is known about differences in sensitivity of larval and pupal stages of different species to the morphogenetic effects of juvenile hormones. However, since all insects appear to be sensitive to a greater or lesser extent, it seems reasonable to assume that ant larvae and pupae do not differ dramatically in this respect. The effect of hormone analogues upon the reproductive capacity of queens is more problematical since effects on female reproduction are by no means universal in insects, even in closely related groups. Laboratory studies have shown that queens of two Myrimicine species - M. pharaonis and Solenopsis invicta can be sterilized by application of juvenile hormone analogues (Edwards, 1975; Troisi and Riddiford, 1974). In the case of S. invicta only temporary sterilization was obtained although this may have been due to the relatively low doses and method of application used. For ants in other subfamilies eg Paratrechina (Formicinae) or Iridomyrmex (Dolichoderinae) it is difficult to speculate on the likely effects of juvenile hormones on the fecundity of queens. However, it is noteworthy that topical application of a juvenile hormone analogue to queens of Plagiolepis pygmaea had only marginal effects on their reproductive capacity (Passera and Suzzoni, 1974). Since sterility in queens is an important factor in the successful use of hormone analogues against ant species, the absence of such effects in a target species would seriously impair the chances of successful control.

In addition to the known effects of hormone analogues on development and reproduction in M. pharaonis, other factors associated with the particular life-style of this species undoubtedly contribute to the susceptibility of Pharaoh's ant to hormone-based control methods. Monomorium pharaonis is omnivorous and workers are attracted to a wide variety of food materials. In our experiments we have used various mixtures of liver (dried and powdered to remove any enzyme activity) honey and sponge cake. This mixture is highly attractive to M. pharaonis workers and provides for the nutritional requirements of the colony (i.e. protein and carbohydrate) which may change emphasis from time to time. The same mixture is attractive to Pheidole megacephala and P.siniatica although this is not the case with the two Paratrechina species. Iridomyrmex humilis or Lasius niger. The problem with these species is not so much that they are unattracted by various bait mixtures but rather that they are inconsistent in their choice of baits from day to day. Since good attractancy is a prerequisite of any successful baiting method,evaluation of the effectiveness of hormone analogues against these and other species will depend upon the development of suitably attractive baits. In this respect, it is probably worth investigating the use of trail or brood pheromones to increase the attractiveness of baits for some species.

When colonies of Pharaoh's ants are exposed to juvenile hormones, adult stages present at the time of treatment are not killed since the compounds have no direct toxic action. The death of adult stages occurs naturally and, in the case of M.pharaonis all adult workers die by about 20 weeks after treatment. Queens of M. pharaonis live longer than workers (sometimes as much as 1 year) but because their numbers are few compared to workers and because they are unable (even when fertile) to found new nests in the absence of brood and workers, their greater longevity is not important. Moreover, there is some evidence from laboratory observations that the longevity of both queens and workers is much reduced when they are deprived of the normal social interactions present in a functioning nest (i.e. one with brood). Short worker life and the inability of queens to found nests are undoubtedly further factors which predispose M. pharaonis to control with hormone baits. Information on the longevity of queens and workers of other pest species is scarce. However, the similarities observed in the rate of decline of post-treatment populations of M. pharaonis and the two Pheidole species suggest that the latter have relatively short-lived adult stages. In contrast, in other subfamilies there are species in which workers and queens are particularly long-lived. In the Formicinae for example, which includes Paratrechina, Lasius and other ecobomically important genera, workers have been recorded as living for over 6 years (Lubbock, 1894) and queens for up to 10 years (Janet, 1904). Because short worker life is important in the absence of toxic effects from hormone analogues, some Formicine species may be unsuitable targets for hormone-based control methods on these grounds alone. In M. pharaonis queens are unable to found new colonies in the absence of workers and brood stages, Thus, after the elimination of brood and workers by the effects of hormone treatment the irreversible sterility of queens is largely academic. However, in other ants where colony foundation is typically accomplished by single queens, it would be essential to ensure that hormone treatment resulted in complete and permanent sterility of queens. It is perhaps fortunate that several important pest species, e.g. Pheidole, Monomorium and Iridomyrmex spp appear to have evolved sociotomy as the major method of colony dissemination.

In M. pharaonis there are only two female castes and hormone treatment does not appear to influence the ratio of these castes to an extent that would contribute to the disruption of ergonomics or normal social functions within the colony. However, we have observed that, in the genus Pheidole, the number of soldiers appear to increase dramatically when the colony is exposed to hormone analogues (Edwards, et al., 1981). Although this phenomenon may simply reflect differences in longevity between the two subcastes (workers and soldiers) it is tempting to speculate that the hormone treatment may have changed the caste bias of some larvae towards soldier development particularly as soldier formation can be induced in Pheidole bicarinata by topical application of methoprene (Wheeler and Nijhout,1981). It is possible in some species, that such an effect on subcaste ratio might be sufficiently disruptive to the ergonomic balance of nest to contribute to the death of the colony.

When foraging M. pharaonis workers find a food source they remove more food than is required to satisfy the immediate require- ments of the colony and store the surplus food in the nest. As a result the nest contains a reservoir of hormone activity and it seems likely that developing stages and queens become contaminated with hormone by contact as well as by feeding. Furthermore, the use of particulate bait material may be important in ensuring that trophal- laxis between workers and larvae occurs with the minium of degradation of the hormone which may result if workers ingest and partially digest the food material. In M. pharaonis solid food particles are often fed direct to developing larvae after pre- liminary mastication by workers whereas liquids are apparently ingested and partially digested before being regurgitated to queens and larvae (Buschinger and Kloft, 1973). The rapid metabolism and poor distribution of the hormone analogue R-20458 by colonies of the fire ant Solenopsis invicta reported by Wendel and Vinson (1978) may have been due to the use of liquid as opposed to solid bait.

In summary, ant species which display an appropriate life style and behaviour in terms of sensitivity to morphogenetic and reproductive disruption by hormones, worker longevity, ease of baiting and mode of colony dissemination are suitable targets for control measures utilizing insect juvenile hormone analogues. Against such species this technique can be highly effective in eliminating infestations and has the advantage of utilizing a type of chemical which is essentially non-toxic and lacks the undesriable effects on the ecosystem associated with more conventional methods of control.

References

Beatson S.H. (1972) Lancet, i, 425 - 427

Beatson S.H. (1973) Lancet, i, 606 - 607

Buschinger A. and Kloft W. (1973) Forchungs. des Landes NRW, No. 2306, Westdeutcher Verlag, Opladen. pp 2 - 28.

Cartwright R.Y. and Clifford C.M. (1973) Lancet, ii, 1455 - 1456.

Donisthorpe H. (1927) British ants - their life history and classification. Routledge and Sons. London. pp 102 - 109.

Edwards J.P. (1975) Bull. ent. Res, 65, 75 - 80.

Edwards J.P. and Clarke B. (1978) International Pest Control, 20, 5 - 10.

Edwards J.P., Pemberton G.W. and Curran P.J. (1981) In, Regulation of insect development and reproduction. (Eds.Sehnal, Zabza, Menn and Cyborowski) Wroclaw Technical Univ. Press. pp 769-779.

Janet C. (1904) Observations sur les Fourmis. Ducortieux et Gout, Limoges. 68pp.

Lubbock J. (1894) Ants,bees and wasps. Appleton and Co. New York. 448pp.

Passera I. and Suzzoni J.P. (1974) C.R. Hebd. Sean.Acad. Sci. (D) 279, 2079 - 2082.

Troisi S. and Riddiford I.M. (1974) Environ. Entomol. 3,112 - 116.

Wheeler D.E. and Nijhout H.F. (1971) Science, N.Y., 213, 331 - 333.

Wendel L.E. and Vinson S.B. (1978) J. econ. Entomol., 71, 561 - 565.

Economic Aspects of the Imported Fire Ant in the United States

C. S. Lofgren and C. T. Adams,
USDA-ARS, Gainesville, Florida

Two species of imported fire ants (IFA), *Solenopsis invicta* and
S. richteri, were accidentally transported to the United States in
the early 1900's. Their spread from the initial point of introduction, (Mobile, Alabama) was rapid and enhanced by concealment of
newly-mated queens or young colonies in nursery stock. Currently,
they are estimated to infest more than 9.3×10^7 ha (2.3×10^8 acres)
in nine southern states and Puerto Rico.

The economic impact of the IFA has been a subject of much discussion. Generally, they have been classified as nuisance pests because of their mound-building and stinging habits (Lofgren et al.
1975); however, documentation of the problems they cause is meagre
and based primarily on survey reports rather than experimentation.
Consequently, we began studies several years ago on the impact of IFA
on man and his crops and domestic animals. We concentrated our efforts on their effect on public health and the production of soybeans.
Other studies included the interrelationship of IFA with sugarcane,
okra and various vegetable crops, but these will not be discussed in
this report.

Public Health Impact

The impact of IFA on man has been the subject of a number of
investigations which have centered on systemic allergic reactions to
the venom. The literature was reviewed by Lockey (1974) who gave a
good description of the sequence of events that result from an imported fire ant sting. The initial reaction is a burning sensation
which is followed by the formation of a wheal which may become as
large as 10 mm in diameter. Typically, a small vesicle containing
clear fluid forms at the sight of the sting after about 4 hours. The
vesicle becomes cloudy and a white pustule forms, usually after about
24 hrs. The pustule may remain from a few days to as much as a week
before rupturing and often, especially in older persons, a small scar
may remain at the sting site for a few weeks to several months.
Secondary infections may occur if the pustule is broken. This is a
special problem for laborers, farmhands and others who may not adequately protect the exposed area.

While the venom of the imported fire ant is composed primarily of alkaloids (2,6 disubstituted piperidines; MacConnell et al. 1970), it does contain a small aqueous component (less than 5 percent) in which Baer et al. (1979) demonstrated the presence of 3 allergenic proteins. Susceptible individuals require immediate medical attention and deaths have been reported.

The only recorded attempt to correlate the incidence of allergic systemic reactions within known populations was made by Rhoades et al. (1977) in Jacksonville, Florida. This city was serviced by only 2 allergists who reported 21 new cases of allergy to fire ant venom in a population area of 560,000, an incidence rate of 3.8 per 100,000 per year. They considered that these figures only "scratched" the surface since some patients probably reported to emergency rooms and others may have had reactions that were not diagnosed.

Using the previous incidence rate, we computed a minimum estimate of the total number of individuals that may become sensitized to IFA venom each year. According to the 1980 census, approximately 38.4 million people live within the 9 infested states. At the rate of 3.8 per 100,000, at least 1460 new cases could be expected yearly. While there is no manner in which we can estimate the total expenses that might be incurred by persons sensitized to IFA venom, we were able to obtain the following data on the cost of hyposensitization therapy at the J. Hillis Miller Health Center, Gainesville, Florida (H. J. Wittig, personal communication) for the year 1978. Cost of the initial patient evaluation was $60.00 while follow-up consultations cost $45.00. Total cost of injection materials was $36.00. Based on these data the cost of desensitizing 1460 new patients each year would be $205,860.

While systemic allergic reactions pose the greatest danger from IFA attacks, many lesser primary or secondary reactions may require medical treatment. Data from several studies are available now from which we can make estimates. Clemmer and Serfling (1975) surveyed a total of 240 households (777 persons) by telephone in Metarie, Lousiana. Sting attacks were reported for 29% of the persons during the months of June-August, 1973 and 1.3% required medical consultation.

We collaborated with health officials in two counties in Georgia to conduct similar surveys. In Lowndes County, Georgia (Yeager, 1978) a sampling of 156 families, including equal numbers of urban and rural residents, revealed that 1 out of every 5 residents could expect to be stung each month and less than 5% of those stung required medical care. The second study (Adams and Lofgren, 1981) was conducted in Sumter County, Georgia in a predominately rural area. A total of 213 sting attacks were reported during one year (1976) on 95 of 272 survey participants (35%). Two individuals (1%) classified their sting reaction as severe, 26 (12%) as moderate and 183 (87%) as mild.

Adams and Lofgren (1982) reviewed data on patients reporting for medical treatment for arthropod sting/bite attacks at Ft. Stewart, Georgia. IFA were responsible for 161 (49%) of a total of 329 patients treated from April 1 to September 30, 1979. This represents 0.7% of the post population including military personnel (12,000) and dependents (11,000). Only 7% of the sting attacks were caused

by bees and wasps. Eight persons (5%) exhibited symtoms of shock
due to IFA stings and 11 (7%) developed secondary infections. Five
patients were hospitalized for 1 day each. Direct medical cost for
outpatient visits was $23.10 while the cost of hospitalization was
$176.45 per day. Total estimated cost attributable to IFA was $5,070.

For purposes of calculating potential medical costs for the 38.4
million people living in the IFA-infested area, we averaged the sting
rates for the studies by Clemmer and Serfling (1975) and Adams and
Lofgren (1981) to obtain an average sting attack rate of 32%. Also,
we averaged the percent of persons requesting medical treatment in
the Clemmer and Serfling (1975) study (1.3%) and the percent of the
Ft. Stewart population reporting to the dispensary for treatment of
IFA stings (0.7%) to obtain a medical treatment rate of 1.0%. Based
on these data, the annual cost for medical treatment (one office
visit per patient at $23.10) would be approximately 2.84 million
dollars (38.4 X 10^6 X 0.32 X 0.01 X $23.10).

Agricultural Impact

In 1949 Wilson and Eads (1949) conducted a survey on the economics of the IFA for the Alabama Department of Conservation in the main
infested areas at that time (Mobile, Baldwin and Washington Counties,
Alabama). They concluded from their systematic poll of 174 farmers
that IFA were a major crop pest. Primary damage was attributed to
feeding on the seeds or seedlings of crops such as corn, peanuts,
beans, potatoes and cabbage. Following their survey, complaints of
damage to crops by IFA decreased and as recently as 1976 a report
published by the Council for Agricultural Science and Technology
(Anonymous, 1976) stated that they were not a major pest of crops.

However, an important development that coincided with the apparent decrease in economic importance of IFA in the 1950's was the ad-
vent and wide-scale use by farmers of chlorinated hydrocarbon pesticides (dieldrin, heptachlor, chlordane) for control of numerous soil
insect pests. In addition federal-state supported programs for control of white-fringed beetles, *Graphognathus* spp., resulted in the
application of these insecticides to many cultivated fields. Since
these pesticides are extremely effective residual chemicals they also
controlled IFA and thus, indirectly reduced the importance of IFA as
crop pests. However, after registrations for use of these chemicals
on cropland were withdrawn by the Environmental Protection Agency
about 1970, we suspected that IFA might become a major crop pest
once again. This conclusion, as well as complaints from farmers in
Georgia in the mid-1970's about IFA interfering with soybean harvest,
prompted us to initiate a series of studies on interactions between
IFA and various crops.

Our initial research was directed toward the impact and interference of IFA mounds with harvesting of soybeans. Soybean plants
develop pods over the entire plant thus the plant must be cut close
to ground level to harvest all of the soybeans. Consequently, IFA
mounds only a few inches high interfere with harvest of soybeans.
If the combine operator raises the header bar to avoid the mound and
protect his equipment, a large amount of beans on the lower portion

of the plant are missed, if the operators choose not to raise the
header bar, plants ahead of the mound are pushed over with soil.
Also, soil taken into the combine increases wear and damage to the
equipment. We conducted two studies to evaluate the impact of IFA
mounds on soybean harvest. The first study near Valdosta, Georgia
(Adams et al. 1976) revealed a loss of about 0.22 hl/ha or $6.00/ha
in a field with an infestation of 109 mounds/ha. The second study
in southeastern North Carolina (Adams et al. 1977) showed a loss of
0.64 hl/ha or $12.35/ha (140 mounds/ha). Considerable undetermined
expenses resulted from damage to harvesting equipment.

While conducting the prior tests we noted that total yield of
soybeans indicated a greater loss than could be attributed to the
incomplete harvest caused by the mounds. Consequently, we conducted
additional tests in which fields were divided and one-half was treated with mirex bait to eliminate IFA while the other half remained
untreated. A summation of all of our data (Lofgren and Adams, 1981)
showed an average decrease in yield for 8 paired fields of 14.5% or
5.1 hl/ha with infestation rates of 49 to 176 mounds per ha.

Because of this significant loss of soybeans we planned further
tests in Florida, Mississippi, and North Carolina. Preliminary evaluation of the data from this research confirms the prior results.
The tests at Gainesville show a reduction in yield of 7 hl/ha (99
mounds/ha) while those in Mississippi (W. A. Banks, unpublished data)
show a loss of 5.2 hl/ha (160 mounds/ha). The North Carolina tests
(C.H. Apperson, unpublished data) revealed that plant stand, plant
height and IFA were negatively correlated with yield and that a regression analysis of IFA against yield indicated a 5.2 to 8.7 hl/ha
reduction where ant activity was high compared to where it was low.

Additional observations in our tests suggested that yield reductions were associated with feeding on germinating seeds, since
plant stand was reduced from 40.7 to 26.2 plants per row meter in
the check plots and IFA-infested plots, respectively. Damage to the
growing plants probably occurred also, since a high percentage of
ants foraging near plants injected with ^{32}P were found to contain
radioactivity.

An estimated 5,557,085 ha were planted with soybeans in the 7
most heavily infested states in 1981. (Southeastern Farm Press, Inc.
Clarksdale MS; Vol. 8(45)) If we assume that (1) 25% of the land in
these states is heavily infested, (2) the average reduction in yield
is 5.2 hl/ha and (3) the sale price of soybeans is $17.00 per hl,
then in 1981 about 7,254,000 hl of soybeans were lost to IFA with a
value of almost $125,000,000.

Our observations confirm the study by Wilson and Eads (1949)
that IFA feed on and destroy germinating soybean seeds or seedlings
and that they are a major pest of soybeans. One reason for the susceptibility of soybeans may be that they are normally planted in
late spring or early summer. At this time of year the IFA are actively producing sexual and worker brood and require large amounts
of food. Also, rainfall in the southeastern U.S. is limited during
April and May or until the summer thundershowers begin. Both of
these conditions could create a critical food stress situation for
IFA in newly cultivated and planted fields and thus germinating

soybean seeds would provide a ready source of food and water. Crops such as cotton, corn and peanuts that are planted earlier in the year (February to April) before ant activity reaches its peak may not be as susceptible to attack. It is obvious that much more research will be required to assess the economic impact of IFA on soybeans as well as numerous other agricultural crops.

References

Adams, C. T. and C. S. Lofgren. 1981. Red imported fire ants (Hymenoptera: Formicidae): Frequency of sting attacks on residents of Sumter County, Georgia. J. Med. Entomol. 18: 376-380.

Adams, C. T. and C. S. Lofgren. (In press).--Incidence of stings or bites of the red imported fire ant and other arthropods among patients requesting medical treatment at Ft. Stewart, Georgia. J. Med. Entomol.

Adams, C. T., J. K. Plumley, W. A. Banks and C. S. Lofgren. 1977. Impact of the red imported fire ant, *Solenopsis invicta* Buren (Hymenoptera: Formicidae), on harvest of soybeans in North Carolina. J. Elisha Mitchell Sci. Soc. 93: 150-2.

Adams, C. T., J. K. Plumley, C. S. Lofgren and W. A. Banks. 1976.-- Economic importance of the red imported fire ant, *Solenopsis invicta* Buren. I. Preliminary investigations of impact on soybean harvest. J. Georgia Entomol. Soc. 11: 165-9.

Anonymous. 1976.--Fire ant control. Council for Agricultural Science and Technology. CAST Report No. 62; 2nd Ed. Rep. No. 65. 24pp. Iowa State Univ.

Baer, H. T. Y. Liu, M. C. Anderson, M. Blum, W. H. Schmid and F. J. James. 1979.--Protein components of fire ant venom (*Solenopsis invicta*). Toxicon 10: 259-71.

Clemmer, D. L. and R. E. Serfling. 1975.--The imported fire ant: Dimensions of the urban problem. South. Med. J. 68: 1133-38.

Lockey, R. L. 1974.--Systemic reactions to stinging ants. J. Allergy Clin. Immunol. 54: 132-46.

Lofgren, C. S. and C. T. Adams. 1981.--Reduced yield of soybeans in fields infested with the red imported fire ant, *Solenopsis invicta* Buren. The Fla. Entomol. 64: 199-202.

Lofgren, C. S., W. A. Banks and B. M. Glancey. 1975. Biology and control of imported fire ants. Annu. Rev. Entomol. 20: 1-30.

MacConnell, J. G., M. S. Blum and H. M. Fales. 1970.--Alkaloid from fire ant venom: Identification and synthesis. Science 168: 840.

Rhoades, R. B., W. L. Shaeffer, M. Newman, R. Lockey, R. M. Dozier, P. F. Wubbena, A. W. Townes, W. H. Schmid. G. Neder, T. Brill and H. J. Wittig. 1977.--Hypersensitivity to the imported fire and in Florida: A report of 104 cases. J. Fla. Med. Assoc. 64: 247-54.

Wilson, E. O. and J. H. Eads. 1949.--A report on the imported fire ant, *Solenopsis saevissima* var. *richteri*, in Alabama. Spec. Rep. Ala. Dep. Conserv. Mimeo. 54pp.

Yeager, W. 1978.--Frequency of fire ant stinging in Lowndes County, Georgia. J. Med. Assoc. Ga. 2: 101-2.

The Impact on Man of *Polistes* Wasps with Special Reference to Caterpillar Suppression

James E. Gillaspy, Texas A & I University

Activity and general effectiveness of *Polistes* in caterpillar predation have been reported in the literature by many observers (see Gillaspy 1979b for a few of many possible examples). Use of augmented populations of *Polistes* in agriculture was first reported by Ballou (1915). Value of such populations for integrated pest management programs has been demonstrated experimentally (Lawson *et al.* 1961), and there is at least 1 functioning program at present utilizing *Polistes* (Bellotti and Arias 1977). However, use of *Polistes* has not become widespread. Needed may be impressive cost/benefit statistics derived from controlled experiments and behavioral studies to reveal more fully the range of hosts taken and the intensity of predation.

Feasible management strategies related to different crops and agronomic practices and possibly the manipulation of wasps for improvement of foraging efficiency need to be demonstrated. Through propagation of appropriate species of *Polistes* or through ability to achieve dense wasp populations, it may be possible to promote taking of concealed, boring, or sparse prey and thus overcome weakness in these regards reported by some investigators. Overall management strategy should permit either leaving colonies permanently in place at field margins or within fields or shifting them between fields or from reserves in holding areas; subduing parasitism; providing material requisites, to include supplementary feeding at times of prey scarcity; taking advantage of pheromonal influences on aggregation, homing, etc.; and, in temperate zones, promoting survival during hibernation.

NESTING RECEPTACLES

Dependence of *Polistes* on availability of suitable nesting sites appears very great (Kirkton 1970; Reed 1979). Factors governing choice of spots for nest initiation need study as a primary management tool. From commonly noted locations of nests, it is inferred that sites chosen by foundresses may combine, to the degree available within the general vicinity, protection from climatic factors, concealment, favorable light intensity, stability of substrate, absence of spider webbing, and proximity to the nest of origin. Both for operational considerations and for elicitation of nest initiation, receptacles used for colonies on a one-to-one basis appear advantageous. Fasteners on

plastic containers used by Gillaspy (1979a, b) created movable units
and friction lids permitted entrapment of wasps for movement or to
keep them out of the field. On wooden shelters to which they were
attached as modular units such containers were not used by foundresses,
which chose instead the tablelike shelter itself, especially a narrow
space between 2 joists. However, in other circumstances caged as
well as free-living wasps have spontaneously chosen the containers,
especially when they were of container natality.

POPULATION CARRYOVER AND CONFINED REARING

In regions where *Polistes* have a non-colonial or hibernating
phase, ability to carry over populations should have relevance. These
may be released to spontaneously populate nesting shelters or receptacles in the field or used as a source of developed colonies for
introduction into the field at the start of the growing season. *Polistes* have been reared both in outdoor walk-in cages and indoors in
boxes, a nest constructed in 1 case in a .005 m^3 box (Gillaspy 1979b).

Since caterpillars are not easily reared in quantity and not
feasibly procured from nature as needed, commercial or formulated
rations as a substitute would greatly further field as well as confined feeding. Phagostimulants would be of aid in use of such artificial
dietetic materials and evidently exist, since dead and even minced
caterpillars are used, implying an olfactory cue (pers. obs. and Y.
Hirose, pers. comm.). As an alternative to fresh or living caterpillars, frozen ones have been used, but with poor results. Caged wasps
in 1980 and 1981 at Kingsville, TX initiated and developed nests, 3
of which attained 30 or more cells and mature, cocooned brood. However, wasps dismantled large portions of these nests and destroyed
all mature brood, discarding the apparently healthy larvae. None of
the latter were noted to be consumed, as often happens in brood
abortion. In 1 dismantled nest with 3 attendant wasps the dismantling
wasp was also the forager and chief nest builder, and extensively re-
built portions of the dismantled nest (Gillaspy, unpublished).

HOSTS OF *POLISTES*

Lawson and Rabb (1957) and Rabb (1960) in North Carolina developed data based on over 2000 interceptions of prey at the nest. Lepi-
dopterous larvae of all instars were taken, although early instars
were noted to often be consumed in the field and not identifiable in
pellets at the nest. Represented were 16 families, 36 genera, and 31
identified species of Lepidoptera and also 1 Coleoptera, the Colorado
potato beetle, *Leptinotarsa decemlineata* (Say), identified in 19 pellets, 21 pellets with Orthoptera, 1 pellet each of Diptera and Hemiptera, and 11 pellets of other Arthropoda. Oliver (1964) found eggs
and all larval instars of the fall webworm, *Hyphantria cunea* (Drury)
taken by 6 species of *Polistes*. Taking of adult Lepidoptera (a butter-
fly) has also been reported (Garcia 1971) and a few adult moths have
been taken by caged wasps (Gillaspy 1979b). Iwata (1976) reported
Homoptera (aphids), Hymenoptera (Tenthredinidae) and Orthoptera (Man-
tidae) and there is a report also of Thysanoptera (Dhaliwal 1977).
Gibo (1974, 1977) used mealworms, *Tenebrio molitor* Linnaeus, to rear

over 100 colonies of *Polistes* in box cages.

Although 7 insect orders and some other Arthropoda have been reported as prey, the taking of prey other than Lepidoptera by free-living wasps would appear exceptional on the basis of present information. Studies involving extensive interceptions and prey identifications similar to those in North Carolina are needed in other areas to clarify host range and host specificity of more of the 150 or so species of *Polistes*. Species with favorable attributes may be chosen for propagation in management programs even though rare in nature or of exotic occurrence.

CONTROL EPISODES AND EXPERIMENTS

Ballou (1915), Kirkton (1970), Gillaspy (1979) and Chinese investigators (Anon. 1978) have reported augmentation of *Polistes* for control of caterpillars on cotton; Lawson *et al.*(1961) and Gallego (1950) for tobacco; A. Alfonso *in* Gallego (1950) and Morimoto (1960, 1961) for cabbage; and Gallego (1950) and Bellotti and Arias (1977) for cassava. Sheds, low shelters and boxlike or canister-like receptacles were used to provide nesting sites.

Kasuya *et al.* (1980) found cannibalism to occur between conspecific *Polistes* colonies in Japan. Ability to achieve dense wasp populations around protected fields could be impaired if this is found to be widespread, but at the same time it could possibly provide a mechanism for survival of wasps during periods of prey scarcity, important in maintaining predator pressure as pest populations fluctuate.

CONSPECTUS OF *POLISTES* IMPACTS ON MAN

Polistes impacts on man might include (1) natural and managed predation of caterpillars and some other insects; (2) pollination of plants (Free 1975); (3) food for humans (Spradbery 1973) and for wildlife provided by brood and by adult wasps; (4) "plant guarding" of plants having extrafloral nectaries (Berkmann and Stucky 1981); (5) value as subjects for research in behavior, especially social behavior; (6) value as source of venom for sting allergy research and therapeutic use in desensitization of hyperallergic individuals (Gillaspy 1979a); and, as debits, (7) the pain and rarely medical bills occasioned by stings; and (8) impedence of activities in agriculture, around residential premises or elsewhere through fear of stings. Since a large proportion of nests tend to be in the open on structures, with ordinary care around bushes, etc. it should be possible to largely avoid stings. There appears a considerable consensus that, as with the honey bee, potential benefits outweigh detriments.

132

References

Anonymous, 1976. -- A preliminary study on the bionomics of hunting
    wasps and their utilization in cotton insect control. *Acta
    Entomologica Sinica*, 19(3), 303-308.
Ballou, H. A., 1915. -- West Indian wasps. *Agric. News*, 14, 298.
Beckmann, R. L. Jr. and Stucky, J. M., 1981. -- Extrafloral nectaries
    and plant guarding in *Ipomoea pandurata* (L.) G. F. W. Mey (Con-
    volvulaceae). *Amer. Jl. Bot.*, 68(1), 72-79.
Bellotti, A. and Arias, B., 1978. -- Biology, ecology and biological
    control of the cassava hornworm, *(Erynnis ello)*, *in* Brekelbaum,
    T. *et al.* Eds. *Proc. Cassava Prot. Workshop. CIAT, Cali, Colom-
    bia*, 7-12 November 1977.
Dhaliwal, J. S., 1975. -- *Polistes hebraeus* preying upon Thysanoptera.
    Curr. Sci., 44, 368.
Gallego, F. L., 1950. -- Estudios entomologicos: el gusano de las
    hojas de la yuca. *Rev. Fac. Nac. Agron. Medellin, Colombia*,
    12, 84-110.
Garcia, C., 1971. -- *Polistes* spp. predatory on adult Lepidoptera.
    Ent. News, 82, 274.
Gibo, D. L., 1974. -- A laboratory study on the selective advantage
    of foundress associations in *Polistes fuscatus* (Hymenoptera:
    Vespidae). *Can. Ent.*, 106, 101-106.
Gillaspy, J. E., 1979a. -- Mass collection of *Polistes* wasp venom by
    electrical stimulation. *Southw. Ent.*, 4, 96-101.
------------ 1979b. -- Management of *Polistes* wasps for cater-
    pillar predation. *Southw. Ent.*, 4, 334-352.
Iwata, K., 1976. -- *Evolution of Instinct. Comparative ethology of
    Hymenoptera*. Amerind, New Delhi (For Smithsonian Inst.), ix +
    535 p.
Kasuya, E., Hibino, Y. and Ito, Y., 1980. -- On "Intercolonial"
    cannibalism in Japanese paper wasps, *Polistes chinensis antenalis*
    Perez and *P. jadwigae* Dalla Torre (Hymenoptera: Vespidae). *Res.
    on Pop. Ecol.*, 22, 255-262.
Kirkton, R. M., 1970. -- Habitat management and its effect on popula-
    tions of *Polistes* and *Iridomyrmex*. *Proc. Tall Timbers Conf.*, 2,
    243-246.
Lawson, F. R., Rabb, R. L., Guthrie, F. E., and T. G. Bowery, 1961. --
    Studies of an integrated control system for hornworms on tobacco.
    Jl. Econ. Ent., 54, 93-97.
Oliver, A. D., 1964. -- Studies on the biological control of the fall
    webworm, *Hyphantria cunea*, in Louisiana. *Jl. Econ. Ent.*, 57,
    314-318.
Rabb, R. L., 1960. -- Biological studies of *Polistes* in North Carol-
    ina (Hymenoptera: Vespidae). *Ann. Ent. Soc. Amer.*, 53, 111-121.
----------, 1971. -- Naturally-occurring biological control in the
    eastern United States, with particular reference to tobacco
    insects. p. 294-311 *in* Huffaker, C. B. (Ed.). *Biological Con-
    trol*. Plenum Press, N. Y. xix + 511 p.
-------- and Lawson, F. R., 1957. -- Some factors influencing the
    predation of *Polistes* wasps on the tobacco hornworm. *Jl. Econ.
    Ent.*, 50, 778-784.
Reed, H. C. and Vinson, S. B., 1979. -- Nesting ecology of paper wasps

(Polistes) in a Texas urban area (Hymenoptera: Vespidae). *Jl. Kans. Ent. Soc.*, 52(4), 673-689.

Spradbery, J. P., 1973. -- *Wasps.* University of Washington Press, Seattle, xvi + 408 p.

The Economic Impact of the Africanized Honey Bee in South America

Lionel Segui Gonçalves,
Universidade de São Paulo

Since its introduction into Brazil (Kerr, 1967), the spread of the pure African bees and later the so-called Africanized bees has been relatively rapid, demanding major adaptations of the traditional beekeeping methodology to this new bee wherever it appears.

Today the occupied areas in South America extend from Argentina in the South to Venezuela in the North. It is reported to be already in Panama (Robinson, 1981). However, no data is available about these bees in that country.

Much of the rapid spread of the Africanized bees in South America can be attributed to its rapid colony growth as well as to its high adaptation to tropical and sub-tropical conditions and its swarming behaviour. According to Nascimento Júnior (1981) the swarming behaviour of the Africanized bees, in contrast to migratory behaviour, is not due to adverse conditions but to a combination of optimum environmental conditions and abundant food. He observed that the frequency of swarms was about 1.5 per year (in European bees the frequency is 0.5 per year), occurring practically throughout the year but especially in July and August in São Paulo State in Brazil, showing that the Africanized bees have an adaptive value at least three times higher than their European counterparts, which explains in part its unexpected rapid spread in South America.

The African bees (<u>Apis mellifera</u> <u>adansonii</u>) were introduced into Brazil in November, 1956. A review of the literature revealed that the <u>adansonii</u> bees had an outstanding productive ability which convinced Prof. W. E. Kerr to import these bees to Brazil and to carry out a selective program to reduce aggressiveness of these bees. Unfortunately the selection program was not carried out in time because of an accident caused by a visiting beekeeper who removed the double queen excluders from the hive entrances which allowed pure <u>adansonii</u> queens to swarm, starting the Africanization of the bees in South America. Very soon the beekeepers were surprised by the unexpected behaviour of the first colonies of the <u>adansonii</u> bees which caused panic and distress among apiculturists and the people in general. The reactions were immediately observed among beekeepers and among laymen who were greatly influenced by the

134

exaggerated "scare stories" printed in the media and shown in horror movies inspired by the fantastic news about the bees introduced into Brazil. Very soon the new bees from Brazil (adansonii) were referred to as "Brazilian-bees" or "killer-bees", which we considered improper names because these bees are not native to Brazil and also because we call our stingless bees "Brazilian bees".

The honey bee now spreading widely in South America is not pure adansonii, but rather the result of a hybridization of races, a hybrid retaining largely adansonii characteristics. This is why we call them Africanized bees. One of the most striking facts about the pure adansonii bees in their native habitat is the great variability in their behaviour, especially concerning aggressiveness, under different climatic conditions. This seems to explain the high variability in aggressiveness shown by the Africanized bees in South America. It is well known that the general nature of the aggressive behaviour (defense) in honey bees, which can be considered as a phenotypic character, is the product of the interaction between their genetic composition (genotype) and environmental factors. In many cases it is difficult to decide which of these two components is the more important. Some responses are caused by internal and some by external factors or by both. According to Stort (Gonçalves and Stort, 1978) the number of genes responsible for the aggressive behaviour of the honey bee is eight. Brandeburgo et al (1976) were able to prove that the aggressive behaviour of several colonies of Africanized bees tested in two regions of Brazil (São Paulo State and Pernambuco State) with completely different climatic conditions was much more influenced by external factors than by the bee's genotypic composition. The colonies tested in Pernambuco were four times more aggressive than those from São Paulo. The results indicated that the main responsibility for the different aggressiveness responses of the bees in the two places can be attributed to climatic conditions. The difference of aggressiveness of the Africanized bees tested decreased when the samples were under the same environmental conditions (Brandeburgo, 1979).

Even though the Africanized bees are more aggressive than the European bees, the adansonii bees brought to Brazilian beekeeping a completely new dimension. Just after the appearance of the first swarms of adansonii bees in Brazil many beekeepers went out of business, especially the hobby beekeepers. However, others adapted themselves to the new requirements imposed by the new bee, mainly the beekeepers from Santa Catarina and Paraná States. These are today the most productive states in honey production. Each of these two states produced between four and five thousand tons of honey in 1981, which represented about 40% of the Brazilian production of 20,000 tons.

The same phenomenon occurring in Brazil is happening in other countries in South America. When the first Africanized bees appear, there is a great divergence of opinion among the apicultural leaders of each country. It is important to take advantage of the knowledge of beekeepers and specialists who already learned how to handle these bees and adapted themselves to the new situation. In order to attack the problems caused by the introduction of the adansonii bees into Brazil, an integrated effort between scientists, beekeepers, and

authorities was developed for about 20 years. Under Prof. W. E.
Kerr's orientation many bee researchers dedicated full time all
these years to the study of the biology and behavior of the
Africanized bees as well as to the improvement of the bees and the
perfection of handling techniques and equipment. An intensive study
of the technology related to the instrumental insemination of queens,
very important for breeding programs, was also developed at the
Genetic Bee Lab in Ribeirão Preto since 1965. Many devices have
been developed since that time as, for example, the models of
Instrumental Insemination Apparatus "Gonçalves & Brites" which
received prizes in the last two International Apicultural Congresses
of Apimondia.

Many extension programs were carried out by extensionists,
contributing also to the renovation of apiculture with Africanized
bees and, consequently to the rebirth of Brazil's bee industry.
Today, 25 years after the introduction of adansonii into Brazil, it
can be said that thanks to these bees the beekeeping industry is now
better organized, the honey, wax and royal jelly production is
increasing each year, and pollination by the bees, especially
Africanized bees, as another source of income to beekeepers, is
already present in this country, mainly for citrus and apple trees.
The increased demand for beekeeping supplies also has affected other
sectors of the industry. Many beekeeping supply factories appeared
in the last few years in Brazil, especially in São Paulo State, to
satisfy the demand for wooden hives. The number of Brazilian
apiculture associations has considerably increased over the last 10
years, but the goal of the Brazilian Beekeeping Confederation to
reach 1 million hives has not yet been met. Even for the hobby
beekeepers who want to avoid excessive stinging by bees there is now
a solution, the split-sting mutation obtained by Soares (1981). The
bees with this mutation are unable to sting. This new characteristic which has a frequency up to 62%, after a selection program, can
be applied in programs demanding gentleness (Soares, 1980, 1981).

Hundreds of letters are answered monthly by the staff of the
Genetic Bee Lab in Ribeirão Preto, most of them related to questions
about beekeeping in general, how to start beekeeping, how to get
mated queens, etc. What surprizes us is the increasing number of
beekeepers who are interested in Africanized queens. We also
received letters from other countries in South and Central America
asking for information on how to obtain Africanized queens or pure
adansonii queens, which shows clearly that the "taboo" about the
"killer-bee" is no longer taken into consideration. Today it is
rare in Brazil in the media to hear of occurrences of accidents with
bees. In most of the states of Brazil apiculture with the
Africanized bees has been growing with handling no longer considered
to be a serious problem as it was in the sixties when these bees
were regarded as totally undesirable.

The improvement of the beekeeping industry in Brazil is very
encouraging for other countries in South America. Unfortunately for
everyone, a serious problem has appeared in the last few years in
South America, the mite Varroa jacobsoni. This mite is present in
Paraguay, Uruguay, Argentina and Brazil, representing a serious
threat to the beekeeping industry of these countries, especially

because there is no treatment until now, proven to be effective in the control of the mite.

References

Brandeburgo M. A. M., 1979. -- Estudo da influência do clima na agressividade da abelha africanizada. Faculdade de Medicina de Ribeirão Preto, São Paulo Brazil: Tese de Mestrado.

Brandeburgo M. A. M., Gonçalves L. S., Kerr W. E., 1976. -- Nota sobre o estudo do efeito das condições climáticas sobre a agressividade das abelhas africanizadas. Cienc. Cult. São Paulo, 28, 276-277.

Gonçalves L. S., Stort A. C. G., 1978. -- Honeybee improvement through behavioral genetics. Ann. Rev. Entomol., 31, 197-213.

Kerr W. E., 1967. -- The history of the introduction of African bees in Brazil. S. Afr. Bee J., 39, 3-5.

Nascimento Júnior A. F., 1981. -- Estudo da influência de fatores ambientais no comportamento enxameatório, migratório e no desenvolvimento de colmeias de abelhas africanizadas. Faculdade de Medicina de Ribeirão Preto, São Paulo Brazil: Tese de Mestrado.

Robinson F. A., 1981. -- Africanized bees a problem that won't go away. Am. Bee J., Sept. 1981, 625-626.

Soares A. E. E., 1980. -- Estudo do carater ferrão aberto em Apis mellifera L. (Hymenoptera: Apoidea). Faculdade de Medicina de Ribeirão Preto, São Paulo Brazil: Tese de Doutoramento.

Soares A. E. E., 1981. -- Split-sting: A new honeybee character. J. Apic. Res., 20, 140-142.

The Economic Importance of Termites in North America

Joe K. Mauldin, USDA-Forest Service,
Gulfport, Mississippi

Termites are beneficial in nature because they help convert dead wood to mineral soil, serve as food for other animals, and are ideal for scientific investigations of insect social systems. Only when termites make their home in, and take their food from, man's wooden possessions should they be controlled. Control in this case does not necessarily mean killing the insects, although currently used termiticides are toxic to termites.

An effective repellent could be as effective as a toxicant. In fact, the γ isomer of chlordane, the most widely used termiticide in North America, is both toxic and repellent to subterranean termites and termites do not necessarily come in contact with the termiticide.

Subterranean termites are found in every one of the United States except Alaska. The predominant subterranean termites, primarily Reticulitermes spp., Coptotermes formosanus Shiraki, and Heterotermes spp., account for about 95 percent of the termite damage to wood and wood products (Moreland, 1981). Although of lesser economic importance, dry-wood termites, primarily Incisitermes spp., Cryptotermes brevis (Walker), Neotermes castaneus (Burmeister), and Kalotermes spp., may cause considerable damage to wood and wood products. Dry-wood termites occur primarily in the Southeastern Coast States, California, and Mexico. Damp-wood termites, Zootermopsis spp., cause damage and economic loss in the northwestern United States and southwestern Canada.

Wooden buildings, marine pilings and docks, furniture, utility poles, fenceposts, paper products, and logs and lumber in storage often are infested and damaged. Termites also damage many noncellulosic items, including golf balls, lead cables, etc. In such situations these items are typically in the termites' path to food or are located adjacent to an infestation. Most subterranean termite damage occurs to buildings in formerly forested areas. Termites simply begin feeding on the new sources of food (a house) because they are conveniently available and the termite's usual food sources, such as dead trees and stumps, are not available.

Despite the availability of technology for minimizing or virtually eliminating infestations (Johnston et al., 1972; Moore, 1979), subterranean termites continue to cause millions of dollars in damage. No data have been published on the total cost of termite damage in Canada, Mexico, or the United States. It is apparent that a well-organized effort is needed to record, report, and publish the extent of losses caused by termites in these three countries.

There have been, however, several estimates of damage cost in the United states. Ebeling (1968) estimated that the expense of termite prevention, control, and repair of damaged wood costs the people of the United States $500 million annually. Ebeling's figure is quoted most often but other estimates range from $100 million (Lund, 1967) to $3.4 billion (USDA, 1974) annually. The only estimate based on data was by Williams and Smythe (1979). They used treatment and cost records to estimate that $168.8 million (in 1976 dollars) was spent for prevention, control, and repair on single family homes in 11 southeastern states. These states are in a region of the United States with the highest termite hazard, but the estimate does not include losses to multifamily dwellings, commercial establishments, public buildings, new houses built between 1970 and 1980, and military structures.

The U.S. Environmental Protection Agency (EPA, 1981) extrapolated the Williams and Smythe data using nationwide survey data on pesticide usage (EPA, 1979) and arrived at an annual loss figure of $470.8 million (1976 dollars) for the entire United States. Using the Bureau of Census' "New One Family Houses Construction Cost Index," the $470.8 million figure was increased to $753.4 million. Again, multifamily dwellings, commercial establishments, public buildings, new homes built between 1970 and 1980, and military structures were not included. As the EPA (1981) noted, the $753.4 million figure has an upward bias because it is extrapolated from 11 states with the highest termite infestation. EPA also justifiably noted that upward bias in the estimate is undoubtedly offset by the downward bias resulting from the exclusion of losses in multifamily dwellings, commercial establishments, and public buildings. Therefore, the figure of $753.4 million for annual termite damage in the United States is probably the best estimate available. No similar economic estimates of termite damage are available from Mexico or Canada.

Prevention of termite infestations involves proper design, construction, pretreatment with an effective termiticide, and at least an annual inspection. The seven chemicals currently registered for termite control in the United States are aldrin, chlordane, chlorpyrifos, dieldrin, heptachlor, lindane, and pentachlorophenol.

Research is now aimed at controlling termites through the use of baits impregnated with a slow-killing toxicant (Esenther and Gray, 1968; Mauldin and Rich, 1980). The bait for Reticulitermes spp. is a small (2.5 X 0.6 X 5 cm) block of sweetgum (Liquidambar styraciflua L.) decayed to about 15 percent weight loss by the fungus Gloeophyllum trabeum (Pers. ex Fr.) Murr. Mirex was somewhat effective in the bait-block method against Reticulitermes spp. (Esenther and Beal, 1978) but is no longer available in the United States. Other chemicals are now being tested.

This publication reports research involving pesticides. It does not contain recommendations for their use, nor does it imply that the uses discussed here have been registered. All uses of pesticides must be registered by appropriate State and/or Federal agencies before they can be recommended.

CAUTION: Pesticides can be injurious to humans, domestic animals, desirable plants, and fish or other wildlife--if they are not handled or applied properly. Use all pesticides selectively and carefully. Follow recommended practices for the disposal of surplus pesticides and pesticide containers.

REFERENCES

EBELING, W., 1968. - Termites: Identification, biology, and control of termites attacking buildings. Univ. Calif., Calif. Agric. Exp. Stn Ext. Serv. Man. 38, 74 p.

ESENTHER, G.R., BEAL, R.H., 1978. - Insecticidal baits on field plot perimeters suppress Reticulitermes. J. Econ. Entomol. 71, 604-607.

ESENTHER, G.R., GRAY, D.E., 1968. - Subterranean termite studies in southern Ontario. Can. Entomol., 100, 827-834.

JOHNSTON, H.R., SMITH, V.K., BEAL, R.H., 1972. - Subterranean termites. Their prevention and control in buildings. U.S. Dep. Agric., Home and Garden Bull. No. 64, 30 p. (Rev. Mar. 1979)

LUND, A.E., 1967. - The laboratory and field study of subterranean termites. Int. Pest Control, 9 (4), 29-33.

MAULDIN, J.K., RICH, N.M., 1980. - Effect of chlortetracycline and other antibiotics on protozoan numbers in the eastern subterranean termite. J. Econ. Entomol. 73 (1), 123-128.

MOORE, H.B., 1979. - Wood-inhabiting insects in houses: Their identification, biology, prevention and control. U.S. Dep. Agric., For. Serv., and Dep. Housing and Urban Dev. (Interagency Agreement IAA-25-75), 133 p.

MORELAND D., 1981. - Subterranean termites. Pest Control Technol. 9 (3), 30-34.

U.S. DEP. AGRIC., 1974. - Insects affecting man and his possessions. Research needs in the southern region. Joint Task Force Rep. of the Southern Region Agric. Exp. Stns. and USDA, 34 p.

U.S. ENVIRONMENTAL PROTECTION AGENCY, 1979. - National household pesticide usage study, 1976-1977. Final Report of a study (Contract No. 68-01-4663) conducted by the Epidemiologic Pesticide Studies Center, Colorado State Univ., 126 p.

U.S. ENVIRONMENTAL PROTECTION AGENCY, 1981. - Comparative benefit
    analysis of seven chemicals registered for use against
    subterranean termites. Draft of part of the Benefits and Field
    Stud. Div., Off. Pestic. Programs, Environ. Prot. Agency's
    cluster analysis, 185 p.

WILLIAMS, L.H., SMYTHE, R.V., 1979. - Estimated losses caused by wood
    products insects during 1970 for single-family dwellings in 11
    southern states. U.S. Dep. Agric., For. Serv. Res. Pap. SO-145,
    10 p.

Economically Important Termites of the Oriental Regions and Their Management

P. K. *Sen-Sarma,* Forest Research
Institute, Dehra Dun, Uttar Pradesh

Termites which form an integral part of extensive fauna in the tropical and subtropical regions are of great economic importance in the Orient. The range of material damaged by these ubiquitous insects is very wide and includes agricultural, horticultural, forestry and plantation crops, cellulosic components in buildings, wooden bridges, poles, posts, underground cables, and hosts of synthetic material, various ordnance stores, etc. Out of approximately 2,500 and odd species known so far from the world, about 550 species have been recorded from the Oriental Region and many still await discovery. Though the precise figures for the monetary value of losses are not available for any country in the Oriental Region, the loss must run into millions of dollars annually.

TERMITES DAMAGING PLANTS

(1) Agriculture crops commonly damaged by termites are wheat, millets, pulses, cotton, sugarcane, vegetables, etc. in South Asia, Indonesia and the Philippines. The common species of termites causing serious damage to these crops include *Coptotermes heimi, Eremotermes paradoxalis, Odontotermes/assmuthi, O. distans, O. indicus, O. microdentatus, O. obesus, Microtermes mycophagus, M. obesi, M. unicolor* and *Trinervitermes biformis* in the Indian subcontinent (Sen-Sarma, 1974); *Heterotermes philippinensis*(damaging only sugar-cane) in the Philippines (Harris, 1961), *Odontotermes formosanus* attacking sugar cane and other crops in Taiwan, Thailand, Vietnam, etc., *Hypotermes obscuriceps* attacking different crops in Ceylon. A loss of 2.5% in the tonnage of the cane and 4.5% of sugar output is caused in Bihar in India (Aggarwal, 1955). The most important termites attacking wheat in seedling stage in South Asia are *Microtermes obesi* and *Odontotermes obesus. Anacanthotermes macrocephalus* attacks wheat grains during storage in arid areas of western India and Sind in Pakistan (Roonwal, 1979). Groundnut plants in India and Pakistan are attacked by *O. obesus* and *Trinervitermes biformis* (Sen-Sarma, 1974).

(2) Horticultural crops attacked by termites include apple, cashew, *Zizyphys mauritiana, Sapota acharis, Citrus* spp., grapevine, guava, mango, peach, pomegranate, etc. Termites recorded attacking these in

South Asia are *Bifiditermes beesoni, Neotermes greeni, Coptotermes heimi, Odontotermes obesus, Microtermes mycophagus, M. obesi, Trinervitermes biformis,* etc.

Taproot and basal part of the stem of young apple plants are damaged by *C. heimi.* Living branches of cashew tree are excavated by *N. greeni.* Seedlings and young plants of *Citrus* spp. are damaged by *O. obesus* and *T. biformis.* Newly planted sets of grape-vines are often attacked by *Odontotermes* spp. which hollow out the entire vine and kill the tender sprouting shoots (Sen-Sarma, 1974). For the countries in South East Asia, very little information is available. Important termite species attacking different horticultural crops are *Coptotermes curvignathus, C. havilandi, Globitermes sulphureus, Macrotermes gilvus,* in Thailand, Malayasia, Indochina Region, Indonesia and the Philippines (Roonwal, 1979).

(3) Plantation crops commonly damaged, often quite seriously, are tea, coffee and cocoa, coconut palm, jute and allied fibre crops, and rubber tree. The major pests of tea bushes in South Asia are *Postelectrotermes militaris, N. greeni, Glyptotermes dilatatus* (all in Ceylon), *Microcerotermes* spp. in N. E. India, *Coptotermes ceylonicus* in Ceylon and S. India, and *Microtermes pakistanicus* in Bangladesh and Malaysia. Coffee is attacked in Ceylon by *G. dilatatus* and in South India by *Dicuspiditermes fletcheri, Grallatotermes grallatoriformis* and *Nasutitermes indicola.* The damage to coconut palm by termites is restricted primarily to the seedlings in nurseries and to young palms in the plantations. *O. obesus* is an important pest in the coconut along the coast-line of India and as many as 30-40% seedlings may be killed in years of drought (Roonwal, 1979).

In Ceylon, damage to coconut palm is done by *O. redemanni* and *Hypotermes obscuriceps* (Fernando, 1962), and latter also attacks living palms in Indochina region. *O. bogoriensis* is a pest of living palms in Indonesia. Cultivated jute in India is subjected to attack by *M. obesi* which damages mature standing crop, the damage, however, remaining obscure until plants start lodging (Dutt, 1962), Sunn hemp *(Crotolaria juncea)* and Sisal *(Agave sislana)* are attacked by *Odontotermes* spp. in India. The rubber trees are seriously attacked by *Coptotermes curvignathus* in Thailand, Malayasia, Indochina Region and Indonesia (Roonwal, 1979). In Ceylon and South India *Glyptotermes dilatatus* and *Coptotermes ceylonicus* are serious pests of rubber (Fernando, 1962).

(4) Forestry crops are damaged primarily during seedling stages particularly when these are one to three years old. The valuable and fast-growing *Eucalyptus* is perhaps the worst sufferer specially in arid areas. Termite species involved in South Asia are *Anacanthotermes macrocephalus, Microcerotermes minor, O. indicus, O. microdentatus* and *O. obesus.* Recently poplars have been damaged severely by *Coptotermes heimi* and *Odontotermes* spp. in India. In Indonesia, plantation teak trees are severely damaged by trunk

inhabiting termite, *Neotermes tectonae*, often killing the plant outright (Kalshoven, 1950)

(5) Cellulosic components in buildings, bridges, etc. Termites have attracted the attention of human beings primarily because of enormous damage which they cause to cellulosic material used in human dwellings. With the advancement of civilization and clearance of areas that formed natural sources of food for termites, termite problems in human dwellings have, of late, assumed serious proportion in the entire Oriental Region. In addition to destruction of wooden doors, window frames, built-in wooden almirahs, decorative panels on walls, termites also damage building contents like carpets, furniture, documents, depositories in archives, etc. In addition, synthetic fibres like nylon, dacron, terelyne, etc. and natural fibres like cotton, silk, jute, etc. are destroyed by termites. Both dry-wood and subterranean termites are responsible for the damage to timber. Dry-wood termites involved are *Cryptotermes bengalensis* (India, Bangladesh), *C. domesticus* (India, Ceylon, Thailand, Malaysia, Indochina Region, Indonesia and Taiwan), *C. dudleyi* (India, Bangladesh, Ceylon, Indonesia and the Philippines). The incidence is, however, severe in coastal areas. The presence is revealed by the fallen hexagonal faecal pellets.

Among the subterranean termites, the most important are the species belonging to the genera *Heterotermes, Coptotermes, Odontotermes,* etc. In addition, *Anacanthotermes vagans* is a very serious pest of wood-work in houses in Baluchistan in Pakistan (Choudhry and Ahmad, 1972). *Heterotermes indicola, Coptotermes heimi, Odontotermes indicus,* etc. cause serious damage in India and Pakistan. In Ceylon, damage is primarily caused by *C. cyeonicus, O. redemanni* and *Hypotermes obscuriceps. C. havilandi* is a very serious pest of buildings in Indonesia, Malayasia and Thailand, while *C. vastator* causes considerable damage to houses in the Philippines (Light, 1934; Kalshoven, 1962). Detailed survey indicates massive termite infestation in buildings in Jakarta, Indonesia (Rentokil, 1980). *C. formosanus* which is a very destructive species in Japan, Taiwan and southern mainland of China has been introduced in Ceylon and Pakistan in the recent past (Choudhry and Ahmad, 1972). In Thailand, Malayasia and Indochina Region, *Globitermes sulphureus* is a serious pest of buildings. *Schedorhinotermes malaccensis* and *S. medioobscurus* regularly attack buildings in Malayasia. *Prorhinotermes libiaoensis* and *Nasutitermes panayensis* attack wood-work in houses in the Philippines.

MANAGEMENT PRACTICES

The management of termites in the Oriental Region can be conveniently grouped into two categories, *viz* management of (1) wood-inhabiting termites and of (2) subterranean termites. Termites inhabiting tea bushes are primarily managed by cultural practices (Fernando, 1962). Attack by *Cryptotermes* spp. is avoided by either using treated wood or preventing swarming adults to invade wood-work in houses by putting off the light. *In situ* treatment by toxic dusts or liquid is, in general, followed for eradication of existing

infestation. Fumigation of the entire building as is commonly done
in advanced countries is not followed due primarily to prohibitive
cost and inconvenience for the dwellers to vacate the premises during
fumigation. Attack by all types of subterranean termites is managed
primarily by soil-poisoning by means of soil insecticides like aldrin,
heptachlor, chlordane and lindane. The dosages used are 0.5% in all
cases except in chlordane where 1.0% is used. This practice is
followed to protect living plants as well as human dwellings. For
village huts, the recommended insecticide is mixed with the mud
thoroughly and then used for the construction of mud-wall. Precon-
struction treatment in buildings made of bricks and mortar involves
poisoning the soil of the foundation trench, floor area before laying
the slabs, etc. with aqueous solution of an appropriate insecticide.
Infestations in existing buildings are eradicated by post-construc-
tion treatment with an insecticide which includes poisoning the soil
around the foundation walls, underneath the joint of the walls and
the floor by drilling holes suitably (Anon., 1973). Some decay
fungi have been recorded to attract termites (Tyagi *et al.*. 1980,
Natawiria *et al.*, 1979), but no work on use of attractant-toxic baits
to suppress termite population has been carried out.

DISCUSSION AND CONCLUSION

Notwithstanding their destructive propensities, termites play no
mean role in the break-down of cellulosic material left in the
forests, thus adding nutrients to the soil which become available to
succeeding generations of plant communities. The continuation of the
biospheric ecosystem, therefore, is greatly dependent, along with
other biodeteriorating agents, on the beneficial activities of
these tiny organisms. In the Oriental Region, the magnitude of
their beneficial role is immense and therefore the salvation lies
in allowing these tiny creatures to contribute their mite in
maintaining the biosphere along with restrictive use of chemicals
to safeguard human properties.

REFERENCES

Anonymous, 1973. - Code of practice for anti-termite measures in buildings. IS 6313, Indian Standard Inst., New Delhi, 9 p.

Aggarwal, S. B. D., 1955. - Control of sugarcane termites. J. Econ. Ent. , 48, 533-537.

Chaudhry, M. I, and Ahmad, M., 1972. - Termites of Pakistan. Pakistan For. Inst., Peshwar, 70 p.

Dutt, N., 1962. - Preliminary observations on the incidence of termites attacking jute. In Termites in the Humid Tropics (Proc. New Delhi Symp., 1960). UNESCO, Paris, 217-218.

Fernando, H. E., 1962. - Termites of economic importance in Ceylon. In Termites in the Humid Tropics, 205-210.

Harris, W. V. 1961. - Termites, their recognition and control. Longman, London, 187 p.

Kalshove, L. G. E., 1950. - De Plagen van de Cultuurgewassen in Indonesia, Vol. I, W. van Hoeve, the Hague, 512 p.

Kalshoven, L. G. E., 1962. - Observations on Coptotermes havilandi Holmg. (javanicus Kemn.) (Isoptera). Beaufortia, 101, 121-137.

Light, S. F.. - The termite fauna of the Philippine Islands and its economic significance. In Termites and termite control, (ed. C. A. Kofoid), California Univ. Press, 319-322.

Natawiria D., Tarumingkang, R. C. et al., 1979. - (The attraction of fungus-invaded wood extracts to subterranean termites). Forum Sekolah Pasca Sarjana, 3, 1-14. (In Indonesian with English summary).

Rentokil, Indonesia, 1980. - Survey reveals massive infestation of termites in Jakarta, Indonesia. Pest Control, 48, 20-22.

Roonwal, M. L., 1979. - Termite life and termite control in tropical South Asia. Scientific Publisher, Jodhpur, 176 p.

Sen-Sarma, P. K., 1974. - Ecology and Biogeography of the termites of India. in Ecology and Biogeography in India (ed. M. S. Mani), W. Junk, The Hague, 421-472.

Tyagi, B. K, Sen-Sarma, P. K., Rehill, P. S., and Pandey, P. C., 1981. Termites-fungi interactions. I. Biossassay of decayed wood to Coptotermes heimi (Was), Neotermes bosei Snyder and Microcerotermes beesoni Snyder. - Proc. 2nd Intern. Sem. on Management of termites in Arid zone, Assuit (Egypt), Nov. 1981.

Abstracts

Structural wood destroying termites
(Isoptera) of Pakistan

Muhammad Saeed Akhtar
Department of Zoology, University of the Punjab,
New Campus, Lahore, Pakistan

Eleven species of termites have been recorded damaging wood-
work in the buildings in Pakistan. These are: <u>Anacanthotermes</u>
<u>vagans</u> (Hagen), <u>Psammotermes</u> <u>rajasthanicus</u> Roonwal and Bose,
<u>Coptotermes</u> <u>heimi</u> (Wasmann), <u>Heterotermes</u> <u>indicola</u> (Wasmann),
<u>Odontotermes</u> <u>obesus</u> (Rambur), <u>Odontotermes</u> <u>parvidens</u> Holmgren and
Holmgren, <u>Microtermes</u> <u>obesi</u> Holmgren, <u>Microtermes</u> <u>unicolor</u> Snyder,
<u>Microcerotermes</u> <u>baluchistanicus</u> Ahmad, <u>Microcerotermes</u> <u>heimi</u>
Wasmann and <u>Microcerotermes</u> <u>tenuignathus</u> Holmgren. A key to all the
species, based on soldier and worker castes, is given. Their distribution and foraging patterns are discussed.

Biological Control of the Termite <u>Nasutitermes</u> <u>exitiosus</u>
with <u>Metarhizium</u> <u>anisopliae</u> - Life Cycle of the Fungus

Heinz Hänel
Desowag Bayer Holzschutz AG - University of Frankfurt

After topical infection of termite workers with conidia of <u>Me-</u>
<u>tarhizium</u> <u>anisopliae</u>, the whole life cycle of the insect pathoge-
nic fungus was investigated. Germination of conidia takes place at
the relative humidity, which occurs in mounds of this termite
species. It can penetrate almost any part of the integument. The
formation of appressoria helps the fungus, not to become groomed off
by the termites. Quickly muliplicating hyphal bodies are distributed throughout the body by the haemolymph stream. At that stage
the insect dies due to various fungal toxins.
Now the pentration of tissues begins in the following sequence:
Fat body, muscle tissue, glands and epithelia, nerval tissue,
digestive system. Before the gut is penetrated, the fungus breaks
through to the outside and up to 10^8 conidia per insect are formed.
They can infect nest mates, which get in close contact to the dead
termites, and in this way spread the disease.

148

Ecopathological aspects in the control of <u>Acormyrmex</u> <u>octospinosus</u>
Reich (<u>Form</u>., <u>Attini</u>) by entomophagous fungi

Alain Kermarrec and Maryvonne Decharme
INR-CRAAG, Station de Zoologie, 97170 Petit-Bourg (Guadeloupe)

The economical importance of leaf-cutting ants in agricultural
systems of neotropical developing countries is generally well underlined. Taking in reference the 1978 EPA report on the human impact
of Mirex[R], new control strategies are urgently needed. Studies on
the use of entomophagous fungi for a biological control of Attine
ants have been undertaken. The following <u>Fungi imperfecti</u>
(moniliales) have shown an excellent in virtro pathogenicity
against <u>A. octospinosus</u>: <u>Beauveria bassiana</u>; <u>Metarhizium anisopliae</u>
and <u>Poecilomyces fumosoroseus</u>. Four strains have been selected for
their pathogenic power, conidiospore production, rusticity and
termal preferenda. However, in vivo preliminar tests on laboratory
and natural nests, put in evidence a certain number of constraints
related to the detection of the conidiospores by the worker ant and
to the lack of epizootical process in the social context under
normal conditions. The main reasons for this strong social homeostasis are studied. At least three conjugated antibiohazard events
occur: - social vigilency (i.e. sensory recognition of a biohaz-
ardous objects) and intense grooming activity operated by several
thousand of ants simultaneously.
- antibiotical excretions by these worker ants
- antibiotical productions by the symbiotic basidiomycetes.

The composition, geographical origin and dispersion
of the termites (Isoptera) of the Islands of
Andaman and Nicobar, Indian Ocean

P.K. Maiti
Zoological Survey of India, Calcutta, India.

The composition, geographical origin and probable mode of
dispersal of termites of the islands of Andaman and Nicobar have
been studied through extensive and intensive field explorations
during the last five years. Out of 30 species studied so far, 16
species are endemic to these islands, while the remaining 14
species have migrated from the adjacent land masses of Malaya and
its Archipelagos, India, Burma and Sri Lanka. Present study further reveals that the Andaman group of islands supports predominantly the Indo-Burmese termite fauna and the Nicobar group contains the Malayan elements. Further, the present faunal structure
is the product of accumulated elements from the adjacent land
masses, as well as, of its own endemic species. The mode of colonization and inter-island distribution pattern of these insects have
been discussed in the light of environmental factors, as well as,
the characteristic features of these insects. It is concluded that
the termites have reached these islands through some direct land
connection during the past geologic time or by some artificial means,
such as, drifted wood, raft, human agency, etc., or by overseas dispersal by flight.

Mound Building Behaviour of Odontotermes wallonensis
Wasmann (Isoptera: Termitidae)

D. Rajagopal
University of Agricultural Sciences, Bangalore, India

This mound building behaviour of Odontotermes wallonensis which
includes design and architecture and construction and expansion of
the mound was studied. This species was observed to construct an
extremely hard and compact dome shaped earthen structure with open
chimney like out-growths at the surface, which help in ventilation.
The internal architecture comprised of ventilation shafts, galleries and a number of fungus combs located in separate earthen
chambers. The main depository was comparatively larger. The royal
chamber was always situated very close to the main depository,
normally below ground level. Incubation cavities were present
around the royal chamber, some of which were utilized for incubating eggs during rainy season. Food chambers were usually found
in large mounds and served as temporary storage structures during
unfavourable months. The active construction and expansion of the
mound was confined to April-May and October-December months.
Mound construction, which was usually attended to during cooler
periods in mornings and evenings were favoured by rainfall and
relative humidity. Mounds increased in size every year, both in
height and diameter, above the ground level during its early growth.

The Detection and Distribution of
Formosan Termites in Southeastern Florida

Catherine Thompson
The University of Florida

The Formosan termite, Coptotermes formosanus, was found in a
Hallandale condominium in July, 1980. Dubbed the "supertermite"
colonies of this species may contain 360,000 individuals, compared
to a Florida species with 60,000, and can adapt to living above
ground within buildings. Following the discovery of four more
infestations in the spring of 1981, a survey was undertaken in
Hallandale with 1/4 mile boundries surrounding the original infestation. One by two inch pine stakes were driven into soil near
shrubs and trees in condominium plantings and checked at two month
intervals. The original infestation site was again positive for
Formosans, not a surprising event since stumps at the corners of
the building, left untreated, contained Formosans. To date, an
additional five condominium grounds have been positive for Formosan
termites. Plans are underway to monitor mating flights with a
light trap survey in the spring.

PRESOCIAL BEHAVIOR
Organized by George C. Eickwort

Introduction

George C. Eickwort, Cornell University

     Presocial insects -- those that exhibit some degree of social behavior short of eusociality (Wilson, 1971) -- occur in most orders. They include insects that occur in aggregations while feeding or ovipositing, subsocial insects whose adults care for nymphs or larvae, and parasocial Hymenoptera in which several adult females of the same generation share a nest. Wilson's monumental "The Insect Societies" (1971) includes an excellent discussion of presocial behavior, and Eickwort (1981) has summarized more recent studies. The most exciting current research behaviorally and chemically analyzes the mechanisms that form and maintain associations, analyzes the genetic relationships of the partipants in order to test models of kin selection, or seeks the selective advantages and disadvantages of presocial associations. Analyses of presocial behavior are also providing significant insights into the conditions under which reproductive division of labor, and thus eusocial behavior, might evolve. The five papers in this symposium will present recent studies across a behavioral spectrum, involving three orders of insects studied in four countries. I shall attempt to provide a brief general background to presocial behavior in this introduction so these papers may be viewed in a common perspective.

     Feeding aggregations typically occur among phytophagous insects that cluster their eggs -- Orthoptera, Psocoptera, Hemiptera, Coleoptera, Lepidoptera, and Symphyta. Commonly only the young instars are gregarious. Individuals respond to pheromones to maintain the aggregations and to disperse when attacked, and in the laboratory exhibit more synchronized moulting, faster development, and less mortality than do their isolated siblings. Mutual defense may be the most important advantage for gregarious insects, which are typically chemically protected.

     Even in those gregarious insects like tent caterpillars and argid sawflies that display sophisticated communication systems and coordinated feeding, migrating, and shelter-building behavior, there is no evidence of altruistic behavior. Indeed, offspring from different egg clusters may form common feeding aggregations when they meet. Aphid aggregations are unusual in that all members of a clone are nearly identical genetically, because of their parthenogenetic origin, so kin selection should be an important factor. In this symposium Aoki reports on his discoveries of non-reproductive soldier aphids and migratory morphs that invade galls inhabited by other clones.

Subsocial insects that care for their young without nests occur among the Blattodea, Orthoptera, Dermaptera, Psocoptera, Embiidina, Hemiptera, Thysanoptera, Coleoptera, and Hymenoptera. Such care is an extension of parental guarding of clustered eggs, and is basically a strategy of enhancing the survival of relatively few offspring. Parental care is usually given only to the youngest instars, when the benefit is greatest and the cost to the parents least. Usually only the mother is involved, but there are a few instances of both parents or even males alone (Smith, 1980) providing care.

Most subsocial insects are long-lived as adults and able to survive without feeding for long periods or to feed at the oviposition site. They are usually relatively immune to predation, often because of chemical protection. The main function of subsocial care may be protection from generalized predators and parasites. The parents may also keep eggs fungus-free, maintain feeding aggregations of the hatching nymphs or larvae, and sometimes even provide nourishment.

The order Hemiptera includes many well-studied examples of subsocial care. Those in Heteroptera have been reviewed by Melber and Schmidt (1977). In the Homoptera, subsocial care occurs frequently in the Membracidae. In this symposium Wood places his extensive studies of treehoppers in an ecological perspective to show the interaction of plant hosts, parental care, predation, mutualism with ants, and geography.

Nesting is a form of parental investment in which the parents invest time and energy in the nest, rather than in prolonged contact with the eggs and nymphs or larvae themselves. I define a nest as a structure in which eggs are deposited and food for the resulting offspring is brought from outside the structure by the parents (Eickwort, 1981). Subsocial nesting is limited to those cases in which the adults remain in the nests until after the immatures hatch, and in Hymenoptera is further (and artificially) restricted to those taxa that practice progressive provisioning (Michener, 1974). True nesting has evolved in only five orders; the Orthoptera, Isoptera, Dermaptera, Coleoptera, and Hymenoptera. All of these orders include species that are subsocial (or eusocial) nesters.

Nesting insects can gather widely scattered or ephemeral food for the later consumption by immatures, so the food resources are more diverse than those used by non-nesting subsocial insects. Both parents may participate in providing for the nest, although there are no instances of males providing the sole care. In subsocial nesters, the adults defend the nest after the immatures hatch and may keep it free from fungal contamination, and in some cases add to or condition the provisions. In this symposium, Halffter considers the subsocial dung beetles (Scarabaeinae), carrion beetles (Silphidae), and passalid beetles in an evolutionary and ecological context, stressing the significance of sexual behavior and adaptations to ephemeral food. The contrast between

subsocial Coleoptera and Hymenoptera should interest all participants in this congress.

It is in the Hymenoptera, of course, that the greatest diversity of nesting behavior is found. The success of the Aculeata is
tied to nesting. In contrast to the Coleoptera, with a few exceptions male Hymenoptera are not involved in building and provisioning
nests. Many species aggregate their nests in close proximity
despite widespread, apparently suitable substrates. In addition,
several females may occur in the same nest. Such parasocial associations can be subdivided into communal societies, in which each
female builds, provisions, and oviposits in her own cells; quasisocial societies, in which several females cooperate in building
and provisioning a cell in which only one oviposits, although all
are potential egglayers; and semisocial societies, in which some
females are the principal egglayers and others lay few or no eggs
(Michener, 1974).

Communal females occur commonly among bees and wasps, but only
recently have studies of marked individuals enabled us to appreciate the dynamics of these associations. In this symposium Evans
and Hook report on the remarkable nests of Australian Cerceris wasps
(Sphecidae), in which egglaying females are consistently either
guards or foragers. They speculate that communal nesting evolved
within Cerceris as a response to ground-dwelling nest predators.

Pleometrotic (multiple foundress) associations of sisters that
become eusocial colonies when their offspring mature are common
among bees and wasps. West Eberhard (1978) has convincingly argued
that such pleometrotic associations are the major setting for the
evolution of castes in wasps. They are frequently semisocial; the
behaviorally dominant female is the principal egglayer and does
little or no foraging. In this symposium Sakagami and Maeta
present their latest experiments in artificially inducing the formation of semisocial and eusocial colonies of normally solitary
Ceratina bees. Why reproductive castes form when these bees are
forced to nest together, and not when sister Cerceris or other
communal Hymenoptera do likewise, should be a topic for discussion.

References

EICKWORT G.C., 1981. - Presocial insects, p. 199-280 in Hermann H.R.
    ed., Social Insects vol. II. Academic Press.
MELBER A., SCHMIDT G.H., 1977. - Sozialphänomene bei Heteropteren.
    Zoologica (Stuttgart), 127, 19-53.
MICHENER C.D., 1974. - The Social Behavior of the Bees. Harvard
    Univ. Press, 404 p.
SMITH R.L., 1980. - Evolution of exclusive postcopulatory paternal
    care in the insects. Fla. Entomol., 63, 65-78.
WEST EBERHARD M.J., 1978. - Polygyny and the evolution of social
    behavior in wasps. J. Kans. Entomol. Soc., 51, 832-856.
WILSON E.O., 1971. - The Insect Societies. Harvard Univ. Press,
    548 p.

Soldiers and Altruistic Dispersal in Aphids

Shigeyuki Aoki, Hokkaido University

Recently some aphid species producing sterile soldiers have been found. Whether these aphid species are "eusocial" or not is merely a matter of definition. As Dawkins (1979) pointed out, it is clear that altruistic behaviour can readily evolve in a clonal colony of the aphid species adopting cyclic parthenogenesis (see also Hamilton et al., 1981, Krebs and Davies, 1981). Also, except non-feeding sexuales of some species, every aphid has stylets which can be used as a weapon. Therefore, the occurrence of soldiers in an aphid colony may not be so surprising provided that the colony is a clone. Other than as a soldier or perhaps also as a worker (cf. Aoki, 1980 a), an aphid can benefit its relatives as a migrant who does not use those resources that its relatives are likely to use, but who disperses to compete with non-relatives for other resources. In the following I exemplify the occurrence of the two kinds of altruism in aphids.

APHID SOLDIERS

Aphid soldiers have the following characteristics: 1) They attack predatory intruders, often in a self-sacrificing manner. 2) They are 1st or 2nd instar and do not moult any more, so that they, of course, do not reproduce at all. 3) They differ morphologically from the conspecific "normal" larvae of the same instar.

Three types of soldiers have hitherto been found (Table 1): 1) pseudoscorpion-like 1st instar soldiers of Colophina spp. (Pemphiginae, Eriosomatini) which clutch at a predatory intruder like a syrphid larva and sting it with their stylets; 2) pseudoscorpion-like 1st instar soldiers of Pseudoregma spp. and Ceratovacuna japonica (Hormaphidinae, Cerataphidini) which clutch at a predatory intruder with their thickened forelegs and pierce it with their acute frontal horns; and 3) second instar soldiers of Astegopteryx styracicola (Hormaphidinae, Cerataphidini) which not only sting insect predators but also sting human skin with their stylets and cause troublesome itch.

These soldier-producing aphids are thought to have passed through a stage where larvae of a certain instar attack predators but no dimorphism occurs in the instar. Colophina spp. on Zelkova

Table 1. -- Aphid species producing soldiers. Information based on unpublished observations is included.

Species	Host on which soldiers are produced	Characteristics of soldiers	References
Colophina clematis C. arma C. sp.	Clematis apiifolia* C. stans* C. floribunda*	First instar, fore and mid legs greatly thickened, stinging a predator with stylets.	Aoki (1977 a, b, 1980 b)
Pseudoregma alexanderi P. bambucicola P. koshunensis P. panicola Ceratovacuna japonica	Dendrocalamus latiflorus* Bamboos* Bamboos* Oplismenus undulatifolius* Bamboo grasses*	First instar, fore legs greatly thicken-ed, piercing a preda-tor with a pair of well-developed frontal horns.	Aoki and Miyazaki (1978), Aoki et al. (1981)
Astegopteryx styracicola	Styrax suberifolia** (making a gall)	Second instar, legs not thickened, sting-ing a predator with stylets, with a habit of readily falling from the gall when disturbed.	Aoki et al. (1977), Aoki (1979 a)

*Secondary host, **primary host.

serrata (Aoki, 1980 b), Colopha sp. on Z. serrata and Pemphigus dorocola on Populus maximowiczii (Aoki, 1978) are extant examples of this stage. Monomorphic 1st instar larvae of Colopha sp., for an example, appear on the surface of their leaf-roll gall and attack a predatory intruder in a self-sacrificing manner.

By the way, are the colonies of these aphid species really clones? No study has yet been carried out to answer the question. In Pseudoregma alexanderi, a species producing bizarre pseudo-scorpion-like soldiers on the ma bamboo Dendrocalamus latiflorus, many non-soldier 1st instar larvae were observed dispersing on the wind. If these larvae reach other colonies of P. alexanderi, or if more than one larva from different clones makes a new colony, then the resultant colonies will not be pure clones (Aoki et al., 1981).

If colonies of an aphid species producing soldiers or other al-truists are usually not pure clones, we can expect the occurrence of "cheaters", who, for example, produce fewer or no altruists when they join other colonies.

INTERGALL MIGRATION IN APHIDS

As a model of altruistic dispersal, consider the following hy-pothetical aphid species:

In spring fundatrices of this aphid species hatch from overwintered eggs, and each makes an open gall on the leaf of a tree. The fundatrix becomes an apterous adult and produces m children in the gall. The second generation larvae can feed and grow only in the gall of this species, and when mature, they become alates which emigrate to secondary host plants. In autumn, progeny of these alates return to the primary host, and there sexual reproduction occurs.

A gall can supply the inhabitants a limited quantity of a resource (e.g. food), R. And one fundatrix can produce enough children for consuming R. The minimum number of inhabitants which can consume R is a (hence, $m \geq a$). When more than a inhabitants are in a gall, the inhabitants gain R, while when not, each of the inhabitants gains R/a. Suppose, then, two types of fundatrices. A fundatrix of Type A produces m children which all remain in the maternal gall and grow there. A fundatrix of Type B, on the other hand, produces a children which remain in the maternal gall, and $m - a$ children which leave the maternal gall and intrude into galls made by other fundatrices and grow there. If there are fundatrices of only the two types in a population, fundatrices of Type B will gain an advantage over those of Type A and will occupy the whole population. Then, we will observe intergall migration in this aphid species.

However, the strategy of Type B fundatrices is not an ESS if other fundatrices can adopt any value of v ($0 \leq v \leq 1$) in the strategy that a fraction v of her children are programmed to be migrants. The ESS value of v in this case is given as follows:

when $(2 - p)a < m$, $v^* = 1/(2 - p)$,

when $a \leq m \leq (2 - p)a$, $v^* = (1 - a/m)/(1 - p)$,

where p ($0 < p < 1$) is the survival rate of migrants during intergall migration. These formulas are adapted from the first model in Hamilton and May (1977). Note that, since a fundatrix and her children are genetically identical, no conflict occurs among them.

The above model is admittedly artificial. But it clearly shows that even if p is very small, the fundatrix producing at least some migrants gains an advantage over the fundatrix producing no migrants, provided that m is larger than a. From a migrant's point of view, as the migrant may die at the probability $1 - p$, it behaves altruistically toward its resident siblings.

Such intergall migration is known to occur in Pachypappa marsupialis (Aoki, 1979 b) and Pemphigus populitransversus (Setzer, 1980), although some uncertainty remains as to my interpretation.

Pachypappa marsupialis is an aphid species making an open gall on the leaf of Populus maximowiczii. In Sapporo and vicinity, northern Japan, fundatrices of this species each makes a gall in May and produces many 2nd generation larvae in the gall. The 1st instar larvae of the 2nd generation are dimorphic; i.e., the fundatrices each parthenogenetically produces 1st instar larvae of two remarkably different forms. What I (1979 b) called "migratory" 1st instar larvae have a very long rostrum and sclerotized plates on the tergites, and no wax plates, whereas "normal" 1st instar larvae have a normal rostrum, no distinct sclerotized plates on the abdominal tergites, and well-developed abdominal wax plates. Normal 1st instar larvae usually grow in the maternal gall and become alates which emigrate to unknown secondary host plants. Migratory 1st instar larvae leave the maternal gall and feed on the twig or petiole of P. maximowiczii, but they never moult there. By marking migratory 1st instar larvae outside galls, it was found that migratory 1st instar larvae intrude into empty conspecific galls or successfully developed conspecific galls, where they grow and become alates which are not apparently distinguished from those originating from normal 1st instar larvae. In short, the migratory 1st instar larva of P. marsupialis is an obligatory intergall migrant, although I have not yet confirmed that no migratory 1st instar larva goes back to the maternal gall.

Intergall migration in the 2nd generation of Pemphigus populitransversus was studied by Setzer (1980). As in Pachypappa marsupialis, fundatrices of P. populitransversus each makes an open gall on the petiole of Populus sargentii or P. deltoides, and produces many 2nd generation larvae in the gall. The 2nd generation larvae grow in the gall and become alates which emigrate to secondary host plants (crucifers).

Setzer found that 2nd generation aphids of P. populitransversus exhibit surprising amounts of within-gall diversity for allozyme phenotypes. He showed that this variation is the result of among-gall dispersal by larvae, by wrapping some galls in cloth bags to prevent the intrusion of aphids into these galls. All of the 36 galls sampled at Hamlin Town, New York, contained 2nd generation aphids showing an allozyme phenotype different from that of the fundatrix, and such intruders were 17.5% of the 2nd generation aphids examined by him. Setzer ascribes this dispersal behaviour to the unpredictable environment, but his explanation is unlikely to be true if the intergall migration is risky.

If intergall migration occurs widely among aphid species making open galls, even a colony in an open gall formed by a single fundatrix may not be a pure clone. Aphids in open galls are vulnerable to predation by syrphids, anthocorids, etc., so that they are likely to need defence, and some gall-making species have actually developed soldiers or monomorphic larvae attacking predators. Is there any species making open galls in which both intergall migration and larvae attacking predators occur? If we find such a species, then it will be interesting to know whether intruders of this species do not engage in defence and behave "selfishly".

DISCUSSION

Social behaviour is likely to evolve in a genetically viscous population like an aphid clone. However, in such a viscous population, altruistic migrants are also likely to evolve. Where we find soldiers, we might expect the occurrence of altruistic migrants. If altruistic migrants very often succeed in intruding into other conspecific colonies, the viscosity of the colony will decrease, resulting that there will be no advantage in producing altruists. In Pachypappa marsupialis and Pemphigus populitransversus, this does not matter because they start a new gall colony annually. But in such perennial colonies as those of Pseudoregma species on bamboos, this might happen. What is actually going on remains to be investigated.

References

Aoki S., 1977 a. -- Colophina clematis (Homoptera, Pemphigidae), an aphid species with "soldiers". Kontyu, 45, 276-282.

Aoki S., 1977 b. -- A new species of Colophina (Homoptera, Aphidoidea) with soldiers. Kontyu, 45, 333-337.

Aoki S., 1978. -- Two pemphigids with first instar larvae attacking predatory intruders (Homoptera, Aphidoidea). New Entomol., 27, 67-72.

Aoki S., 1979 a. -- Further observations on Astegopteryx styracicola (Homoptera: Pemphigidae), an aphid species with soldiers biting man. Kontyu, 47, 99-104.

Aoki S., 1979 b. -- Dimorphic first instar larvae produced by the fundatrix of Pachypappa marsupialis (Homoptera: Aphidoidea). Kontyu, 47, 390-398.

Aoki S., 1980 a. -- Occurrence of a simple labor in a gall aphid, Pemphigus dorocola (Homoptera, Pemphigidae). Kontyu, 48, 71-73.

Aoki S., 1980 b. -- Life cycles of two Colophina aphids (Homoptera, Pemphigidae) producing soldiers. Kontyu, 48, 464-476.

Aoki S., Akimoto S., Yamane Sk., 1981. -- Observations on Pseudoregma alexanderi (Homoptera, Pemphigidae), an aphid species producing pseudoscorpion-like soldiers on bamboos. Kontyu, 49, 355-366.

Aoki S., Miyazaki M., 1978. -- Notes on the pseudoscorpion-like larvae of Pseudoregma alexanderi (Homoptera, Aphidoidea). Kontyu, 46, 433-438.

Aoki S., Yamane Sk., Kiuchi M., 1977. -- On the biters of Astegopteryx styracicola (Homoptera, Aphidoidea). Kontyu, 45, 563-570.

Dawkins R., 1979. -- Twelve misunderstandings of kin selection. Z. Tierpsychol., 51, 184-200.

Hamilton W.D., Henderson P.A., Moran N.A., 1981. -- Fluctuation of environment and coevolved antagonist polymorphism as factors in the maintenance of sex. In Alexander and Tinkle: Natural Selection and Social Behavior. Blackwell Sci. Publ., Oxford, pp. 363-381.

Hamilton W.D., May R.M., 1977. -- Dispersal in stable habitats. Nature, 269, 578-581.

Krebs J.R., Davies N.B., 1981. -- An Introduction to Behavioural Ecology. Blackwell Sci. Publ., Oxford, 292 p.

Setzer R.W., 1980. -- Intergall migration in the aphid genus Pemphigus. Ann. Entomol. Soc. Amer., 73, 327-331.

Communal Nesting in Australian *Cerceris* Digger Wasps

Howard E. Evans and Allan W. Hook,
Colorado State University

There is growing evidence that many tropical and subtropical species of the sphecid genus <u>Cerceris</u> nest communally, that is, that several females live together in a nest and display no noticeable aggression toward one another. Nests are maintained by several successive generations, and cooperating females are believed to be products of individual nests, that is, sisters or mother and daughters (Grandi, 1961; Salbert and Elliott, 1979; Alcock, 1980; Evans and Hook, in press). Nests may ultimately become deep and complex. Prey consists of paralyzed beetles which are first deposited in the burrow, then later removed to freshly prepared cells, where oviposition occurs.

MATERIALS AND METHODS

In the summer of 1979-80, we studied 16 species of <u>Cerceris</u> in eastern Australia. Not all were studied in equal detail, and in the case of 5 species our sample size is too small to permit meaningful conclusions. Nests were located in areas of bare, firm sand or clay by means of the characteristic tumulus at the entrance. Females were marked with paint, and their comings and goings at the entrance recorded by placing plastic cups over the entrance to delay them. Selected nests were excavated and contents recorded.

RESULTS

The 11 species studied in some detail appear to fall into 3 groups (Table 1). Two relatively large species, <u>froggatti</u> and <u>megacantha</u>, appear to be solitary species, like most species of the genus that have been studied in temperate regions. Three other species, of medium size, appear to be mostly solitary, but occasional nests were found to have two females active at the same time. In each case about 20% of the nests were polygynous, with a mean of 1.2 females per nest.

The final group, of 6 species (from small to fairly large size) consisted of communal nesters, with typically more than two females active per nest at the same time; nests with single females were usually newly formed nests with only a few cells. Overall, 40 to

Table 1.-- Nest data for 11 species of Australian Cerceris.

Species	Mean head width (mm)	No.nests studied	No. females/ nest	Mean no. females/ nest	% poly- gynous nests	Max.no. cells/ nest
GROUP I						
froggatti	5.15	5	1	1	0	25
megacantha	4.25	3	1	1	0	29
GROUP II						
xanthura	2.84	18	1-2	1.2	22	30
unispinosa	2.74	6	1-2	1.2	17	5
windorum	2.38	11	1-2	1.2	18	18
GROUP III						
australis	3.59	32	1-8	2.8	68	179
goddardi	3.30	20	1-7	3.3	89	231
antipodes	2.85	31	1-7	2.8	77	196
armigera	2.85	5	1-5	2.6	40	37
minuscula	2.06	7	1-4	2.0	43	12
anthicivora	1.85	4	1-4	2.0	50	20

89% of the nests we found were polygynous, but in the case of the 3 species for which we have the largest sample size, the percentage was from 68 to 89.

Of these 3 species, some information has already been published on antipodes (Alcock, 1980) and we have a paper in press on australis. We therefore focus this report on goddardi, a common species of subtropical and tropical eastern Australia that has not been studied previously. These 3 species are closely related taxonomically and differ but little in their nesting behavior. It was common to find two of them, and occasionally all 3, nesting together, their nests more or less intermingled.

We studied goddardi at 3 sites in Queensland and one in northeastern New South Wales. Twenty nests were studied in some detail, and 16 of these were eventually excavated. Two contained single females and relatively few cells (12-28) and were evidently recently founded nests which would have eventually contained additional females. Others contained groups of older cells, containing fragments of beetles, as well as groups of cells deeper in the soil containing viable contents or recently evacuated cocoons. These nests had many cells (20-89) and had obviously been occupied for 2 seasons. One further nest had 161 cells and another 231; in the latter the burrow measured 1.8 m in length. Nests of these last two groups contained 2-7 females active at the same time.

Of the females active at a given nest, 1-3 were normally "provisioners", bringing in beetles over much of the day. Provisioners generally remained in the same role day after day (maximum recorded 27 days). From 1 to 3 females in each nest were "non-provisioners", leaving the nest each day for up to 60 minutes, presumably to obtain nourishment from flowers, returning without prey. Non-provisioners also tended to maintain the same role (maximum recorded 23 days). Evidently non-provisioners play an important role as guards within

the nest entrance. We observed guards actively driving away mutill-
id wasps on 2 occasions. On another occasion about 20 ants (<u>Pheidole
megacephala</u>) approached an entrance as a group. Over a period of
about 15 minutes the guard succeeded in driving them away, making
considerable use of her mandibles; 12 dead ants were later counted
outside the entrance.

As in other species of <u>Cerceris</u>, prey is deposited in the
burrow and not removed to a cell until later. This is suggested
by the short time spent in the nest by provisioners between hunting
trips. For example, one female brought in 23 beetles between 0805
and 1042 hours, staying in the nest each time from 5 to 50 seconds
(mean 18.4 seconds). The total number of beetles brought in per day
(by several females) is commonly enough to provision more than one
cell.

We collected all the females from several nests and found that,
in general, both provisioners and non-provisioners may be in egg-
laying condition (Fig. 1). As a general rule, larger females, in
either role, display the most fully developed ovaries. The size of
newly laid eggs varies little, approximating 2.5 mm. Oocytes found
by dissection to be appreciably smaller than this are either not
fully formed or are undergoing resorption. The frequency and recognition of oosorption in female <u>Cerceris</u>, both provisioners and
non-provisioners, has been discussed elsewhere (Evans and Hook, in
press). We compared 6 provisioners and 6 non-provisioners with
respect to head width and size of the largest oocyte and found no
significant difference. Thus the evidence suggests that the distinct
division of labor apparent in these wasps is not accompanied by a
reproductive division of labor.

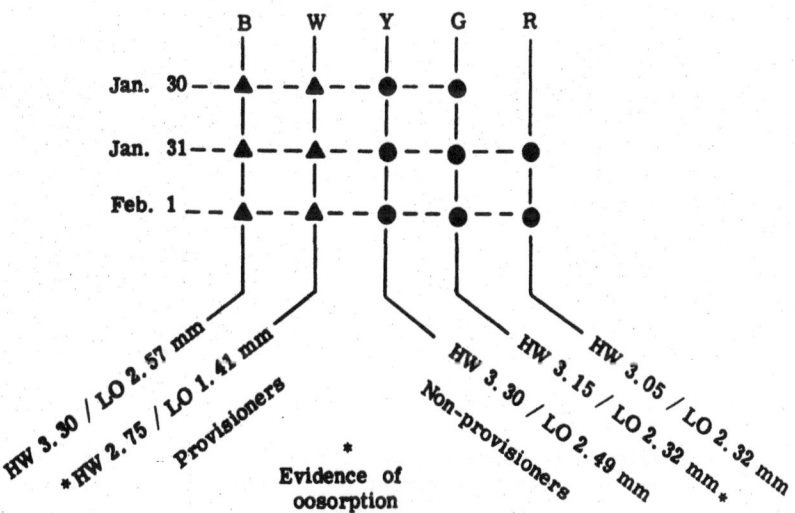

Fig. 1.-- Condition of oocytes in a nest containing 5 females,
observed over 3 days. HW= head width (mm); LO= length of largest
oocyte (mm). Letters at the top identify individual females (e.g.,
B= blue, etc.). Provisioners shown as triangles, non-provisioners
as circles.

162

DISCUSSION

We believe that in C. goddardi and in other communal nesters we studied, one or more females overwinter in the nest and begin to expand and provision the nest in the spring. This is suggested by the fact that females with badly worn mandibles are commonly found alone or with one or two freshly emerged females early in spring. Over time, other females appear and assume a role as provisioners or non-provisioners, or disperse. In the variable and often rigorous climate of Australia, and with the abundance of ants, mutillids, and other predators, cooperative nesting may well be a more successful strategy than nest founding, at least at times and places.

Why do we find that some species are solitary nesters and some exhibit only occasional evidence of communality? There appears to be no strong correlation between degree of nest-sharing and type pf substrate in which they nest, and no strong correlation with size of the wasp. We believe the degree of cooperative behavior to be related to past selection pressures, chiefly from natural enemies. While guards are distinctly advantageous with respect to predators that try to enter the nest, in the case of miltogrammine flies guards may actually increase the opportunities for successful attacks. These flies deposit small maggots on the prey as it is being taken into the nest, and slight delays caused by a guard may result in increased success of the predator. We frequently saw miltogrammine flies at nest entrances, found maggots in several cells, and occasionally reared adult flies from cells.

Since females are, in general, able to plunge into the nest more quickly when no guard is present, it would seem likely that solitary nesting is more successful under strong selection pressure from miltogrammines. But when ants and mutillids are the major natural enemies, communal nesting would appear to be more adaptive. We believe that the many species of Cerceris in Australia evolved in refugia during past periods of extreme drought, as is believed to be the case in other groups of animals. We hypothesize that it was the relative importance of various nest predators during isolation that determined the degree of communal nesting characteristic of individual species. Under conditions of extreme drought, it may also be advantageous for a female to be a joiner rather than dig her own nest, since the soil surface may be difficult to break through under these conditions. Even today, it is probable that degree of communal nesting varies within species depending upon climatic factors and the local abundance of certain natural enemies.

Acknowledgments

These studies were conducted while the senior author held a research fellowship in the Department of Entomology, University of Queensland, Brisbane, Australia, and a travel grant from the National Geographic Society, U.S.A. The junior author held a grant for dissertation research from the National Science Foundation, U.S.A., no. BNS79-12602.

References

ALCOCK, J., 1980.- Communal nesting in an Australian solitary wasp, Cerceris antipodes Smith (Hymenoptera, Sphecidae). J. Aust. Ent. Soc., 19, 223-228.

EVANS, H.E., HOOK, A.W. (in press).- Communal nesting in the digger wasp Cerceris australis (Hymenoptera, Sphecidae). Aust. J. Zool.

GRANDI, G., 1961.- Studi di un entomologo sugli imenotteri superiori. Boll. Ist. Ent. Univ. Bologna, 25, 1-659.

SALBERT, P., ELLIOTT, N., 1979.- Observations on the nesting behavior of Cerceris watlingensis (Hymenoptera: Sphecidae, Philanthinae). Ann. Ent. Soc. Amer., 72, 591-595.

Evolved Relations Between Reproductive and Subsocial Behaviors in Coleoptera

Gonzalo Halffter,
Instituto de Ecología, México

The dominant concepts about social and subsocial behaviors in insects were derived from the well known societies of Hymenoptera and termites. Most scientists that study Hymenoptera consider that the basis of the social behavior of the group is found in the haplodiploid system of sex determination. A consequence of haploiploidy is that females are more similar to their sisters than to their daughters, so sister care is a way of increasing fitness. Another consequence is a social organization based in the existence of many sterile and few fertile females, a typical structure for Hymenoptera societies.

Perhaps Wilson's (1975) profound knowledge of ants' societies makes him state that from the point of view of evolution sexuality is an antisocial phenomena. That is, sexual relations, when confronting two individuals, genetically and phenotypically different, cause conflict and consume great quantities of energy, two factors that go against the sociability process. This social effect extends to the parent-offspring relation, which, after breeding, goes into a conflict stage. For Wilson, social behavior dependent on family structure is limited to a few groups of mammals, such as canids and higher primates which have a sufficiently developed intelligence to remember detailed patterns of behavior with which to build barriers or ties in the social structure. Even so, these societies are relatively unstable and their behavior includes an important anti-social component of aggression and self-satisfaction. As Wilson points out, according to observations on Hymenoptera, the sociability increase in insects seems to be based on a restriction of the desegregating forces of sexuality; that is, sexual reproduction restricts and shapes social evolution but does not promote it. Wilson's point of view has very strong foundations. The three theories that explain eusocial insect evolution, group selection, kin selection and offspring breeding (see Starr, 1979) do not include the evolution of sexual behavior, but a reduction of its importance instead.

In previous work (Halffter, Halffter and López, 1974, Halffter, 1977, Halffter and Edmonds, 1981, Halffter and Edmonds, 1982), we have described the existing relations in Scarabaeinae (Coleoptera: Scarabaeidae) in regards to evolution of nesting, bisexual relations and the initiation of patterns that tend to be subsocial. In this study we extend our analysis to two more groups of Coleoptera: Necrophorus (Silphidae) and the Passalidae.

The central hypothesis is that there is an evolution towards subsocial forms of behavior through development of patterns based on sexuality.

In the three groups studied, the evolutionary process begins with elaborate responses to the special situations in which food is found. That is, food is discontinuously distributed and ephemeral, and motivates an intense intra- and interspecific competition. Because two provide better than one in different species a male-female cooperation exists during the provisioning. During food storage for the offspring, still more competitive conditions take place. This favors even more the development of male-female cooperation in nesting patterns, which in its simpler forms derive from feeding behavior. In many cases, from the collaboration and evolution of nesting, offspring care appears performed by the female or by either sex. For this type of care to appear, and be sustained for long periods of time, the existence of communication mechanisms, acoustical or chemical, is necessary. It is under these conditions that the intersection between social and subsocial is found, when the family nucleus is established with generation superposition, acoustical and chemical mechanisms of communication, shared home and food, and even parent-offspring trophallaxis. In these Coleoptera, in contrast with Hymenoptera, the evolved process is derived from very clear expressions of sexual behavior. In other words, sexuality not only does not restrain the development of social behavior, but promotes it.

Ecological situation of food and evolution of behavior. The Scarabaeinae (Coleoptera: Scarabaeidae), adults as well as larvae, feed themselves from excrement of the large mammals. Specially in tropical America many species use the corpses of small vertebrates, and they manipulate them in the same way as excrement, being in most cases copronecrophagous. In the tropics, some species are saprophagous, feeding themselves with ripe fruit. Other species are mycetophagous. But the group's usual food, the one to which their behavior and morphology is adapted to, is the large mammals' excrement (see Halffter and Matthews, 1966). This food is the basis of behavior's evolution because it determines an initial situation of ecological stress: the food is relatively abundant, but ephemeral and highly localized. There is great inter- and intra-specific competition about it. Its desiccation and subsequent hardening, when staying exposed at ground level, is very quick, which makes it non-usable for most of the Scarabaeinae, for feeding as well as for nesting. The process is quick in savannahs, and in isolated prairies. In humid pastures, in the forests, and especially in the humid tropical ones, desiccation is no longer important. Under these conditions, another factor limits the Scarabaeinae's use of excrement, the quick laying of eggs and further development of fly larvae. The pressure of these two factors, desiccation and competition with flies and the spatial distribution of excrement are decisive in the evolution of feeding behavior as well as in this derivate: the nidification Scarabaeinae's response has been to make the excrement disappear from the surface, as quickly as possible, protect it from desiccation and flies, either to feed the adult, or to feed the larvae (nesting).

Necrophorus uses for feeding and nesting small vertebrates' corpses, which are buried completely. The situation is totally similar to that described for Scarabaeinae.

Passalidae feed from damp trunks and stumps in process of decomposition. In forest ecosystems, these elements constitute an abundant resource, but its distribution, however, is not uniform but localized. On the other hand, within the long process of decomposition and degradation, the food is usable by Passalidae during a relatively short time. The stress derived from the ecological situation of food still exists, although not as pronounced as in Scarabaeinae, because food is less ephemeral and less competed for.

On the face of their food and environmental pressures, Scarabaeinae and Necrophorus have not reacted multiplying their offspring, but they have looked for specialization and an increase in the efficiency of their feeding behavior and as a last result in reproductive behavior. It is clearly a K strategy, evident in the scarce number of eggs per female, that reaches extreme reductions. In Scarabaeinae, the reduction of fertility goes along with the reduction of the female reproductive apparatus to a single ovary, with a single ovariole, a phenomenon that is exceptional in the animal world.

The convergence around an ephemeral food facilitates the meeting of the sexes. This has great evolutionary importance, because it frees all the energy of sexual behavior to cooperation for nesting. Two create a nest quicker and better (by means of a division of labor) than one. This has been a factor that has favored the development of bisexual cooperation, and from it, certain forms of subsocial behavior. Cooperation is a form of altruistic behavior that increases rearing possibilities and, above all, its own fitness.

In Passalidae, food is not protected or changed from one place to another. The dead stump, besides being a feeding source, is a nesting site. In it, the animal spends the greatest part of its long life (adults live more than a year), leaving it only to colonize new stumps. In contrast to Scarabaeinae and Necrophorus, Passalidae do present elaborated patterns of copulation and courtship. In Passalidae, the existence of family groups is reinforced by a feeding system that requires the conditioning of wood by adults, the transmission from one generation to another of the microorganisms responsible for the transformation of cellulose into assimilable carbohydrates, as well as the woods' enrichment in nitrogen due to the excrement of adults and larvae.

In Passalidae, as in Necrophorus, as well as in those Scarabaeinae that reach more complex levels of nesting, the nest is made by a bisexual couple. One nest of Passalidae includes more juveniles than those of Necrophorus or evolved Scarabaeinae. Unlike Scarabaeinae, the egg and later the larvae are not separated from adults by physical barriers. In Scarabaeinae, even in those that make multiple nests, each egg occupies a brood-ball, from which the larva never goes out.

Evolution of bisexual cooperation. Scarabaeinae are an exceptional group for studying bisexual cooperation because they present a whole sequence; within the great evolutionary line represented by the burrowing species, from nesting pattern I (for a description of the patterns see Halffter, 1977, Halffter and Edmonds 1982) in which cooperation first appears, to pattern II, where it is more elaborated but not compelling, to pattern III where collaboration is always present. In all cases, the female digs the gallery which is going to be transformed into a nest and manipulates the larva's food. In patterns I and II, the male role consists of helping the female as provider, at the gallery's surface and entrance. In pattern III, the male can participate in the first stages of cake conditioning, from which the female will later make the brood-balls. He stays with the female during the long caring period, until his death, but he does not actively participate in it   Only in Cephalodesmius providing is sustained, during the larvae development, by addition of prepared food to the exterior of the brood-balls. The creation and increase of the brood-balls is the female's job, but surface recollection is performed by the male.

In the other great line of Scarabaeinae alimentary-reproductive behavior, in the species that prepare and roll a ball on the surface, with only a known exception male-female cooperation is always present, although practically restricted to the rolling process. As in the burrowings' pattern III, some rolling species reach a more complex level of nesting, thanks to the formation of a multiple nest cared for by the female accompanied by the male, although he does not perform a specific task.

In all Scarabaeinae, the couple is established in relation to the nest, be it during the excavation and its providing, or during the rolling of a ball or fragment of excrement. Copulation is an incidental act. Although in the rollers it is always part of the nesting process, it is not present in each sequence of nesting of the burrowing species when the female has been fertilized. In any case, this takes a short time and little energy and it does not include elaborated patterns of courtship.

The same thing happens in Necrophorus. The couple is established as a response to the emergency situation: to bury the little corpse. Male and female work on it simultaneously or alternatively. Copulation comes, only after the burial, without any special courtship. The couple stays together until the larvae's pupation when the caring of the offspring is no longer necessary. The male's part is greater in the burial and the female's part is more important in the caring of larvae.

In Passalidae the union of the bisexual couple is done on the site where the food is found. Any of the sexes may begin a gallery and attract the opposite sex. The encounter of sexes and the copulation require much more energy expenditure and display than in the other two groups. The behavioral display includes a complicated courtship (studied mainly by Jack Schuster); there are movements, contacts and different auditory signals for about an hour to twelve continuous hours. We suppose that the elaborate and complex sonorous communication system that characterize relations in the bisexual couple has an ecological meaning in securing conspecificity, for a group of insects that is very uniform in its ways and behavior and lives in the single environment of the rotten stump, which propitiates the encounter in a reduced space, of individuals from different species. The couple collaborates during their whole life in digging the nest galleries, and in providing food for the larvae.

Care of the offspring - communication. The care of the offspring appears in various groups of Scarabaeinae not taxonomically related. Mainly in patterns V (rollers) and III (burrowers); also in the genera Eurysternus and Oniticellus, the female stays in the nest until the emergence of the offspring, accompanied by the male in patterns III and V. During this time, she continuously mends the brood-balls, avoiding the development of fungi and closing any crevice from the outside. For this she uses the larvae's fresh excrement that may leak from the interior.

In the genus Cephalodesmius a cake is prepared in the underground chamber made of leaves, petals, and small excrement fragments that the female, and perhaps the male, contaminates with its own faeces. Microbial fermentation transforms this cake into something very similar to excrement. The female builds the brood-balls with this material, providing each one of them with an egg. The difference in respect to the other Scarabaeinae with known nesting, is that the cake (a real fermenter) continues being provided by the male, and the female incorporates new fermented material to the balls that contain developing larvae. On the other hand, both parents lick the liquids that drain from the cake. This continuous providing establishes a closer contact between parents and offspring. The stridulation of the larvae has been clearly perceived in Cephalodesmius. The stridulatory system in adults and larvae is known in other Scarabaeinae, even though it is still not well known in what way auditory signals contribute to maintain group cohesion. The chemical mechanisms -apparently very important in Scarabaeinae- that play a role in the union of the couple as well as in the nest's continuity are being studied right now.

In Necrophorus, the process of nidification is very complex (Milne and Milne, 1976; original observations). The couple buries a small corpse, from which parents and larvae are going to feed. Occasionally, the male abandons the nest at this moment, and the inseminated female continues, but usually both parents stay. Once the corpse is buried, it is made compact until it reaches a spheric form. The beetles' movements around the sphere create a real underground chamber, in whose upper part the female digs a little enlargement in which she deposits the eggs. On the corpse, combining ingestion and digging, the female prepares a conical depression, exactly under the egg chamber. Both parents regurgitate tissues partially digested in this depression, together with liquid drains from the corpse. In this stage, one of the parents settles on the depression's edge and begins to stridulate. The sound attracts the newly born larvae beside their parents. They transfer mouth to mouth liquid nourishment to all of them. The task is mainly done by the female, but the male can also participate. Mature larvae able to eat directly from the pond or from small fragments of carrion renew their petition of regurgitated food approaching either of the parents that is close to the depression and pressing with their mouth pieces the adult's mandibles and palpus. Parents can then prepare an horizontal gallery for pupation. Only at this moment do they abandon the nest.

In Passalidae we find a real social structure with a complex system of auditory and possible chemical signals, as well as overlapping generations. As in the previous examples, caring is always from the parents and never among sisters. The female can deposit eggs at different times. The result is that a family group is constituted by a) the parent couple, that condition the environment and dig the nest; b) eggs, placed in very finely triturated wood; c) larvae from the I, II and III stage, moving freely within the galleries and keeping themselves in groups that follow close to the adults; d) prepupae and pupae that are being transformed within a cocoon which isolates them from the rest of the family; e) adults recently emerged, that stay within the group for a certain length of time, maturating and waiting the period of dispersion.

Larvae depend for food on the triturated wood that adults prepare when they enlarge the nest and its gallery system. The wood that they can scratch directly from the walls seems to be insufficient to complete their development. There is also a close collaboration from parents to larvae in the third stadium or stage to build the pupal cocoon, which is repaired by the adults when it suffers any damage. The generation coexistence is also important to enrich the wood that is separated and triturated, with microorganisms that help the digestion of cellulose, and with the excrement that increases the nitrogenated content. The parents specially, but also the imago-offspring and the larvae, contribute to create adequate conditions for feeding on wood.

When the colony grows, density increases and with it aggression, which finally provokes the dispersion of the new imagos and the constitution of new family units.

Note. - Most part of the information about Passalidae has not been published. It comes from my dear friend Pedro Reyes-Castillo, from the Instituto de Ecología, México.

REFERENCES

Halffter G., 1977.-Evolution of nidification in the Scarabaeinae (Coleoptera, Scarabaeidae). Quaest. Ent., 13, 231-253.
Halffter, G., Edmonds W.D., 1981.-Evolución de la nidificación y de la cooperación bisexual en Scarabaeinae (Ins., Col). An. Esc. nac. Cienc. biol., México, 25, 117-144.

Halffter G., Edmonds W.D., 1982.-The nesting behavior of dung
    beetles. Instituto de Ecología, México.
Halffter G., Halffter V., Lopez I., 1974.-Phanaeus behavior: food
    transportation and bisexual cooperation. Environmental Ent.,
    3. 341-345

Halffter G., Matthews E.G., 1966.-The Natural History of dung
    beetles of the subfamily Scarabaeinae (Coleoptera, Scarabaeidae)
    Folia Entom. Mex., 12-14, 1-312.
Milne L.J., Milne M., 1976.-The social behavior of burying beetles.
    Sci. Amer., 235, 84-89.
Starr C.K., 1979.-Origin and evolution of insect sociality: a
    review of modern theory. In Hermann H.R., Social Insects,
    vol. 1, Academic Press, 437 p.
Wilson E.O., 1975.-Sociobiology. The new synthesis. Harvard Univ.
    Press, 697 p.

Further Experiments on the Artificial Induction of Multifemale Associations in the Principally Solitary Bee Genus *Ceratina*

Shoichi F. Sakagami, Hokkaido University

Yasuo Maeta, Shimane University

Artificial induction of MFA[1] was attempted with the principally solitary bee Ceratina japonica (Sakagami, Maeta, in press). In successful MFA, a tendency was confirmed for the larger ♀ (L) to serve as guard and main egg layer while the smaller ♀ (S) is pollen forager and subsidiary layer. This paper reports two further experiments: 1) In C. japonica some natural MFA have been found. Induction of MFA was tried with C. flavipes for which no such have been discovered. 2) All induced MFA of C. japonica were either "semisocial" (of ♀♀ of the same generation) or "delayedly eusocial" (of ♀♀ emerged in 2 successive years). Induction of "truly eusocial" MFA (of ♀♀ of 2 successive generations but born in the same year) was attempted with a multivoltine species C. okinawana.

MATERIALS AND METHODS

Methods are essentially like those adopted previously. But instead of liberating 20 - 50 ♀♀ in each cage (method I, ratio ♀♀/nest tubes = 2.0) only 1 or 2 nest tubes were fixed at a side of a smaller cage and only 2 or 4 ♀♀ were liberated (method II). C. flavipes was collected in Morioka, N. Japan and C. okinawana in Naha, S. Japan.

RESULTS AND DISCUSSION

1. Induction of MFA in C. flavipes.
   In C. japonica natural MFA occupied 1.2% of newly built nests (n = 412) and 31.0% of nests surviving from the last year (n = 203). In this species 12 stable (= rearing brood) MFA were induced with

---

[1] Abbreviations used in this paper:
   MFA - Multifemale associations
   L   - Larger female
   S   - Smaller female
   P   - Pollen foraging
   FT - Food transfer
   G   - Guard or guarding
   D   - Daughter
   M   - Mother

method I and 257 ♀♀. In C. flavipes no MFA have been found among
more than 3,000 natural nests and none were induced by method I
using 130 ♀♀. In 1981 13 MFA were induced by method II (all 2 ♀♀,
period of living together 7 - 88 days, $\overline{X}$ = 34.6 $\pm$ 27.9), but 10
were unstable: 1) No brood cells were formed. 2) In 6 cases a
basal partition was not built.   In 6 cases a sequence of building
(mostly by L) and destruction (mostly by S) was repeated. 3) Pollen
foraging (P) and food transfer between ♀♀ (FT) were not seen.
4) Associations were ended by deaths of one or both ♀♀ or drifting
out by S. 5) In all cases guarding (G) was mainly performed by L.
6) All cases were semisocial. The chronicles of 3 other nests are
summarized below (all ended with death of L):
     N-1. Delayedly eusocial (L = 2 years old), lasting 71 days.
Generally as above but FT L → S and P by L observed.
     N-9: Semisocial, lasting 88 days. Ten cells provisioned and
oviposited, mostly by L even after start of P by S in the later 22
days (n observed, L 20, S 7). Both L and S served as G and cleaned
immatures. FT L → S observed but not the opposite. After death
of L, 3 cells provisioned and oviposited by S.
     N-3: Delayedly eusocial (L = 2 years old), lasting 76 days.
Among 7 cells produced, first 3 were provisioned and oviposited by
L. During P by L (n = 15), S served as G and occasionally modeled
the pollen loaf. After cell 4, P was performed only by S (for 22
days, n = 12), and G and probably ovipositions (confirmed only once)
by L. Since P by S, FT S → L was observed several times but not
the opposite. After death of L, 6 cells were provisioned and
oviposited by S.
     The absence of nest reuse in C. flavipes certainly should make
formation of MFA difficult. But the above results show also a
difficulty at the behavioral level. In all MFA excepting the last
stage of N-9 and N-3, intranidal passing of L and S was much less
smooth than in C. japonica, though no distinct agonism was expressed. Nevertheless, the results show as a whole a gradual approach
to sociality: From instability in most cases except G by L to
appearance of P and FT in N-1, somewhat "quasisocial" brood rearing
in N-9, and finally a rudiment of the tendency found in a C.
japonica MFA in the later stage of N-3 (G by L and P by S). This
threshold sociality may depend on the behavior found in the
solitary part of the life cycle, particularly the association of
mother and juvenile daughters as in C. japonica.

2. Induction of eusocial MFA in C. okinawana.
     In C. japonica the mother deprived of immatures can oviposit
again. Hoping for the formation of a truly eusocial MFA, 1 or 2
young ♀♀ were introduced in the bottom of a nest with another ♀
whose brood was removed except for an egg and a young larva. In
all 5 cases, the older ♀ fed the stepdaughter. The latter stayed
in the nest some days but never performing P, and the stepmother
ceased oviposition. This result was assumed to have been caused by
the reproductive diapause in this univoltine species.
     In the multivoltine C. okinawana, a rudiment of eusociality
was found in a nest in a cage: June 25. One daughter (D_2)

performed P and prepared a pollen loaf. Mother (M) laid on it
(= eusocial behavior); June 27-28. D_2 fed emerged sibs; June 29,
D_2 closed the cell (= eusocial behavior), June 30, D_2 fed younger
sibs. July 1, M left nest, July 2, D_2 left nest. In other nests,
too, eusociality did not develop further and daughters left the
natal nests.

In this species, natural MFA were found in 12.6% of reused
nests ($\underline{n}$ = 103). Eusocial MFA were artificially induced as follows:
4 delayed eusocial and 1 truly eusocial MFA by method I (1978-'79,
with 35 M and 73 D) and 5 truly eusocial MFA by method II (1980-'81,
with 5 M and 5 D). The result suggests that method II is better
to induce truly eusocial MFA, probably by decreasing the chance of
nest abandonment and increasing that of homing. Besides, several
semisocial MFA were also induced by method II.

Among 10 eusocial MFA induced, 6 consisted of real M and D and
4 (all delayedly eusocial) also probably so. In all cases but one,
M was larger than D. And in all cases, M was G and the principal
egg layer while D was a pollen forager and subsidiary layer, being
comparable to L and S in semisocial or delayedly eusocial MFA in $\underline{C}$.
japonica, even in the unique case in which D was larger than M.

The behaviors in these eusocial (and also semisocial) MFA were
essentially as in semisocial or delayedly eusocial MFA in $\underline{C}$.
japonica, including elaborated mutual passing and lack of overt
agonism among nestmates. Exceptions are: 1) In all cases, FT
L → S was rare in $\underline{C}$. japonica. In both species FT D → M (or S → L)
prevailed when D(S) began P. 2) In many cases, the tendency M = G,
D = P, was preceded by M = G + P and D without tasks. This sequence
was rare in $\underline{C}$. japonica.

It is difficult to explain why truly eusocial MFA were produced
in $\underline{C}$. okinawana but not in $\underline{C}$. japonica. A possible explanation is
that $\underline{C}$. japonica DD can forage pollen to feed their MM or younger
adult sibs, but, at reproductive diapause, cannot prepare pollen
loaves, at least under the temperature/photoperiod regime adopted
for the experiments.

$\underline{D}$. okinawana possesses elaborated mother-daughter interactions
in the post-reproductive association as in $\underline{C}$. japonica and $\underline{C}$.
flavipes.

CONCLUSION

The genus $\underline{Ceratina}$ has long been considered to consist of
solitary bees (Michener, 1974). Apparently the majority of female
offsprings may leave maternal nests and found their own, solitarily.
However, they can circumstancially develop natural MFA, in most
cases probably by remaining together in old nests and reusing them.
Moreover, in $\underline{C}$. japonica and $\underline{C}$. okinawana semisocial and even
eusocial MFA can be induced when nest substrates are scarce. Even
in $\underline{C}$. flavipes, natural MFA of which have not been discovered, a
rudiment of such social organization is formed. To produce such
MFA's with rudimentary polyergism, they need few behavioral
innovations. They are exceptional among solitary bees by having
elaborate inter-adult coactions in the post-reproductive nest; these

can be adopted with little change in the occasional reproductive
associations.

REFERENCES

Michener, C. D. 1974. -- The social behavior of the bees. A
    comparative study. Harvard U. Press, xii+404pp.
Sakagami, Sh. F., Maeta, Y. 1977. -- Some presumably presocial
    habits of Japanese _Ceratina_ bees, with notes on various social
    types in Hymenoptera. _Insectes Sociaux_, _24_, 319 - 343.
Sakagami, Sh. F., Maeta, Y. (in press). -- Multifemale nests and
    rudimentary castes found in the basically solitary bee
    Ceratina japonica.

Selective Factors Associated with the Evolution of Membracid Sociality

Thomas K. Wood, University of Delaware

Membracids (treehoppers) exhibit a variety of social patterns which are consistent within generic and, in most cases, tribal lines. Solitary species may or may not interact with ants, while those which form nymphal and adult aggregations are usually associated with mutualistic ants. Aggregations may be composed of siblings or offspring from several females, but parent females are not involved in the maturation of offspring. In some species parent-offspring associations occur in which the parent female plays an active role in the maturation of offspring.

Parent-offspring interactions occur in at least 50 species of Membracids (Wood, Unpublished). Among these species female investment in offspring and their interaction with ants are variable. Females of some species desert offspring after they reach the 1st instar leaving ants to provide protection to offspring. In others, females remain with and protect offspring until they mature but do not interact with ants (Wood, 1974; 1976a, b; 1977a; 1978; 1979). Although parental care occurs in a number of orders including the Homoptera (Review by Eickwort, Unpublished; Wilson, 1971), only in the Membracidae does it involve ant mutualism. The influence of ant mutualism on Membracid life histories is poorly understood.

I propose here that the following interactive causal agents (Fig. 1) are responsible for the variety of social organization in the Membracidae: 1) predation, 2) ant mutualism, 3) host plant utilization, and 4) geography. These factors and predictions of how they influence Membracid life histories are presented below.

1. <u>Predation</u> - Egg mortality can be postulated as a primary factor influencing whether eggs are deposited singly or in clusters. A female which deposits eggs in small groups throughout a single host plant or among several plants may spread out the risk of predation for any single egg. This appears to be the case for many solitary species. Other Membracids deposit eggs into masses which contain a female's entire reproductive output. This behavior increases the consequences of predation unless there is some countering selective advantage in nymphal survival.

Active or passive egg guarding by parent females is mechanism to reduce predation on clusters of eggs (Wood, 1974; 1976a, 1977a, 1978, 1979) and may establish aggregations of nymphs which in turn

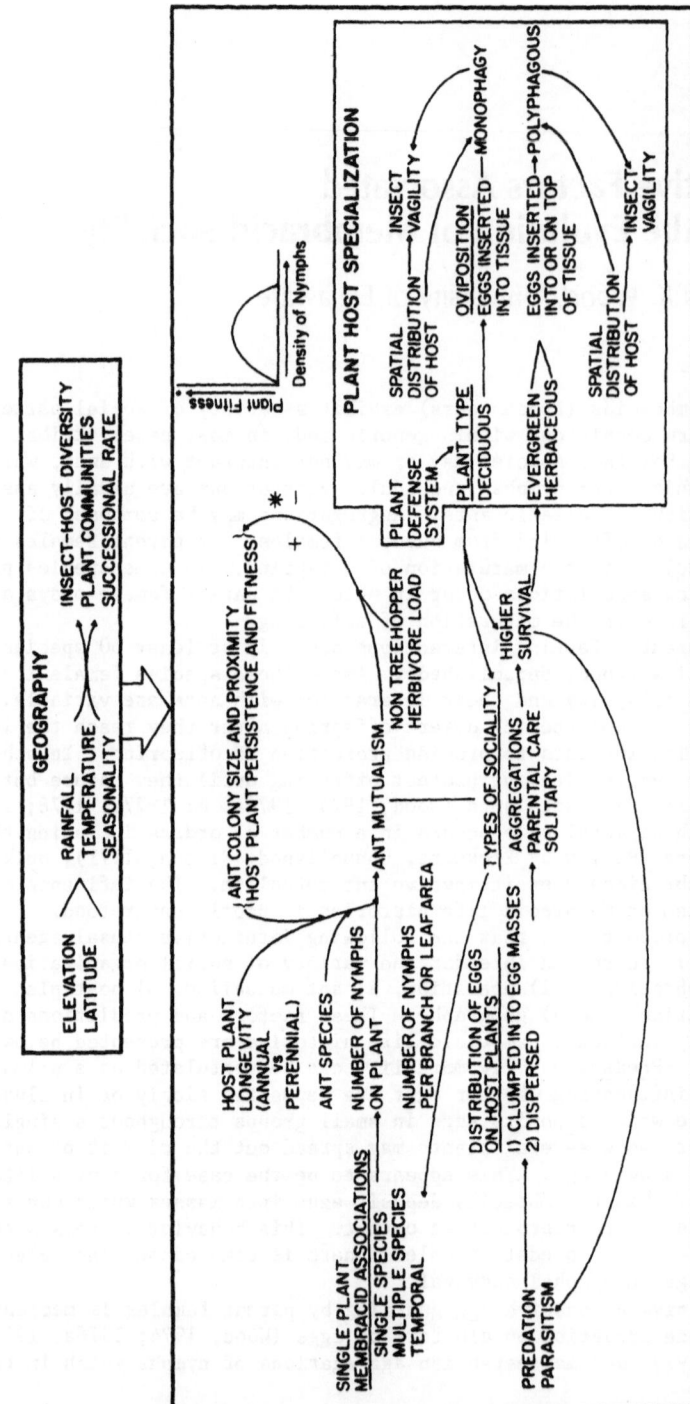

Fig.1.– EVOLUTION OF MEMBRACID LIFE HISTORIES

attract mutualistic ants. In some tropical habitats predation by
ants is heavy (Jeanne, 1979; Leston, 1973a, b) but Membracids which
release honeydew to ants become an energy resource to these ants.
In turn, ants are a resource to Membracids because they reduce predation to nymphs (Wood, 1977a).

Many tropical Membracids in wet, lowland forests interact with
ants. Females of some species deposit a clutch of eggs and then
leave so that location of nymphs by ants is a fortuitous event. In
some species the duration of egg guarding by females is increased by
ant attendance. When ant attended the survival of both females and
eggs is greater in addition to increasing the chance offspring will
be attended by ants (Wood, 1977). In other species females cover
eggs with a frothy substance which not only protects eggs but also
serves as an attractant to other ovipositing females. The resulting
clumping of eggs produces a large aggregation of nymphs at egg hatch
which facilitates attraction of ants but does not involve parental
care (Wood, Unpublished).

2. Ant mutualism - How effective ant mutualists are in promoting Membracid survival appears to depend on a number of factors.
Early location and consistent ant attendance appears to depend upon:
a) the overall number of nymphs on a plant, b) the number of nymphs
on a branch or leaf, c) the proximity of the ant colony and its
size, and d) the ant species. Longevity of the host plant also
influences ant mutualism. Annual herbaceous plants may be less
predictable to Membracids in terms of their distribution over time
as compared to relatively long lived woody perennials. Ant attendance on annuals may also be more variable from year to year since
ants may have to forage longer distances from nests before locating
nymphs. Long lived perennials provide predictable oviposition sites
to Membracid females and the resulting nymphal aggregation an energy
resource to ants. The establishment of ant colonies close to such
Membracid host plants would not only promote nymphal survival but
also the fitness of the ant colony. Location of nymphs by ants does
not necessarily enhance nymphal survival. Some ant species are more
aggressive and thus more effective in protecting nymphs.

Membracids may benefit the host in a manner analogous to extrafloral nectaries (Messina, 1981). Ants attending extrafloral
nectaries or associated with trees (Janzen, 1966; Bentley, 1976,
1977; Inouye and Taylor, 1979) appear to reduce herbivore damage
particularly those eating leaves or interfering with seed set.
Nymphs which attract ants may not stress a plant at low densities
but rather reduce herbivore damage through the action of ants. At
high treehopper densities predation on nymphs may be high because
of the inability of ants to protect all aggregations.

3. Host plant utilization - The degree of host specialization,
the predictability of the resource in space and time, and resource
quality appear to influence Membracid life histories (Wood, 1980;
Wood and Guttman, 1982). Some treehopper species on long lived
evergreen legumes have low vagility because of host predictability
which results in a high degree of relatedness on individual trees.
The quality of the host resource changes as treehopper population
density increases, which in turn appears to have a profound influence on female vagility, sex ratios of offspring, and body size of

offspring (Wood and Dowell, Unpublished). Whether host plants are herbaceous or woody, annual or perennial, deciduous or evergreen appears to influence the degree of resource specialization and the life histories of Membracids.

4. <u>Geography</u> - The influence of predation, ant mutualism and host plant specialization is modified by constraints imposed by geography. Evevation and latitude influence rainfall, temperature, and seasonality which in turn affect insect and host plant diversity as well as community sturcture. In view of these geographical considerations, different types of Membracid life histories are predicted in varying geographical regions and habitats.

In the tropics, there maybe considerable differences in life histories, ant mutualism, and host specialization among diverse habitats. In lowland wet forests with high ant and plant diversity, the majority of Membracid species appear to interact with ants and are polyphagous. In higher elevations ant diversity and abundance is reduced and hosts are evergreen. I expect fewer treehoppers which are ant attended, solitary species more common and parental care which is not dependent on ant mutualism. In contrast seasonality is pronounced in deciduous dry forests and imposes limitations on life history. I expect Membracids to be seasonally abundant and show some degree of host plant specificity as a result of coordination of life history to host phenology. Seasonal ant activity should affect the number of species which are ant attended. Membracids with parental care should be on evergreen hosts and be polyphagous with restricted dependence on ants.

In North temperate regions such as the northeast United States, the majority of species deposit eggs on deciduous woody hosts. Thus there is a high degree of host plant specialization leading to monophagy (Wood, 1980; Wood, Unpublished; Wood and Guttman, 1982; Guttman et al 1981). Polyphagous species appear to feed on herbaceous plants. Some species interact with ants but not as commonly as in the tropics. Solitary life histories are common although on some host plants such as oak, multiple species associations may be formed to attract ants. Parental care is rare or highly modified.

References

Bentley, B. L. 1976. Plants bearing extrafloral nectaries and the associated ant community: Interhabitat differences in the reduction of herbivore damage. Ecology 57:815-820.

Bentley, B. L. 1977. Extrafloral nectaries and protection by pugnacious bodyguards. Ann. Rev. of Ecology and Syst. 8:407-427.

Eickwort, G. C. Pre-social insects. <u>In</u> Herman, H. R. Social Insects. Vol. 2. Academic Press, New York. In press.

Guttman, S., T. K. Wood and A. Karlin. 1981 Genetic differentiation along host plant lines in the sympatric <u>Enchenopa binotata</u> Say complex (Homoptera: Membracidae). Evolution. 35:205-217.

Inouye, D. W. and O. R. Taylor. 1979. A temperate region plant-ant Seed Predator System: Consequences of extra-floral nectar secretion by <u>Helianthella quinquenervis</u>. Ecology. 60:1-7.

Janzen, D. H. 1966. Coevolution of mutualism between ants and
    Acacias in Central America. Evolution 20:249-75.
Jeanne, R. L. 1979. A latitudinal gradient in rates of ant pred-
    ation. Ecology. 60:1211-1224.
Leston, D. 1973a. Ecological consequences of the tropical ant
    mosaic. Pro. VII Congr. I. U.S.S.I., London. pp. 235-42.
Leston, D. 1973b. The ant mosaic-tropical tree crops and the
    limiting of pests and diseases. Pans. 19(3):311-41.
Messina, F. J. 1981. Plant protection as a consequence of an ant-
    membracid mutualism: Interactions on goldenrol (Solidago sp.).
    Ecology. 62:1433-1440.
Wood, T. K. 1974. Aggregating behavior of Umbonia crassicornis
    (Homoptera: Membracidae). Can. Ent. 106:169-73.
Wood, T. K. 1976a. Alarm behavior of brooding female Umbonia
    crassicornis (Membracidae: Homoptera). Ann. Entomol. Soc.
    Amer. 69:340-44.
Wood, T. K. 1976b. Biology and presocial behavior of Platycotis
    vittata F. (Homoptera: Membracidae). Ann. Entomol. Soc.
    Amer. 69:807-811.
Wood, T. K. 1977a. Role of parent females and attendant ants in
    the maturation of the treehopper, Entylia bactriana (Homoptera:
    Membracidae). Sociobiology. 2(4):257-272.
Wood, T. K. 1978. Parental care in Guayaguila compressa. Psyche.
    85:135-145.
Wood, T. K. 1979. Sociality in the Membracidae. Misc. Pub.
    Entomol. Soc. Amer. 11:15-22.
Wood, T. K. 1980. Divergence in the Enchenopa binotata Say
    Complex (Homoptera: Membracidae) Effected by Host Plant
    Adaptation. Evolution. 34(1):147-160.
Wood, T. K. and S. Guttman. 1982. The Ecological and Behavioral
    basis for the development of reproduction isolation in the
    sympatric, Enchenopa binotata complex. Evolution. In press.

Abstracts

Nest Sharing in Solitary Wasps:
Mutualism or Parasitism?

H. Jane Brockmann
University of Florida

Nest sharing is normally thought to be one of the earliest stages in the evolution of social behavior. I have studied the nesting behavior of two species of sphecid wasps which are normally solitary, the ground-nesting golden digger wasp *Sphex ichneumoneus* and the pipe-organ mud-daubing wasp *Trypargilum politum*. In both species females usually build and mass provision their own nests, but they also use the abandoned nests of conspecifics. In dense nesting aggregations, I have observed two females jointly provisioning the same brood cell within the same nest. They are highly aggressive toward intruders and defend their nests fiercely. However, in both species the females spend most of their time hunting away from the nest, so that two females can sometimes jointly occupy the same nest for several days without encountering one another. Long-term studies on both species reveal that on the occasions when females nest together, they are less successful than when nesting alone. In fact, one female effectively steals from another by laying an egg on the jointly accumulated provisions before the other female has a chance to do so. Some female mud-daubers also show a specific pattern of brood parasitism, opening a nest, removing the egg and replacing it with one of their own.

Maternal Care Behavior Among Sawflies (Hym., Symphyta)

Braulio F. S. Dias
Departamento Regional de Pesquisas Ecológicas,
Fundação IBGE, Brasília, Brazil

Maternal care is a rare phenomenon among Holarctic Tenthredinoid sawflies, being known only for Pachynematus itoi of Eurasia. Females of the Nearctic Neodiprion lecontei and N. nanulus remain at the base of the needle where they have laid there last eggs, suggesting an

180

incipient maternal care. The phenomenon, however, is not rare among tropical and subtropical sawflies of the families Argidae and Pergidae, where it evolved independently at least six times. At least 15 species in eight genera (Dielocerus, Digelasinus, Pachylota, Themos and Sericoceros in Argidae, and Pseudoperga, Philomastix and Syzygonia in Pergidae) from South America, West Indies and Australia are known to exhibit elaborate maternal care toward their eggs and young larvae. Studies on Themos olfersii and Dielocerus diasi in Brazil and a revision of the literature suggest that maternal care in sawflies is present only in species with extreme semelparity (all eggs in one compact egg cluster) as a protection against predators, but failing against parasitoids.

Colony Composition of the Woodroach
Cryptocercus punctulatus

Christine A. Nalepa
North Carolina State University

The composition of colonies of the woodroach Cryptocercus punctulatus was studied in spring 1981 by field dissection of logs at Mountain Lake Biological Station, Virginia. Four collection trips were made and approximately 100 social groups were sampled each trip. A social group was defined as all roaches found in one gallery system. Families, consisting of a group of nymphs together with either a male-female pair of adults or a single adult, were between 36 and 48% of each sample and are considered the basic social unit. Adult pairs were consistently about one-fourth of each sample. The remainder included lone adults, lone nymphs and groups of nymphs with no adult present. No gallery system contained more than two adults. In 96.8% of the family groups only one age class of nymphs from the past three years was present, i.e., adult Cryptocercus do not reproduce annually, but remain with a brood at least three years, a period during which additional offspring are not produced. The occurance of an extensive period of parental care in Cryptocercus supports the hypothesis that eusociality in termites (Isoptera) may have emerged in close association with the extension of parental care.

Population dynamics of the quasisocial spider Anelosimus
eximius (Aranea: Theridiidae).

William L. Overal & Pe. Romeu Ferreira da Silva
Museu Paraense Emílio Goeldi, Belém, Pará, Brazil

Nine colonies of Anelosimus eximius (Keyserling) remained intact in a Primary forest reserve ("Macambo") near Belém for over 24 months, whereas 23 other colonies emigrated or became extinct within 12 months after discovery in the same 5 hectare area. Large colonies (>1000 spiders) intermittantly produced daughter colonies of up to 70 spiders which dispersed to nearby trees. On two occasions daughter colonies dominated insect fly-ways near the parent colony which consequently declined. Reproduction in A. eximius occured throughout the year, although colonies were reduced in size during the rainy season (Dec. to March) with correspondingly fewer oothecae. Males, which composed 5 to 22 percent ($\bar{x}$=8.1) of the adult populations of 6 colonies of 100 to 200 spiders, reached maturity in approximately 12 weeks, whereas females required 15 weeks. Sixteen oothecae held an average of 38.3$\pm$2.54 eggs. The ratio of oothecae to adult females varied from 0.05 to 0.42, and the time to eclosion of eggs was 22.3$\pm$ 1.77 days (10 oothecae). In two colonies of 50 to 100 spiders, several adult females (ca. 10 percent) apparently never produced oothecae, but this could not be confirmed in larger colonies. No evidence for natural recruitment or fusion of colonies was found, nor were solitary spiders found to disperse. Colonies of A. eximius are probably highly in-breed as a consequence of their derivation from another colony and the small proportion of males.

Social Parasitism and the Evolution of
Reciprocal Altruism in Lace Bugs

Douglas W. Tallamy
The University of Delaware

Field and laboratory studies suggest that sociality in the Lace Bug Gargaphia solani (Hemiptera: Tingidae) has advanced beyond subsocial parental care. Females oviposit into communal egg masses containing eggs from two-several females. A variety of interactions ensue. Initiators of egg masses either accept eggs from other females or aggressively prevent other females from adding eggs. Females that leave eggs for other females to guard are social parasites. Large proportions of the population cooperate in brood care, i.e. "take turns" guarding against predators until eggs hatch. Following egg hatch only one female remains with nymphs until they mature. Sociality in this species may be at an evolutionary interface between social parasitism and cooperative brood care.

Advanced Parental Care and Mate Selection
in a Tropical Tortoise Beetle

Donald M. Windsor
Smithsonian Tropical Research Institute

Achromis bisparsa (Chrysomelidae: Cassidinae) is one of six tortoise beetles that commonly feed on the vine, Merremia umbellata (Convolvulaceae), in second-growth habitats in the Republic of Panama. As the only one of these species to exhibit parental care, females actively use broad elytra to shield eggs, larvae and pupae from invertebrate predators over the 40-day developmental period. Females avoid competition with larvae of other Cassidinae by ovipositing beneath leaves on fast-growing vine shoots. Achromis males provide no care for offspring and as a result are present in the mating population in far greater numbers than females. Males precede females to new vine shoots where they lock elytra with and flip smaller sexual competitors from the plant. Females mate repeatedly while searching for oviposition sites; however, their final mating usually occurs at the oviposition site with the largest male on the plant. Unisexual parental care is discussed as a factor predisposing the evolution of elytra specialized for combat as occurs in larger males of this species.

THE EVOLUTION AND ONTOGENY OF EUSOCIALITY
Organized by Mary Jane West-Eberhard

Introduction

Mary Jane West-Eberhard,
Smithsonian Tropical Research Institute

The organisms featured in this symposium represent all of the major groups of social insects (wasps, ants, bees, and termites) and many regions of the world (Europe, England, Malaysia, India, Central America, and North America). They were chosen to illustrate diverse aspects of a particular phenomenon: reproductive competition among the members of the same colony. There is evidence of such competition in a wide range of social insects, from the most primitively social species (e.g., the stenogastrine wasp studied by Hansell et al, in which group life is optional and a worker-queen functional dimorphism poorly developed) to the most highly specialized social species (e.g., the termites studied by Thorne, in which group reproduction is obligatory and queens so highly specialized that they are incapable of performing worker tasks). These studies suggest that all elements of a colony -- workers (van Honk), queens (Elmes, Thorne), and even larvae (Hunt) -- may be implicated in competitive interactions.

Why is intracolony reproductive competition an appropriate theme for a symposium on the Ontogeny and Evolution of Eusociality? As the colony histories of more and more species are scrutinized, evidence of intracolony competition (worker oviposition, multiple queens, ritualized dominance interactions, and suppressive pheromones) increases, and the stereotyped image of a social insect colony as a harmonious "supraorganism" declines. The revised vision of the colony requires a new view of social insect evolution. This is not to say that the supraorganismic qualities of social insect colonies are not important: certainly insofar as the group is vital to individual reproductive success, inherited tendancies promoting the survival and well-being (or "homeostasis") of the group must be favored under natural selection at the individual and genic levels. And individually advantageous characters helpful to the colony rather than detrimental to it must persist longer through evolutionary time.

However, if more than one female can lay eggs in a colony, a colony-selection explanation of cooperation and communication is incomplete and can be misleading. If, as van Honk finds in Bombus (below), workers regularly lay eggs late in the ontogeny of colonies, then their helping behavior (or lack of it) cannot automatically be viewed as good for the colony, or even as a product of kin selection

(helping relatives). Instead, as Michener long ago suggested, it
may represent investment in the "worker's" own offspring. Similarly,
the mutual tolerance of cohabiting termite queens (Thorne) may be
favored due to their (temporary) need for each other in maintaining
a colony which each one has a certain probability (or "hope") of
later monopolizing. In polygynous colonies (Gadagkar and Joshi;
Elmes) and in laying-worker colonies (van Honk) much communication,
and caste determination itself, must be regarded as evidence of
intracolony strife, not simply a mechanism of coordination analagous
to that effected via the central nervous system of a multicellular
organism.

A few highly specialized social insects have colonies approaching a supraorganismic level of integration, and it is of great
interest to discern the ecological and phylogenetic circumstances
under which they occur.

The most "supraorganismic" social insect colonies -- those with
morphologically specialized worker subcastes and highly differentiated physogastric queens -- have long-lived monogynous queens. Thus
they are the social insects which most closely approach multicellular organisms in having a high degree of genetic uniformity of
component parts and, consequently, a low degree of internal conflict
of interests. The ontogeny and evolution of monogyny is therefore
.of special interest. It is being approached through studies of
colony development in species having both polygynous and monogynous
stages (Thorne); and through comparative studies of genera containing
both monogynous and polygynous species (Gadagkar and Joshi).

"Laying workers" are promising subjects for future research.
"Worker" may often be a misnomer for an unmated female, since idle-
ness and waiting can be part of the reproductive tactic of a potential or occasional egg layer. Some females, especially unmated ones,
may contribute neither work nor worker-producing eggs to the colony
in which they oviposit. "Parasitic" oviposition by unmated females
and polygynous queens (see especially Elmes) can lead to interspecific social parasitism. In special circumstances it may even
foster the origin of reproductive isolation of a parasitic lineage
via assortative mating within colonies, as I discuss elsewhere. In
any case, it is clear that socially parasitic behavior can evolve as
an alternative strategy within species, and this can have far
reaching consequences for the social organization of colonies.

Whatever evolutionary interpretations we may choose to invent
or debate, they must ultimately be evaluated in terms of detailed
studies of individual and colony histories like those presented in
this symposium.

A Comparative Study of Social Structure in Colonies of *Ropalidia*

Raghavendra Gadagkar and N. V. Joshi,
Indian Institute of Science

Ropalidia is a large old world genus in the family Vespidae
whose systematics has been studied by van der Vecht (1962) and
Richards (1978). There is very little information on the biology of
Ropalidia (summarised by Richards, 1978) and even less regarding
social organization (summarised below). However, the genus Ropalidia is considered to be of special interest primarily because of
its diversity in social organization and nest architecture. The
genus contains both independent- and swarm-founding species with
both open as well as enveloped nests (Jeanne, 1980; van der Vecht,
1962).

Some information on social organization is available for 2 African and 2 Indian species. In Africa, Roubaud (1916) reported that
R.guttatipennis colonies consist of a mixture of morphologically
indistinguishable females, some with functional ovaries and some
without. The colonies of R.cincta studied by Darchen (1976) were
monogynous and there was a dominance hierarchy with the queen at the
top of the hierarchy and the males at the bottom.

In India, R.marginata has colonies which may be either monogynous or polygynous with the adults all morphologically identical.
New colonies are founded by one or a group of females (Gadagkar et
al. 1978). Gadgil and Mahabal (1974) showed that females with well
developed ovaries tend to be among the heaviest individuals in a
colony and hypothesized that workers spend more energy in food
gathering but receive a disproportionately smaller share of the food
and thus suffer from 'nutritional castration'. Gadagkar (1980)
demonstrated that there is a dominance hierarchy in the colonies
which influences division of labour in such a fashion that the dominant individuals including the queens spend most of their time sitting on the nest and at best show alarm reactions while the subordinate individuals spend a great deal of time making trips to places away
from the nest to bring back food, building material, water etc.
R.cyathiformis appears to be similar except that it is conjectured
to be at a more primitive level of social organization because
several individuals lay eggs as well as forage and therefore combine
the roles of queen and worker (Gadagkar and Joshi, submitted).

Here we analyse time-activity budgets of individually identified wasps of R.marginata and R.cyathiformis by multivariate

statistical techniques such as principal components analysis and hierarchical cluster analysis and show that wasps of both species fall into three distinct clusters or behavioural castes which we call sitters, fighters and foragers.

MATERIALS AND METHODS

Observations were made on two colonies of R.marginata for 165 hours and six colonies of R.cyathiformis for 170 hours in Bangalore (13°00'N and 77°32'E). Individual wasps were identified by marking with a spot of paint.

Ad libitum sampling was used to describe the behavioural repertoire, instantaneous scanning to estimate the proportion of time spent in different behaviours and all occurrences of rare behaviours were recorded in separate sessions to calculate the frequencies with which these were performed (Altmann, 1974).

Time-activity budgets for 20 animals of R.marginata and 32 animals of R.cyathiformis were used in principal components analysis (Anderberg, 1973; Frey and Pimental, 1978) and hierarchical cluster analysis (using the single linkage algorithm, and Pearson product moment correlation as an index of similarity between animals, De Ghett, 1978).

RESULTS

The time-activity budgets (data not shown) reveal that every animal spends most of its time (85-100%) in the six activities, sitting plus grooming, sitting alert (antennae raised), mild alarm reaction (antennae and wings raised), walking, inspection of cells and temporary absence from nest. However, the manner in which an animal allocates its time between these activities is highly variable. For example, the time spent by an animal in sitting varied between 7-56% in R.marginata and between 0-67% in R.cyathiformis; the time spent in temporary absence from the nest varied between 0-69% in R.marginata and between 0-88% in R.cyathiformis. This suggests that most individuals are capable of, and do perform most activities and, the differences between individuals are likely to be quantitative rather than qualitative. For this reason we have subjected the time-activity budgets to multivariate analysis. The results of principal components analysis show that temporary absence from the nest and sitting have the highest weightage in the first two principal components respectively in R.marginata and temporary absence from the nest and sitting alert have the highest weightages in the first two principal components in R.cyathiformis respectively. 92.5% and 98.1% of the total variance are accounted for by the first two principal components in R.marginata and R.cyathiformis respectively. Wasps of both species fall into three distinct clusters when each animal is represented as a point in the coordinate space of the first two principal components (Figs.1 and 2). The distinctness of the three clusters was confirmed by the method of nearest centroid. The method of hierarchical cluster analysis gives identical clusters for both species (not shown). In both the species members of one cluster rank highest in sitting (see Figs.3 and 4)

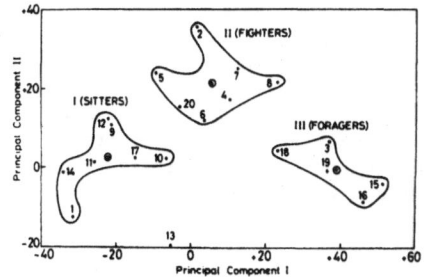

Figure 1. Behavioural castes in _R. marginata_ obtained by principal component analysis. Each point represents an animal. ⊙ =centroid of the cluster

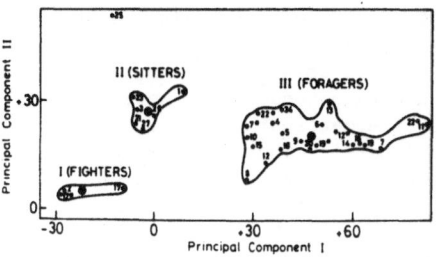

Figure 2. Behavioural castes in _R. cyathiformis_ obtained by principal components analysis. Each point represents an animal. ⊙ =centroid of the cluster.

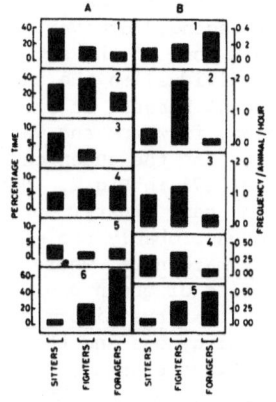

Figure 3. Mean behavioural profiles for the three clusters obtained in Fig. 1 for _R. marginata_. A, percentage time spent in the six activities used in the analysis: 1, sitting plus grooming; 2, sitting alert; 3, mild alarm reaction; 4, walking; 5, inspection of cells; 6, temporary absence from nest. B, mean frequency per hour of 5 other activities not used in the analysis: 1, bringing food loads; 2, attacking; 3, being attacked; 4, snatching food; 5, losing food.

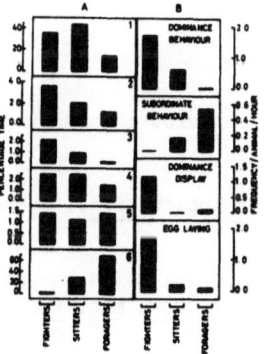

Figure 4. Mean behavioural profiles for the three clusters obtained in Fig. 2 for _R. cyathiformis._ A, percentage time spent in the six activities used in the analysis. Numbering as in Fig. 3. B, mean frequency per hour of 4 other activities not used in the analysis.

and we call these 'sitters'. Members of a second cluster rank highest in temporary absence from the nest. Wasps temporarily absent from the nest often returned with food, building material or water and hence we call these 'foragers'. The third cluster has the highest rank for sitting alert, an activity that is positively correlated with fighting (labelled as attacking in R.marginata and dominance behaviour in R.cyathiformis) and we have therefore labelled this cluster as 'fighters'.

DISCUSSION

Multivariate analysis of time-activity budgets reveals the presence of three clusters or behavioural castes namely sitters, fighters and foragers in R.marginata as well as R.cyathiformis. It is interesting to note that although data on egg-laying were not used in the analysis, both the queens of R.marginata (individuals 1 and 13) fall into the same cluster, 'sitters'. Thus R.marginata queens seem to do little other than egg-laying and spend most of their time sitting - perhaps the best strategy to develop their ovaries and maximise their egg-laying capacity. The rest of the sitters in R.marginata could either be hopeful queens or naive workers yet to be recruited into the worker force. This is being investigated by queen removal experiments. The situation in R.cyathiformis is somewhat different. One of the colonies was polygynous and some animals did both egg laying and foraging. However, individual 2 was the most dominant one which did most of the egg laying. Individual 17 was the next most dominant individual on the same colony which may be called a 'potential queen' because it later left this colony and went on to found a new colony in which she was the queen. The data on this animal after she became the queen on the new colony are treated separately in form of 17* in Fig.2. Note that individuals 2, 17 and 17* fall into the same cluster but, in this species they are the fighters. Although probably too early to generalise, it is very suggestive that in R.marginata, where the colonies studied were monogynous and the roles of queen and worker rather distinct, the queens are 'sitters'. In R.cyathiformis on the other hand, where one of the colonies was polygynous and the distinction between queen and worker was not always present, the 'queens' are 'fighters'. Reproductive competition must obviously be more intense in R.cyathiformis than in R.marginata. Details are being published elsewhere.

REFERENCES

ALTMANN J., 1974 - Observational study of behaviour: sampling methods. Behaviour, 49, 227-265.

ANDERBERG M.R., 1973 - Cluster Analysis for Applications. Academic Press, 359 p.

DARCHEN R., 1976 - Ropalidia cincta, guepe social de la savane de lamto (Cote-D'ivoire) (Hym.Vespidae). Ann. Soc. ent. Fr. (N.S.), 12, 579-601.

De GHETT V.J., 1978 - Hierarchical cluster analysis. In Quantitative Ethology ed. COLGAN P.W. John Wiley and Sons, pp.115-144.

FREY D.F., PIMENTAL R.A., 1978 - Principal components analysis and factor analysis In. Quantitative Ethology ed. COLGAN P.W., John Wiley and Sons, pp.219-245.

GADAGKAR R., 1980 - Dominance hierarchy and division of labour in the social wasp Ropalidia marginata (Lep.) (Hymenoptera: Vespidae). Curr. Sci. 49, 772-775.

GADAGKAR R., GADGIL M., MAHABAL A.S., 1978 - Observations on population ecology and sociobiology of the paper wasp Ropalidia marginata (Lep.)(Family Vespidae). Paper presented at the Symposium on Ecology of Animal Populations. Zool. Survey of India, Calcutta.

GADAGKAR R., JOSHI N.V. (submitted)- Behaviour of the Indian social wasp Ropalidia cyathiformis (Fab.)(Hymanoptera: Vespidae).

GADGIL M., MAHABAL A.S., 1974 - Caste differentiation in the paper wasp Ropalidia marginata (Lep.). Curr. Sci., 43, 482.

JEANNE R.L., 1980 - Evolution of social behaviour in the Vespidae. Ann. Rev. Entomol., 25, 371-396.

RICHARDS O.W., 1978 - The Australian social wasps (Hymenoptera: Vespidae). Aust. J. Zool. Suppl. Ser. No.61.

ROUBAUD E., 1916 - Recherches biologiques sus les quepes solitaires et socialis d'Afrique La genese de la vie sociale et l'evolution de l'instinct maternal chez les Vespides. Ann. Sci. Nat. Zool., 1, 1-160.

VAN DER VECHT, 1962 - The Indo-Australian species of the genus Ropalidia (Icaria)(Hymenoptera, Vespidae)(Second part). Zool. Verh. Rijks. Natl. Hist. Leiden., 57, 1-72.

Liostenogaster flavolineata:
Social Life in the Small Colonies
of an Asian Tropical Wasp

M. H. Hansell, Glasgow University

Charlotte Samuel and J. I. Furtado,
University of Malaya, Kuala Lumpur

The Stenogastrinae have until recently been regarded as a subfamily of the Vespidae but recent evidence suggests that they should be considered as a subfamily of the Eumenidae (Van der Vecht, 1977). Their special importance lies in the smallness of their colony sizes so that, although their biology is known to be unique in some respects, they provide one of the best available systems for obtaining insights into the early stages of evolution of social behaviour in the Vespoid wasps.

Stenogastrinae are confined to the Asian tropics. Of the seven genera, five are found in Malaysia. Stenogastrinae vary greatly in their nest architecture, many nests being species typical. Colony sizes vary between species from one or two to perhaps as high as ten females per nest.

This paper is based upon an ongoing project to study Liostenogaster flavolineata, which is known to have colony sizes among the largest of any stenogastrine species.

RESULTS

1. Colony sizes

Spot checks of the number of females per nest gave figures ranging from one to seven. A sample of 49 nests on one day gave the following distribution:

Females per nest :	1	2	3	4	5	6	7
No. of nests in that category :	17	13	10	5	1	2	1

Extended periods of observation on a small number of nests show that actual numbers of colony females will be slightly larger because a few females spend little time on the nest.

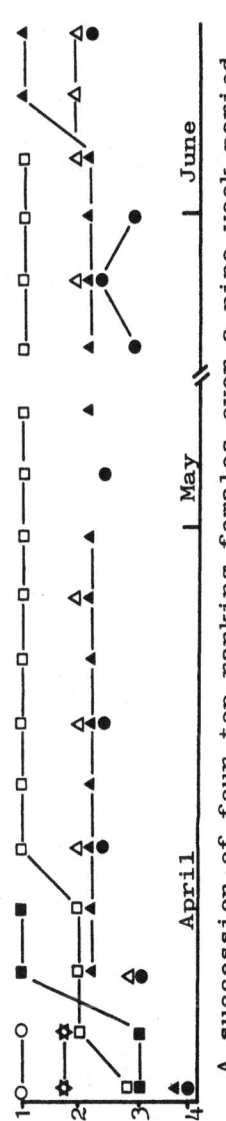

2. Social relationships

a) When two or more females are present on the nest, distinct dominant and subordinate behaviour can be seen. Dominance takes the form of a direct <u>advance</u> <u>towards</u>, which in its extreme form is prolonged harassment, and <u>attack</u>, which takes the form of biting legs and wings.

Subordinate behaviour takes the form of <u>retreat</u>, or <u>crouch</u> in which the motionless insect lowers head and antennae. Other kinds of social interaction occur, notably head butting and antennal fencing but these are not clearly associated with dominance and were not used to assess rank.

b) All multiple female colonies (N = 23) were observed to have a single top ranking female. Below this it was rare to be able to identify a linear rank order, instead it was usually only possible to distinguish a further two levels of rank not least because dominance behaviour was rather rare between females who were not of the top rank.

Extend studies on nests showed that females can rise through the ranks to take over the dominant position. One nest observed for seven months and maintaining from three to eight females on it, had a succession of seven top ranking (dominant) females who retained their position for periods varying from one to 10 weeks (Fig. 1). All these 'reigns' were terminated by the disappearance of the wasp concerned and the accession of another resident from among the second level females.

3. Correlates of dominance

a) The dominant female spends more time on the nest (98%) when compared with all other females (68%), P < 0.001.

b) Second level females do not spend significantly more time on the nest (79%) than third level females (62%) P > 0.1. Both groups are rather variable in this respect:

Fig. 1
Female
rank

A succession of four top ranking females over a nine week period

Interquartile range for level 2 = 100 - 78%
"   "   " level 3 = 100 - 14%

c) The dominant female has significantly larger ovaries than second level females (P < 0.002), who have significantly larger ovaries than third level females (P < 0.05).

d) Females seem to become inseminated before the ovaries are fully developed and females other than the dominant female do lay some eggs.

e) It is not yet clear what advantage the dominant female derives from her position, but the expectation is that through it she enjoys superior reproductive performance to her colleagues. This might result from any one or a combination of the following:

her diminished energy expenditure on foraging,
her selective cannibalism of eggs,
her behavioural inhibition of the ovary development
    of others.

4.   Getting to the top
a) Of 18 wasps disappearing from one nest over a seven month period, 11 (61%) did not become dominant female before their departure.

b) There are two ways for nest leavers to become top ranking females themselves:

    i) Found their own nests. Eight nests were obser-
       ved being founded, all by a single female.

    ii) Join other colonies. Seven cases of joining
       have been observed; all these joined single
       foundresses early in the life of the new
       colony. Single females seem therefore seldom
       to remain single. None of these joiners
       joined as the top ranking female although they
       may well achieve this by succession in the
       same way as any resident. It is not yet known
       whether joiners are admitted to multiple
       female nests, but alien intruders to such
       colonies are reguarly observed being repelled.

5.   The future
The future direction of the work is towards determining the factors influencing the tendency of females to stay or leave the home colony, since this is fundamental to the determination of the small colony sizes seen in the Stenogastrinae.

CONCLUSIONS

L. flavolineata lives in small polygynous colonies consisting probably of both related and unrelated individuals. Behaviour differences are apparent between

females, which are correlated with age. There is also a
partially developed dominance system. This kind of
social organisation corresponds to the 'rudimentary-caste-
containing stage (III)' of West Eberhard (1978) and lends
support to her polygynous family hypothesis for the
evolution of sterile castes in the social wasps, although
this stage may never have been reached in the Stenogastrinae.[1]

References

VAN DER VECHT, J., 1977. - Studies of Oriental Stenogas-
    trinae (Hymenoptera, Vespoidea). Tijdschr. Ent.,
    120, 55-75.

WEST-EBERHARD, M.J., 1978. - Polygyny and the evolution
    of social behaviour in wasps. J. Kans. ent. Soc.,
    51, 832-856.

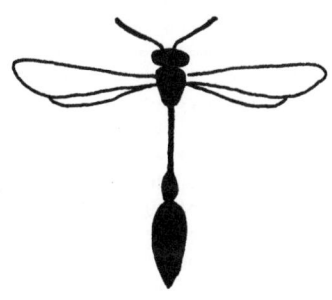

[1]This work is being conducted at the Field Station of the
University of Malaya at Ulu Gombak near Kuala Lumpur and
is supported by a grant from the Leverhulme Trust.

The Social Structure of
Bombus terrestris Colonies: A Review

Cor G. J. van Honk,
Utrecht State University

Within the group of Apidae we find a large variety of levels of social organization ranging from the rather primitive communal type found in some Euglossini (Dodson,1966) to the highly eusocial type found in Meliponini and Apini (Sakagami,1971; Michener,1974). In Eulaema nigrita (Zucchi et al.,1969) we find quite an incomplete division of labour between females, apparently based upon the outcome of agonistic interactions, whereas in Apis and in some Meliponini the division of labour is almost complete and castes have evolved. In the latter two groups one caste succeeds in monopolizing the reproduction, either by inhibiting the egg-laying behaviour of the others as in Apis, or by eating their (trophic) eggs as in some Meliponini. The honeybee queen maintains her dominance over her workers by producing pheromones which inhibit oogenesis in workers (De Groot and Voogd, 1959; Verheyen-Voogd,1959; Butler and Fairey, 1969). These pheromones are distributed in the manifold exchanges of food in the worker population (Nixon and Ribbands,1952; Butler,1954; Verheyen-Voogd,1959). Compared to the situation in which one female has to dominate others by means of agonistic behaviour, the production of pheromones significantly extends the reach of the producer, and colonies headed by a pheromone-producing female actually are bigger. Although the phenomenon of caste determination was studied in extenso in both Meliponini and Apini (Wirtz and Beetsma,1972; Kerr,1974; Velthuis and Velthuis-Kluppell,1975; Engel,1979) the mechanism is still obscure. In Bombus terrestris however, it is known to be regulated by a pheromone of the queen (Röseler,1970, 1974): yet another mechanism of the dominant female to extend her reach.

In bumblebees the level of social organization changes during the lifetime of a colony, which is started by a solitary queen who is able to dominate the firsts of her offspring, but who loses her dominance after some time when many workers start laying eggs. Bumblebees hold an intermediate position in the group of Apidae, mainly because of morphological parameters, colony size, and life-cycle characteristics. The question arises whether the mechanisms which regulate the dominance of the queen are intermediate as well. Is this dominance imposed in agonistic behaviour, by means of pheromones, or both ?

Velthuis (1976,1977) postulated that queen pheromones probably

evolved from sexpheromones. In the honeybee queen both functions are
ascribed to the 9-oxo-decenoic acid produced in her mandibular
glands. We investigated the presence of sexpheromones in bumblebees
and showed (Van Honk et al.,1978) that bumblebee queens produce
species-specific sexpheromones in their mandibular glands which
play a predominant role in mating biology. Although chemical
analysis is not yet completed the structure of the pheromones was
shown to be similar to that of the honeybee queen and the pheromones
of the different species of bumblebees were shown to be differing in
composition.

The intermediate character of dominance in Bombus terrestris
was shown in a number of experiments. The queen produces pheromones
which inhibit JH production in worker corpora allata (Röseler et al.
1981) and prevents worker ovipositions agonistically after the
pheromone production has become insufficient by force of numbers
of workers rather than of senescence of the queen (Röseler et al.,
1981; Van Honk et al.,1981; Van Honk and Hogeweg,1981). An intact
queen is able to maintain her dominance longer than a queen whose
mandibular glands are removed, whereas a sham-operated queen is as
capable as an intact queen. Colonies with an operated queen remain
smaller and the workers have developed ovaries at an earlier age.
The remaining timespan to oviposition (24 days, Van Honk et al.,
1980) however, is still much longer than the "autonomous" development (5 days, Röseler, 1974).

There is no distribution of pheromones in food exchanges like
in the honeybee. The mouthparts of the queen are antennated by
workers, after which the latter withdraw in most cases (Van Honk et
al.,1980; Van Honk and Hogeweg, 1981). The pheromone is volatile to
some extent (Van der Kerk, pers.comm.) and we suppose that the
queen is recognized chemoperceptively. In a later stage of development also workers with well-developed ovaries are antennated and
then again the lower ranking animals withdraw. Laying workers prove
to be able to inhibit oogenesis in younger workers (Röseler et al.,
1981; Van Honk et al.,1981).

Part of the worker population does not participate in foraging
activities and constantly follows the queen on the comb. These are
first workers to develop ovaries. They are often engaged in agonistic interactions with the queen and among each other and in many
colonies they eventually drive the queen off the comb (Van Honk et
al.,1980). Although it appears to be rather contradictory that
workers who are exposed to the queen's pheromones most often should
be the first to develop ovaries, we think this paradox can be easily explained in terms of the handicap principle. (Figure).

The above mentioned withdrawal behaviour was studied in
extenso im one colony. We gathered information on some 25,000 interactions (Van Honk and Hogeweg,1981). The lifecycle of the colony
was divided in six different stages during which of each the number
of individuals hardly changed. This was due to the fact that young
bumblebees emerge in groups. The data were analyzed in both Principal Coordinate Analysis and Cluster Analysis. The results of both
analyses were in perfect keeping.

The results permitted for establishing a hierarchy in the population which proved to coincide with the reproductive success of

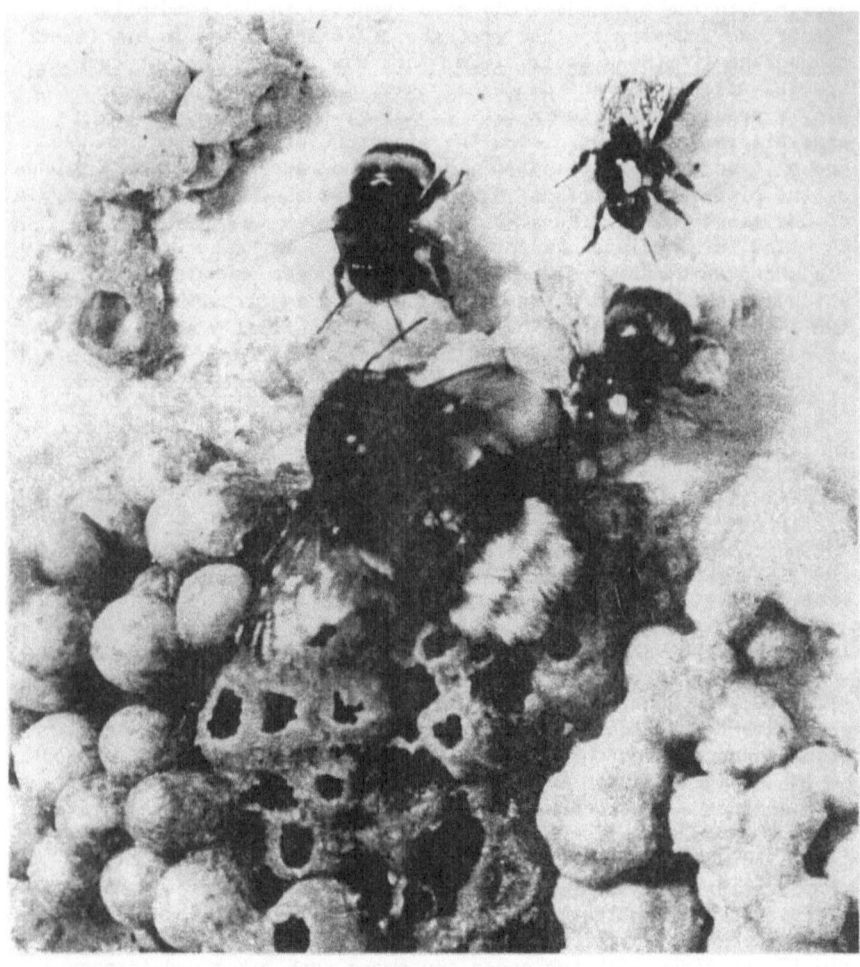

Figure: Workers of the "elite group" antennating the queen and displaying agonistic behaviour towards the end of colony development.

the individuals. High ranking workers in the analysis of the withdrawal behaviour all became laying workers. In the first periods the
queen is the most high ranking animal. After some time a group of
workers emerged of which most of the workers started laying eggs
towards the end of colony development. Within this "elite-group"
workers withdrew at Random for each other, towards workers outside
the group they appeared to be fully dominant. Shortly before an
animal entered the "elite group" it showed an enhanced activity
towards the queen or some members of the group.

After the oldest members of the "elite group" stopped laying
they maintained their status within the group although they lost the
ability to maintain their dominance towards many of the younger
workers. They were consequently used as a "stepping stone" for young
workers to enter the "elite group".

After the queen was forced by the "elite group" to leave the
comb, one worker emerged and took the queen's position in the
hierarchy. By this time the behaviour displayed in the interactions
had become much more openly agonistic. At this time the highest
ranking animals not only produced offspring of their own, almost all
workers started to help the last of the queen's offspring to develop
into new queens, both by reducing the number of larvae and by
adequately feeding the remainder of the larvae. A comparison with the
honeybee at the time of swarming is very tempting.

The dominance mechanisms in a bumblebee colony are intermediate
between dominance based on purely agonistic behaviour and dominance
completely through pheromones: the queen produces pheromones and
additionally displays agonistic behaviour. A dominance hierarchy is
established and there is very strong competition among the workers
to acquire offspring of their own.

When we consider all reproductive options of both workers and
queens in a Bombus terrestris colony we see that the aspirations of
both castes coincide perfectly with the strategies they should
follow in order to maximalize theircontribution to the genome of the
next generation of bumblebees.

+ Author's present address: Royal Netherlands Embassy, P.O. Box
21822, Safat, State of Kuwait.

References

Butler C.G., Fairey E.M., 1963. -- The role of the queen in preven-
     ting oogenesis in worker Honeybees. J. apicult. Res., 2, 14-18.
Dodson C.H., 1966. -- Ethology of some bees of the tribe Euglossini.
     J. Kansas entomol. Soc., 39, 607-629.
Engel M.S., 1979. -- Is caste determination in Melipona quadrifasci-
     ata, a stingless bee, influenced by 9-oxo-decenoic acid?
     Insectes Sociaux, 26, 273-278.
Groot A.P. de, Voogd C., 1954. -- On the ovary development in queen-
     less worker bees (Apismell L.). Experientia, 10, 384.
Honk C.G.J. van, in press. -- Queen pheromones and agonistic behaviour
     in Bombus terrestris. A view on the sociobiology of social
     apidae based on intra-specific competition. Apidologie.

200

Honk C.G.J. van, Velthuis H.H.W., Röseler P.-F., 1978. -- A sex pheromone from the mandibular glands in bumblebee queens. Experientia, 34, 838-839.

Honk C.G.J. van, Velthuis H.H.W., Röseler P.-F., Malotaux M.E., 1980. -- The mandibular glands of Bombus terrestris queens as a source of queen pheromones. Ent. exp. et appl., 28, 191-198.

Honk C.G.J. van, Röseler P.-F., Velthuis H.H.W., Hoogeveen J.C., 1981. -- Factors influencing the egg laying of workers in a captive Bombus terrestris colony. Beh. Ecol. Sociobiol., 9, 9-14.

Kerr W.E., 1974. -- Geschlechts- und Kastendetermination bei stachellosen Bienen. In: Sozialpolymorphismus bei Insekten, G.H. Schmidt, Ed. Wiss. Verlagsges., Stuttgart, p. 336-349.

Michener C.D., 1974. -- The social behaviour of the bees. Belknap Press, Cambridge, Mass.

Nixon H.L., Ribbands C.R., 1952. -- Foodtransmission within the honeybee community. Proc. R. ent. Soc. London, B 140, 43-50.

Röseler P.-F., 1970. -- Unterschiede in der Kastendetermination zwischen den Hummelarten Bombus hypnorum und Bombus terrestris. Z. Naturf., 256, 543-548.

Röseler P.-F., 1974. -- Vergleichende Untersuchungen zur oogenese bei weiselrichtigen und weisellosen Arbeiterinnen der Hummelart Bombus terrestris. Insectes Sociaux, 24, 249-274.

Röseler P.-F., Röseler I., Honk C.G.J. van, 1981. -- Evidence for inhibition of corpora allata activity in workers of Bombus terrestris by a pheromone from the queen's mandibular glands. Experientia, 37, 348-351.

Röseler P.-F., Honk C.G.J. van, in press. -- Castes and reproduction in bumblebees. In: W. Engels ed. Developmental physiology of Social Insects reproduction. Springer, Berlin, 1982.

Sakagami S.F., 1971. --Ethosoziologischer Vergleich zwischen Honigbienen und stachellosen Bienen. Z. Tierpsychol., 28, 337-350.

Velthuis H.H.W., 1976. -- Egg laying, aggression and dominance in bees. Proc. XV Int. Congr. Entomol., Washington, p. 436-449.

Velthuis H.H.W., 1977. -- The evolution of honeybee queen pheromones (from sex attractant to chemosensory signal for workers). Proc. VIII Int. Congr. IUSSI, Wageningen, p. 220-222.

Velthuis H.H.W., Velthuis-Kluppell F.M., 1975. -- Caste differentiation in a stingless bee, Melipona quadrifasciata Lep., influenced by juvenile hormone application. Proc. Kon. Ned. Akad. Wet. Ser. C 78, 81-94.

Verheijen-Voogd C., 1959. -- How worker bees perceive the presence of their queen. Z. vergl. Physiol., 41, 527-582.

Wirtz P., Beetsma J., 1972. -- Induction of caste differentiation in the honeybee (Apis mellifera) by juvenile hormone. Ent. exp. et appl., 15, 517-520.

Zucchi R., Sakagami S.F., Camargo J.M.F. de, 1969. -- Biological observations on a neotropical parasocial bee, Eulaema nigrita, with a review of the biology of Euglossinae. A comparative study. J. Fac. Sci. Hokkaido Univ. (VI, Zool 17, 271-380).

Trophallaxis and the Evolution of Eusocial Hymenoptera

James H. Hunt,
University of Missouri, St. Louis

The exchange of alimentary liquid among individuals of a social species has been recognized for more than two centuries as a prominent aspect of hymenopteran biology. Wheeler (1918) coined the term trophallaxis, which is now widely and generally used in reference to this behavior. Wilson (1971), Spradbery (1973), and Jeanne (1980) have given recent and useful reviews of it. The term trophallaxis is at present usually applied to two patterns of liquid food exchange; in one, larval saliva is imbibed by attendant adults, and in the second, ingluvial liquid is passed between adults. The first of these, adult ingestion of larval saliva, has figured prominently in writings on origins of social behavior. Roubaud (1916) presented the extreme position that the behavior is the cornerstone of vespid sociality. West-Eberhard (1978) has taken the opposite position that this form of trophallaxis need not be invoked at all to explain group life in wasps. My own position, to be presented here, is intermediate between these poles. I will argue that trophallaxis may be the key to our understanding of a more basic and more general phenomenon that has been central to the evolution of hymenopteran sociality.

The distinction between the two patterns of trophallaxis needs to be clearly drawn. The two habits have probably evolved independently, and each is probably polyphyletic. Both patterns of trophallaxis, however, exemplify the more general phenomenon to be presented in this paper.

LARVA-ADULT TROPHALLAXIS

Roubaud (1916) discussed this behavior in the context of a reciprocal food exchange, with larval provisioning as the complementary behavior. Other functions ascribed to the larval saliva include a role in regulation of nest temperature and humidity (Weyrauch, 1936), larval excretion (Brian and Brian, 1952), and as an aid in the ingestion and digestion of solid foods (Spradbery, 1965). The weight of recent evidence and opinion heavily favors a nutritional role in adults for the larval saliva that they ingest (Wilson, 1971; Spradbery, 1973; Jeanne, 1980).

Hunt, Baker and Baker (in press) have argued that though the production of a larval salivary exudate might have originated for any of the above three non-trophic functions (or for another as yet

unidentified function), its evolution and full expression as a pathway of adult nourishment may be reasonably explained. Hunt, Baker
and Baker analyzed the free amino acid content of wasp larval saliva,
and they noted a strong similarity between the nutrient composition
of larval saliva and of floral nectars, which are the typical food
for most adult aculeate solitary Hymenoptera. This similarity, they
argue, reflects selection for a nutritional role for larval saliva,
in those species having larva-adult trophallaxis, that is analogous
to the role played by floral nectars in non-social Aculeata.
Furthermore, they argue, the larval saliva may have facilitated the
evolution of sociality in any or all of three ways: 1) by fostering
more efficient foraging in social vs. solitary wasps; 2) by facilitating reproductive longevity in social vs. solitary wasps; and 3) by
contributing to a trophic mechanism for the production of individuals
with reduced fecundity.

The behavior of larva-adult trophallaxis has evolved at least
twice. It is present in at least some subfamilies of Formicidae
(Wilson, 1971, p.289) and in most if not all eusocial Vespidae.

ADULT-ADULT TROPHALLAXIS

The exchange by adults of liquids regurgitated from the crop is
a conspicuous behavior in many social Hymenoptera. Such trophallaxis
in honeybees is especially well documented, and this pattern of food
exchange serves many integrative functions in advanced social species,
including the efficient and equitable distribution of food among
colony members. In primitively social species a major role for such
trophallaxis, which I feel is not fully appreciated by most researchers, is that an unequal distribution of food is achieved. Pardi
(1948) described dominance hierarchies in primitively eusocial
Polistes wasps, and he noted trophic advantage to dominant individuals as an important result of the dominance order. Though ingluvial
exchange is rarely, if ever, unidirectional in Polistes, dominant
individuals recieve liquids more often than they are donors. Such
inequitable food distribution could serve to foster social evolution
in varied ways analogous to those suggested in the preceding section:
1) dominant individuals could sustain their nourishment yet forage
very little, with consequent low risk of predation or accidental
death; 2) dominant individuals could be sufficiently well nourished
to engender sustained reproduction as compared to otherwise equivalent species that lack such nourishment; and 3) sub-dominant individuals could experience a diminishment of their own nourishment sufficient to contribute to lessened reproductive capacity.

Adult-adult trophallaxis is polyphyletic. It is probably
present in all social Vespidae, and it is variably present in both
Formicidae and in social bees (Wilson, 1971).

ENHANCED PROTEINACEOUS NOURISHMENT AS A GENERAL PHENOMENON

I believe that the two patterns of trophallaxis just described
are significant components of social evolution in those Hymenoptera
that exhibit the behaviors. The most significant aspect of trophallaxis, in my opinion, is that reproductive individuals are

supplied with proteinaceous nourishment that is adequate and appropriate to support relatively high levels of egg laying as compared to
nestmates, be they co-foundresses or offspring. Many social Hymenoptera, however, do not exhibit trophallaxis at all; others show one
form but not the other. Clearly, trophallaxis per se cannot be
essential to the evolution of sociality. It is my position that the
underlying phenomenon, i.e. a pathway of reliable and relatively
abundant proteinaceous nourishment to reproductive females, is a
basic aspect of social evolution in Hymenoptera. In those species
lacking trophallaxis, other behaviors may serve this purpose.
Examples of alternative behaviors that may serve the same ultimate
function include (but are not necessarily limited to) ingestion of
liquids from malaxated larval provision loads, oophagy, and feeding
on pollen.

1. Ingestion of liquids from malaxated larval provision loads
    Several workers have shown that social vespids take liquid from
malaxated provision loads into their crop. In a series of studies on
Polistes metricus I have used radiotracers to show that the ingested
liquids are typically regurgitated to larvae but that a variable
quantity may be retained and so may be utilized as adult nourishment
(Hunt, MS). The potential significance of this nutritional pathway
is suggested by Evans' (1958) analysis of vespid social evolution,
where prey malaxation is the penultimate adaptation preceding
eusociality.

2. Oophagy
    Oophagy has been reviewed by Wilson (1971) and reported to occur
in varied patterns and at varied frequencies in many taxa of social
Hymenoptera. The potential nutritional significance of oophagy is
intimated by Wilson's (1971, p.281) observation that oophagy and
adult-adult trophallaxis are largely complementary behaviors; i.e.,
when a social species conspicuously exhibits one of these behaviors
the other is usually infrequent.

3. Pollen feeding
    The nutritional value of pollen has been widely reported and is
well known. Pollen is the primary larval provision of most bee taxa,
and most adult bees ingest pollen together with nectar. Grogan and
Hunt (1979) have shown in addition that pollens contain proteases of
appropriate activity and adequate quantity for in vivo digestion. My
opinion at present is that pollen in nectar may in general provide to
bees a suitably rich nourishment source for sustained egg laying,
with ecological variables (e.g., degree of polylecty) being primary
determinants of potential reproductive longevity.

ESSENTIAL EXTRINSIC FACTORS IN EUSOCIAL EVOLUTION

    Evans (1977) has drawn an important distinction between extrin-
sic and intrinsic factors that may be significant affectors of eusocial evolution. One extrinsic factor that has been effectively
shown to be requisite to eusocial evolution is complex nesting habits.
No hymenopteran is social in the absence of a nesting habit that

directly facilitates the conmingling of conspecifics. Though complete supporting evidence remains to be assembled, I would like to propose here that a pathway of proteinaceous nourishment that can augment, supplement, or replace entirely the nourishment pathways of solitary species be viewed as a second extrinsic factor that is requisite to hymenopteran eusocial evolution and equal in importance to complex nesting habits. Such pathways of proteinaceous nourishment are, like complex nesting, varied in expression and evolutionary history in Hymenoptera. Trophallaxis encompasses only two specific nourishment patterns that exemplify this more basic and general phenomenon.

References

Brian M.D., Brian A.D., 1952. -- The wasp, Vespula sylvestris Scopoli: feeding, foraging and colony development. Trans. Roy. Ent. Soc. London, 103, 1-26.

Evans H.E., 1958. -- The evolution of social life in wasps. Proc. 10th Intern. Cong. Entomol., Montreal, 1956, 2, 449-457.

Evans H.E., 1977. -- Extrinsic versus intrinsic factors in the evolution of insect sociality. BioScience, 27, 613-617.

Grogan D.E., Hunt J.H., 1979. -- Pollen proteases: their potential role in insect digestion. Insect Biochem., 9, 309-313.

Hunt J.H., in manuscript. -- Adult nourishment during larval provisioning in a paper wasp, Polistes metricus.

Hunt J.H., Baker I., Baker H.G., in press. -- Similarity of amino acids in larval saliva and nectar: the nutritional basis for trophallaxis in social wasps. Evolution.

Jeanne R.L., 1980. -- Evolution of social behavior in the Vespidae. Ann. Rev. Entomol., 25, 371-396.

Pardi L., 1948. -- Dominance order in Polistes wasps. Physiol. Zool., 21, 1-13.

Roubaud E., 1916. -- Recherches biologiques sur les guepes solitaires et sociales d'Afrique. La genese de la vie sociale et l'evolution de l'instinct maternel chez les vespides. Ann. Sci. Nat., (10)1, 1-160.

Spradbery J.P., 1965. -- The social organization of wasp communities. Symp. Zool. Soc. London, 14, 61-96.

Spradbery J.P., 1973. -- Wasps: An account of the biology and natural history of solitary and social wasps. Univ. Washington Press, Seattle, xvi+408pp.

West-Eberhard M.J., 1978. -- Polygyny and the evolution of social behavior in wasps. Jour. Kansas Ent. Soc., 51, 832-856.

Weyrauch W.K., 1936. -- Das Verhalten sozialer Wespen bei Nestuberhitzung. Zeitsch. Vergl. Physiol., 23, 51-63.

Wheeler W.M., 1918. -- A study of some ant larvae with a consideration of the origin and meaning of social habits among insects. Proc. Am. Phil. Soc., 57, 293-343.

Wilson E.O., 1971. -- The Insect Societies. Belknap Press of Harvard Univ. Press, Cambridge, Mass., x+548pp.

Multiple Primary Queens in Termites: Phyletic Distribution, Ecological Context, and a Comparison to Polygyny in Hymenoptera

Barbara L. Thorne, Harvard University

Multiple primary (first-form, alate-derived) queens within termite colonies have been found in a relatively few genera, all within the phylogenetically advanced family Termitidae (reviewed in THORNE 1981). Polygyny of primary forms is distinct from the presence of multiple supplementary (neotenic or ergatoid) reproductives because a] supplementary reproductives are added secondarily (often following death of the initial royal pair); b] supplementary reproductives develop and remain within the parental nest, hence they are always close relatives; and c] differences in morphology and development result in first form queens having a much higher fecundity than second-form (brachypterous) or third-form (ergatoid) reproductives. For these reasons, multiple <u>primary</u> queens in termites is most comparable to polygyny in social Hymenoptera. This paper explores that comparison and concludes with a brief discussion of factors influencing the evolution of polygyny in the two groups.

List of Comparisons and Contrasts: Polygyny in Isoptera vs Hymenoptera
The brevity of this paper precludes elaboration of detail and citation of many pertinent examples. General patterns are presented: discussion of variations and exceptions will be included in a future treatment. The list is topical, with similarities and differences presented within each subject entry.

1. Polygyny in the Termitidae appears to be a derived condition, as it does within many tribes of ants (HOLLDOBLER & WILSON 1977). Foundress associations are found in a number of genera of relatively primitive bees and wasps, as well as in the evolutionarily advanced polybiine wasps.
2. Although data is lacking, pleometrosis is apparently not obligate for any termite species. The frequency of polygynous colonies in a population may be relatively low. In contrast, many primitively social wasps are rarely monogynous (WEST-EBERHARD 1978a) and a variety of ants, polybiine wasps, halictid bees and bumblebees are frequently characterized by multiple foundresses (reviewed in WEST-EBERHARD 1973; HOLLDOBLER & WILSON 1977; MICHENER 1974, respectively).
3. Co-foundresses in many pleometrotic Hymenoptera may be closely related (e.g. HELDMANN 1936; RAU 1940; ORDWAY 1965; WEST 1967; LITTE 1977). Associated termite queens (and kings) are unquestionably sibs

if they are replacement (adultoid) reproductives (COATON 1949, DARLINGTON 1978) or if the nest is a "bud" seeded with progeny from the parent colony (THORNE 1982). However, it appears that in at least some cases polygynous termite colonies are the result of multiple foundresses (DARLINGTON, pers. comm.; THORNE 1982). In these cases alates "digging in" together may or may not be close relatives.

4. Polygyny in termites may be secondary: queen removal results in replacement by multiple primary daughters (adultoid reproductives) in colonies of Macrotermes and Odontotermes (COATON 1949; DARLINGTON 1978). Multiple replacement queens have been found in some species of ants (HOLLDOBLER & WILSON 1977) and in polybiine wasps (WEST-EBERHARD 1978b).

5. Groups of polygynous termite queens remain large for a longer period of time than do associations of fertile Hymenoptera females. I found 33 physogastric, egg-laying primary queens in a Nasutitermes corniger colony in Panama, and associations of 2-8 physogastric primary queens are known from species of Macrotermes and Odontotermes (reviewed in THORNE 1981). WEST-EBERHARD (1973) located one nest of Metapolybia with 36 foundresses, but the number of functional queens is quickly reduced to ≤3 reproductives in these and other Hymenoptera.

6. In Nasutitermes corniger data suggest that colonies eventually become monogynous, although the transition time can take years - much longer than in many Hymenoptera colonies which become functionally monogynous before or shortly after production of the first worker brood (e.g. WALOFF 1957; ORDWAY 1965; WEST-EBERHARD 1973; GAMBOA 1978).

7. A subordinate female in a Polistes foundress group can leave one nest to try to found her own colony, or to attempt to usurp a queen's position on another nest (WEST-EBERHARD 1969; GAMBOA 1978). Such options do not exist for supernumerary termite queens. Wings are lost shortly after the nuptial flight, and gonad development (physogastry) prevents extensive movement. Thus a female alate's decision to found singly or to join a group is largely irreversible.

8. A primary queen termite either continues to lay eggs or else she dies: there is no option to revert to a non-reproductive "helper" (as occurs in all major Hymenoptera groups).

9. Worker termites develop into functional (ergatoid) reproductives only under particular circumstances following queen death (NOIROT 1969). Thus, workers do not compete with queens by laying occasional eggs. Many Hymenoptera workers are capable of producing haploid (male) eggs even while a functional queen is present.

10. With two exceptions, termite queens within a colony are usually of similar size (reviewed in THORNE 1981). If the queens appear to be mother/daughter associations, size and coloration differences among generations are distinct (COATON 1949). Secondly, evidence from N. corniger demonstrates that as queens grow and age, variance in weight within the group can increase (THORNE, unpub. data). Egg production increases linearly with queen weight (THORNE 1981). In Hymenoptera, size differences (especially in degree of gonad development) can be present from the beginning of a polygynous association.

11. There is currently no evidence that behavioral dominance exists among primary queens in termite colonies. All females are found in one queen cell (often within the same chamber), and if separated will crawl back together. They are occasionally seen grooming one another. Dominance heirarchies are common in wasps (PARDI 1942; WEST 1967;

JEANNE 1972; GAMBOA 1978; WEST-EBERHARD 1978a), and weak dominance relations have been observed in bumblebees (BRIAN 1952; FREE 1955).

12. Some Hymenoptera queens guard their eggs in attempts to protect them from destruction by competing queens (WEST-EBERHARD 1969). Such behavior is impossible among termite queens because the physogastric females are confined to the royal chamber and eggs are transported to "nurseries" outside the queen cell. Once an egg is laid its care is dependent upon the workers. In $\underline{N}$. corniger, eggs laid by all queens are placed on the same egg cluster (laboratory observations).

13. In Macrotermes michaelseni (DARLINGTON 1978) and several species of Nasutitermes (THORNE 1981; NMNH Collection) more than one functionally reproductive female can be present when alate-producing eggs are laid. Presumably, the queens share this reproductive output. In many cases of polygyny in the Hymenoptera, queen associations are relatively short-lived, and functional monogyny is accomplished by the time alates are produced (exceptions among eusocial species are some ants with long-term polygyny, e.g. Monomorium pharaonis (PEACOCK 1950), Pseudomyrmex venefica (JANZEN 1973), Leptothorax curvispinosus (WILSON 1974)).

14. Multiple reproductive females give both Isoptera and Hymenoptera colonies a significant boost in colony growth rate (as compared with monogynous conspecifics of similar age) (WALOFF 1957; LITTE 1977; GIBO 1978; THORNE, unpub. data).

15. The probability of survival of colonies initiated by multiple ant, wasp, and termite foundresses appears to be higher than that of haplometrotic colonies (WALOFF 1957; WILSON 1974; LITTE 1977; GAMBOA 1978; GIBO 1978; THORNE, unpub. data).

16. Hymenoptera females are probably each inseminated by different males. Thus offspring in polygynous colonies that have different mothers also have different fathers. Polygynous Isoptera colonies usually contain only one king (DARLINGTON 1978, pers comm.; THORNE 1981, unpub. data). In $\underline{N}$. corniger multiple fertile males have never been found in mature colonies (those producing alates) except during the season of king replacement (asynchronous with the production of alate eggs). Thus, progeny within a mature polygynous termite colony always have the same father even though they may have different mothers.

Such intra-colony variation in parentage results in a range of relatedness among progeny. Among offspring in a polygynous Hymenoptera colony in which all queens are sibs, full sisters bear a coefficient of relatedness of 3/4, while the coefficient is 3/16 among female cousins. Workers promote their own genes 4 times more effectively by tending sisters over cousins. In contrast, termites within polygynous societies are always at least half-sibs (through the single king). If the primary reproductives are unrelated, full sibs have a coefficient of relatedness of 1/2, while half-sibs are related by 1/4. Workers tending siblings over half-sibs increase their inclusive fitness by a factor of 2. If termite queens are sisters, and further, if the king is their brother, variance in relatedness among offspring decreases substantially. Therefore, in comparison with Hymenoptera, this reasoning predicts a smaller premium on the ability of polygynous termite colony members to discriminate sibs, and a concomitant decrease in "conflict of interest" within the colony.

DISCUSSION

The evolution of polygyny is a perplexing enigma nested within the already complex puzzle of the ontogeny of eusociality. Inseminated, fertile females are expected to seek propagation of their own genes; whereas queens in polygynous associations contribute to a single effort and often succeed, at best, in only sharing in the reproductive output. Polygyny has evolved independently and repeatedly within both the Isoptera and Hymenoptera. Convergences and contrasts among groups assist in exploring the evolution of polygyny.

WEST-EBERHARD (1973, 1978a) has argued that polygyny in eusocial Hymenoptera is likely due to a medley of interacting forces including kin selection, maternal control and ecological mutualism. Such compound factors may also be involved in the evolution of multiple primary queens in termite colonies, although our current knowledge of polygynous Isoptera life histories and ecological dynamics is limited. To date neither agonistic behavior among termite queens nor preferential tending of individual queens by workers has been observed, yet, at least in Nasutitermes corniger, colonies eventually become monogynous. Intra-group competition (social competition among queens) is expected in light of individual selection, but it is to all queens' advantage to delay intense competition until after mutualistic benefits.

ACKNOWLEDGMENTS

I thank S.H. Bartz and E.O. Wilson for commenting on earlier versions of this paper. Research was supported by NSF Dissertation Improvement Grant DEB-80-16415 and with Predoctoral fellowships from the Smithsonian Tropical Research Institute and the American Association of University Women.

Thirty-three primary queens and 17 primary kings of Nasutitermes corniger found within the royal cell of a colony collected in Frijoles, Panama on 26 March 1981. Average wet weight of the queens = 0.0306 ± 0.0017 gms. Scale = 0.5 cm.

REFERENCES

Brian, A. D., 1952. - Division of labour and foraging in Bombus
   agrorum Fabricius. J. Anim. Ecol. 21: 223-240.
Coaton, W. G. H., 1949. - Queen removal in termite control. Farming
   in South Africa 24: 355-358.
Darlington, J. P. E. C., 1978. - Populations of nests of Macrotermes
   species in Kajiado and Bissell. Annual Reprot, I.C.I.P.E.
   6: 22-23.
Free, J. B., 1955. - The behaviour of egg laying workers of Bumblebee
   colonies. Br. J. Anim. Behav. 3(4): 147-153.
Gamboa, G. J., 1978. - Intra-specific defense; Advantage of social
   cooperation among paper wasp foundresses. Science 199: 1463-
   1465.
Gibo, E. L., 1978. - The selective advantage of foundress associations in Polistes fuscatus (Hymenoptera; Vespidae); A field
   study of the effect of predation on productivity. Can. Ent.
   110: 519-540.
Heldmann, G., 1936. - Ueber die Entwicklung der polygynen Wabe von
   Polistes gallica. L. Arb. Physiol. angew. Ent. Ber. 3: 257-259.
Holldobler, B., Wilson, E. O., 1977. - The number of queesn: an
   important train in ant evolution. Naturwissenschaften 64: 8-15.
Janzen, D. H., 1973. - Evolution of polygynous obligate acacia-ants
   in western Mexico. J. Anim. Ecol. 41: 727-750.
Jeanne, R. L., 1972. - Social biology of the neotropical wasp
   Mischocyttarus drewseni. Bull. Museum Comp. Zool. 144(3):
   63-250.
Litte, M., 1977. - Behavioral ecology of the social wasp Mischocyt-
   tarus mexicanus. Behav. Ecol. & Sociobiol. 2: 229-246.
Michener, C., 1974. - The Social Behavior of the Bees. The Belknap
   Press of Harvard Univ. Press, Cambridge, Mass. 404 pp.
Noirot, C., 1969. - Formation of castes in the higher termites. In:
   Krishna, K. and F. M. Weesner (eds.). Biology of Termites, Vol.
   I, pp. 311-350. Academic Press, N. Y.
Ordway, E., 1965. - Castal differentiation in Augochlorella (Hymenop-
   tera, Halictidae). Insectes Sociaux 12 (4): 291-398.
Pardi, L., 1942. - Ricerche sui Polistini. V. La poliginia iniziale
   in Polistes gallicus (L.). Boll. Inst. Ent. Univ. Bologna
   14: 1-106.
Peacock, A. D., 1950. - Studies in Pharaoh's Ant, Monomorium pharaonis.
   L. Ent. Mon. Mag. 86: 294-298.
Rau, P., 1940. - Cooperative nest founding by the wasp Polistes
   annularis Limm. Ann. Ent. Soc. Am. 33: 617-620.
Thorne, B. L., 1981. - Polygyny in termites: multiple primary queens
   in colonies of Nasutitermes corniger (Motschulsky) (Isoptera:
   Termitidae). Insectes Sociaux, in press.
Thorne, B. L., 1982. - Reproductive plasticity in the Neotropical
   termite Nasutitermes corniger. In: Jaisson, P. (ed.). Social
   Insects in the Tropics. Proc. I..S.S.I. Symp., Mexico 1980.,
   in press.

Waloff, N., 1947. – The effect of the number of queens of the ant
     Lasius flavus (Fab.) (Hym., Formicidae) on their survival and
     on the rate of development of the first brood. Insectes
     Sociaux 4: 391–408.
West, M. J., 1967. – Foundress associations in polistine wasps:
     dominance hierarchies and the evolution of social behavior.
     Science 157: 1584–1585.
West-Eberhard, M. J., 1969. – The social biology of polistine wasps.
     Miscellaneous Pubs. Mus. of Zoology, Univ. of Michigan. #140:
     1–101.
West-Eberhard, M. J., 1973. – Monogyny in "polygynous" social wasps.
     Proc. VII Int. Congr. I.U.S.S.I., London. 396–403.
West-Eberhard, M. J., 1978a. – Polygyny and the evolution of social
     behavior in wasps. J. Kansas Ent. Soc. 51(4): 832–856.
West-Eberhard, M.S. 1978b. – Temporary queens in Metapolypia wasps:
     Non-reproductive helpers without altruism? Science 200: 441–443.
Wilson, E.O., 1974. – The population consequences of polygny in the ant
     Leptothorax curvispinosus. Ann. Ent. Soc. Am. 67: 781–786.

Intra-Colonial Competition in Ants, with Special Reference to the Genus *Myrmica*

G. W. Elmes, The Institute of Terrestrial
Ecology, Wareham, Dorset

This paper stems from a round-table discussion at the 8th Int. Congr. IUSSI where I proposed that the term 'intra-colonial competition' could be used to embrace the interactions within an ant colony (Elmes 1977). Opposition to this view says that this is not true competition, merely a trade-off of energy within a super-organism (Vepsalainen 1978), in which case it follows that selection must operate mostly at the colony level. I still prefer to consider an ant society as a collection of individuals and expect a certain amount of true competition between them.

My interest is in the how and what of speciation in the morphologically-uniform ant genus *Myrmica*, especially what factors determine the success of the various species in widely differing habitats. I have extended the detailed physiological studies of *M.rubra* (done by M.V.Brian) to other common European *Myrmica* species and have combined these with field studies of the same species. Therefore, taken together with the vast amount of other work on this genus (eg behaviour, anatomy, chemistry), it is probably true to say that more is known about *Myrmica* than any other ant genus. Some of this knowledge is summarised below and this forms the background to my suggestions, some of which will be further explored in my talk.

1. The genus is very widespread and successful in the temperate holarctic, often dominating particular habitats.
2. All the species are morphologically similar. Despite the large number of described varieties, these can be grouped to form a much smaller number of good species which form 3 or 4 natural species groups (eg see Weber 1947).
3. Workers do not show polymorphism but do show polyethism in the division of labour (eg Weir 1958).
4. Colonies vary in size from 50-5,000 individuals. Colony size seems to be a species trait, despite the large intra-specific variability due to habitat (eg Elmes 1973 and Elmes & Wardlaw 1982).
5. Queen number is a species trait and a function of colony size. Contrary to tradition, all the species I have studied so far are sometimes polygynous (eg Elmes 1973 and Elmes 1980).
6. All the species probably recruit queens periodically, so that colonies are potentially immortal. Daughters have been observed to be recruited into a colony but other evidence suggests that the joiners are not always closely related to the parent colony (Elmes 1973,1980).

7. A queen can lay between 500 and 1,000 eggs per year under supposedly optimum laboratory conditions (Brian 1965).

8. Queens affect workers in many ways - 'queen effect'. They suppress worker reproductive egg laying, stimulate worker trophic egg production, induce workers to share food more evenly between the larvae and to work more efficiently. They cause workers to neglect and sometimes bite gyne-potential larvae (eg Brian 1979). 'Queen effect' has a common interchangeable action between all the species - any permutation of queen, workers and larvae for different species shows the same effect (Elmes & Wardlaw, unpublished manuscript).

9. Queen efficiency as a group is less than the sum of the individual components, both in terms of egg-laying and 'queen effect' (eg Brian 1965).

10. In *Myrmica rubra* there is some evidence of aggression and a dominance hierarchy between queens in polygynous colonies (Evesham, unpublished).

11. Workers only work when they are hungry. The queen and larvae both act as a pump, removing food from the worker population (Brian & Abbott 1977).

12. The queens of all the species lay eggs which develop into larvae that either metamorphose during the same season (rapid brood) or hibernate. The former all become workers whereas the latter retain the option of becoming workers or gynes during the following spring. It seems that a period of hibernation is essential for gyne production but optional for male production (Brian 1965, 1979).

13. In *Myrmica rubra* queen eggs are biased to become rapid brood or hibernate at different times during the egg-laying season (Brian 1965). Large queens tend to lay eggs that are more likely to be rapid brood whereas small queens lay eggs that become dormant larvae (Elmes 1976).

14. Gynes tend to be produced from large hibernated larvae but an exception is *Myrmica scabrinodis* Nylander, which can produce gynes from any size of hibernated larvae (Elmes & Wardlaw 1981).

15. Two species, *Myrmica sulcinodis* Nylander and *Myrmica schencki* Em., have some exceedingly large larvae in their nests during winter. These have no more sexual bias and a much worse survival compared to normal large larvae and they may act as food reservoirs (Elmes & Wardlaw 1981).

16. Each different species has a differing basal physiological rate (Q10 ≃ 3) but they do not work and grow at the same rate for any given temperature. The most efficient species normally live in colder habitats and the least efficient species is associated with the hottest habitats. This is a trait of both larvae and workers (Elmes & Wardlaw, unpublished).

17. It is not known whether queens contribute regularly to male production. Workers of all the species examined can lay viable eggs but the numbers of laying workers and quantity of worker eggs vary enormously between the species (Elmes & Wardlaw, unpublished).

18. In queen-right culture, males mostly develop from the eggs of newly-eclosed workers, which seem to be able to evade queen control whilst metamorphosing (Smeeton 1980).

19. There is no obvious pattern to sexual production by wild colonies of *Myrmica*, although some years are more favourable for sexual production. In any species, small colonies rarely produce sexuals. Males are produced by most of the larger colonies, some of which also produce

gynes (eg Elmes & Wardlaw 1982).
20. The more polygynous species frequently possess miniature queens.
It is not clear whether these are strictly microgynes that may establish themselves as separate breeding entities, that may give rise to
true parasites (Elmes 1978), or whether they are separate parasitic
species that have evolved in isolation from the host species (Pearson
1981).

SOCIAL INTERACTIONS

1. Interactions between castes
     In *Myrmica* this falls into 2 categories: (i) the interaction be-
tween the queen(s) and the worker pool and (ii) the interaction with
the next generation, the larvae, which are themselves potentially one
of the 3 adult castes. The queen exerts a profound influence over
the behaviour of the workers, yet, in many ways, the control and
manipulation of caste ratios seems to belong to the workers. It is
tempting to think of the worker population as an independent entity,
regardless of the queen(s). This makes some considerable sense if
queens are short-lived and if the occupation of an existing nest pays
a very high dividend. It would then seem to be profitable for an established colony to periodically recruit queens, even unrelated ones.
     I suggest the following: (i) At any time the worker population
is sure that it has a moderately high degree of internal relatedness.
By a process of mutual tolerance the workers all lay reproductive
eggs that become males. They cannot be sure that the queen(s) have
any relatedness to them, yet, in order to ensure the successful rearing and launching into nuptials of their sons, they must turn some of
the queen-eggs into workers. (ii) The queen(s) likewise cannot be
sure whether she is related to the existing female population, so
that she would like to produce gynes and males (equally) at the expense of the other females. The hibernated larvae may not be related
to her; therefore she tries to prevent them becoming gynes (queen
effect). She can either make an immediate investment in the future
success of the worker pool by laying worker-biased rapid eggs or take
a chance on the following year and bias them towards queens. Older
queens have a greater chance of being related to the workers; therefore this choice may modify with age. The queen(s) may also lay male
eggs. If she suppressed males directly there would be no advantage in
her laying male eggs or investing in the future worker pool. So she
inhibits laying by the workers but tolerates male larvae.
     The larvae themselves have a much more passive role in their
interactions with the adult castes. At best they can counter 'queen
effect' by trying either to make themselves more attractive to the
workers or disguise their sexuality. Young male larvae hide among the
more common sizes of the 2 female castes.

2. Interactions within castes
     Myrmica workers show a distinct division of labour according to
age. The difficulties of ageing workers so confounds the observed
interactions between individuals that it is not possible to be sure
whether they are 'competitive' or some sort of division of labour.
This aspect of *Myrmica* societies deserves more attention.

Interactions between the larvae are also little studied. In both laboratory and field, far more is known about the numbers and fate of hibernated larvae than is known about rapid brood. (Petal (1967 and unpublished) has the best field data on production of rapid workers to date.) Indirect evidence suggests that there is some competition between the categories of hibernated larvae. Colonies that have many larvae tend to have small larvae (Elmes & Wardlaw 1981). Generally, the size of gyne-potential larvae is positively correlated with the number of worker larvae and negatively with the number of male-likely larvae (males can only be guessed by size during early development). This suggests that when gynes are produced they are not in contest with worker larvae but are in some sort of contest with male larvae. This supports the field observations, which suggest that although males are produced regularly, gynes are only reared when the colony resources exceed those required to maintain the worker pool. Therefore, in a good season there are many larvae and the gyne-potential larvae tend to be bigger and are more likely to become gynes. Worker larval number has a negative effect on male larval size, indicating that whilst males can attract resources from the gynes, they are in competition with most female larvae. Worker-potential larvae are smaller when they are numerous, indicating that they interact. These observations are confounded by species variation and more work is needed before definite conclusions can be made.

If *Myrmica* queens are short-lived and merely attempt a short-term control of a persistent worker pool, the queens can be thought of as parasites of the workers (Elmes 1973). Individual queens should strive to monopolise the worker pool: within the polygynous *M.rubra*, Elizabeth Evesham (pers. comm.) has demonstrated aggressive behaviour and a dominance hierarchy among queens. Some queens, attended by a few workers, monopolise the best egg-laying sites in the nest; the others run about meeting many workers. The former contribute most to the next generation but the latter might exert the most 'queen effect'. Thus, in polygynous societies each queen strives to obtain a long-term advantage and tries to ensure that no other attains a short-term one. Also, the workers, being less related, should each tend to invest in their own eggs rather than any other female's.

Once a queen has control of a worker group she should attempt to prevent others joining, but this can only be successful if she is long-lived and can make a long-term contribution to the worker pool. An alternative is for a queen to specialise in laying queen-biased eggs, gambling the short-term advantage of a possibly good season against the more certain advantage of investing in the worker pool. Microgynes could be the product of this strategy, which could easily degenerate into true social parasitism, the larvae evading 'queen effect' by their small size. Contrary to this idea, *M.ruginodis* microgynes produce rapid workers.

CONCLUSION

It seems to me that in *Myrmica* species there is a considerable conflict of interests and competition for resources within the nest. Egg-laying workers may have arisen by a sort of alternation of generations in sexual reproduction through the males, whereas gyne production is primarily environmentally-determined. It seems reasonable to assume that originally queens might have been not much longer-lived than their sterile counterparts. If a nest site and an established worker pool are at a premium, a polygynous exploitation of the worker pool, balanced by the advantage of reproductive workers in maintaining the pool, might arise. In stable environments with abundant nest sites the selection could then favour an increased queen size, monogyny and direct control of reproduction wrested from the worker caste.

References

Brian M.V., 1965. -- *Social Insect Populations*. Academic Press, 135pp

Brian M.V., 1979. -- Caste differentiation and division of labour. In: *Social Insects*, ed. H. Hermann, Vol. 1, 121-222, Academic Press.

Brian M.V., Abbott A., 1977. -- The control of food flow in a society of the ant *Myrmica rubra* L. *Anim. Behav.*, 25, 1047-1055.

Elmes G.W., 1973. -- Observations on the density of queens in natural colonies of *Myrmica rubra* L. (Hym. Form.). *J. Anim. Ecol.*, 42, 761-771.

Elmes G.W., 1976. -- Some observations on the microgyne form of *Myrmica rubra* L. (Hym.Form). *Insectes soc.*, 23, 3-22.

Elmes G.W., 1977. -- Intra- and interspecific competition in ants. *Proc. 8th Int. Congr. IUSSI*, Wageningen, Netherlands, 277-279.

Elmes G.W., 1978. -- A morphometric comparison of 3 closely related species of *Myrmica* (Formicidae), including a new species from England. *Syst. Ent.*, 3, 131-145.

Elmes G.W., 1980. -- Queen numbers in colonies of ants of the genus *Myrmica*. *Insectes soc.*, 27, 43-60.

Elmes G.W., Wardlaw J.C., 1981. -- The quantity and quality of over-wintered larvae in 5 species of *Myrmica* (Hym. Form.). *J. Zool., Lond.*, 193, 429-446.

Elmes G.W., Wardlaw J.C., 1982. -- A population study of the ants *Myrmica sabuleti* and *Myrmica scabrinodis* living in 2 sites in the South of England. 1.A comparison of colony populations. 2.Effect of above-nest vegetation. *J. Anim. Ecol.*, in press.

Pearson B., 1981. -- The electrophoretic determination of *Myrmica rubra* microgynes as a social parasite: possible significance in the evolution of ant social parasites. *Systematics Assoc. Special vol.19 'Biosystematics of Social Insects'*, ed. P.E.Howse & J.L. Clement, Academic Press, 75-84.

Petal J., 1967. -- Productivity and the consumption of food in the *Myrmica laevinodis* Nyl. population. In: *Secondary Productivity of Terrestrial Ecosystems*, ed. K.K.Petruscewicz, Warsaw, 841-857.

Smeeton L., 1980. -- *Male production in the ant Myrmica rubra* L. Unpublished PhD thesis, University of Southampton, England.

Vepsalainen K., 1978. -- Modes of competition in ants. *Proc. 8th Symp. Social Insect Section, Polish Ent. Soc., Pulawy*, 25-29.

Weber N.A., 1947. -- A revision of the North American ants of the genus *Myrmica* Latreille with a synopsis of the palaearctic species. *Ann. Ent. Soc. Amer.* 40(3), 437-474.

Weir J.S., 1958. -- Polyethism in workers of the ant *Myrmica*. *Insectes soc.*, 5, 97-128.

Abstracts

Effects of Queen Removal on Intracolony
Dynamics of <u>Polistes</u> <u>metricus</u>
(Hymenoptera: Vespidae)

Heather E. Dew
The University of Kansas

Queens were permanently removed from three different age
groups of laboratory colonies of <u>Polistes</u> <u>metricus</u>. Queens of
control colonies were removed and immediately replaced on the
nest. In the youngest colonies the first emerged worker was
7 days old; in the oldest colonies the first emerged worker was
29 days old.
After queen removal, the oldest workers of young and old
colonies often assumed the role of queen. Dominant-subordinate
social interactions between females indicated that in young
colonies, after queen removal, the frequency and duration of
interactions increased markedly. In these colonies, workers
also showed reduced activity in foraging and larval care. In
older colonies, after queen removal, the frequency and duration
of dominant-subordinate interactions did not increase as markedly
and workers showed little reduction in foraging and larval care
activities.

When a wasp colony splits

Raghavendra Gadagkar and N.V. Joshi
Indian Institute of Science

On 30 June 1981, a colony of <u>Ropalida</u> <u>cyathiformis</u> (Fab)
(Hymenoptera:Vespidae), a common Indian social wasp, split into two
separate colonies as a result of 5 of the 11 adults leaving the
parent combs to build a new comb about a meter away. The numbers
of eggs, larvae and pupae on the parent combs kept decreasing before the split and reached the lowest value on the day of the split.
After the split, the brood on both the parent as well as new combs
increased rapidly. The split thus increased the genetic fitness
of the queens of both the parent and the new colony. The inclusive
fitness of the subordinate individuals associated with each colony

217

also increased provided they were at least marginally related to
the queens. Quantitative data on the behaviour of the wasps before
and after the split showed that the group that left to found a new
colony was the one that had spent significantly more time foraging
while the group that stayed on at the parent combs had shown a
significantly higher frequency of dominance behaviour. Even before
the split, wasps belonging to the group that eventually left the
colony were highly coordinated in their behaviour. They were away
from the nest site, rested or sat on a particular comb more often
together than two randomly picked individuals.

Social Integration in a Primitively Eusocial Wasp (Polistes fuscatus): The Role of the Queen

George J. Gamboa, Hudson K. Reeve and David W. Pfennig
Oakland University

The queen's role in social integration was investigated in the
paper wasp, Polistes fuscatus by making detailed field observations
(139 h) of post-emergence colonies. In the first study, 2 h observations were made on each of 12 colonies (24 h). The queen was
then removed, and 2 h observations were made on the now queenless
colonies (24 h). Control observations (10 colonies, 40 h) were
made simultaneously on undisturbed colonies. In the second study,
behavioral observations of 1.25 h/colony (9 colonies, 11.25 h) were
followed by the removal and cooling of the queen. The queenless
colony was observed during the 30 min cooling period, whereupon
the relatively inactive, cooled queen was returned to the nest.
Observations then resumed for another 1.25 h (11.25 h total).
In one control, workers of other colonies were removed, cooled,
and replaced instead of queens (4 colonies, 12 h). In a second
control, undistrubed colonies were observed for 3 h (4 colonies,
12 h). The queen is the most active individual on the nest. The
absence or decreased activity of the queen (but not of a single
worker) depresses both overall colony activity and foraging rate.
Furthermore, the queen aligns worker activities in time and her
removal results in the uncoupling of worker activities. We con-
conclude that the queen acts as both a *pacemaker* and a
coordinator of colony activity.

Studies on neotenic reproductive development in a primitive termite.

Sharon Greenberg
Cornell University

The development of neotenic reproductives was studied in labor-
tory colonies of the lower termite Zootermopsis angusticollis (Hodotermitidae). In experimental groups of larvae without reproductives,
neotenic molts began 3-4 weeks after isolation of the larvae from
the parent colony and continued, for 10 weeks, until inhibition was
re-established. Neotenic development could be initiated by compe-
tent larvae after removal of the royal pair for only 48 hours.

Female neotenics were found to develop earlier and in greater numbers than male neotenics. In the absence of males, female neotenics were parthenogenic. At 25° C, vitellogenin was present in the hemolymph of female neotenics on day 7 and eggs were laid on days 25-30 after the molt. Vitellin of neotenic queens was immunologically identical to that of adult queens. In general, reproductives of the lower termites control neotenic development by producing sex-specific, and as yet unidentified, inhibitory pheromones. In Z. angusticollis, the putative pheromones are virtually non-volatile and were not transmitted to larvae in faeces of reproductives. In fact, more neotenics developed in groups of larvae fed faeces compared to controls. Varnishing the anal or genital opening of reproductives did not disrupt inhibition. Inhibition did occur in isolated experimental groups of larvae fed 70% ethanol washes of whole queens and kings.

Early Foundress Associations in
Polistes

Colin Hughes
Rice University

A comparison of foundress associations is made between Polistes exclamans and Polistes annularis, two closely related species whose ranges overlap greatly. P. exclamans typically founds in groups of 1-5 and only occasionally near the parental nest whereas P. annularis founds in groups of 1-26 within 3 meters of the parental nest. P. exclamans never reuses old nests and seldom have more than one egg layer per nest. P. annularis however reuses 10% of old nests and commonly have more than one egg layer. P. exclamans foundresses frequently switch between new nests but P. annularis females will only do so if their own nest is destroyed. These differences may be explained by the higher frequency of nest failure in P. exclamans.

Foundress Associations and Early
Colony Failure in Polistes exclamans

Alan Mac Cormack
The University of North Carolina
at Chapel Hill

The behavior of Polistes exclamans foundresses following nest destruction was studied. Wasps evicted during the middle of the pre-emergence period did not refound their nests. They were observed attempting to join surviving colonies. These attempts were resisted, often successfully, by the resident foundresses. Successful joiners were almost always subordinate to the original foundresses. Joiners stayed on the nest most of the time and rarely foraged. Shortly after the emergence of the first workers, falling fights were observed between foundresses. Original foundresses were frequently deposed by former subordinates and by

newly arrived foundresses during this period. The results are
consistent with a hypothesis of selfish joining.

Social Evolution in the Evylaeus malachurus Species Group

Laurence Packer and Gerd Knerer
The University of Toronto

Evylaeus laticeps, E. pauxillus, E. linearis and E. malachurus
are four very closely related halictine bees which show an increase
in social level from the former to the latter. A preliminary
principal components analysis indicates that the following characteristics are correlated with increased sociality: the number of
workers and the suppression of males in the first brood, a reduction in the frequency of mated and egg-laying workers in the
first brood, an increase in the degree of morphological caste
differentiation and the opening of the brood cells during larval
development. The second component axis reflects the number of
worker broods produced and the characteristics of nest defense.
    In addition to varying in characters which may easily be
related to social advancement there are other behavioural traits
which are not so easily graded. For example, E. linearis exhibits
polygyny in the northern parts of its range where environmental
conditions are less predictable. E.malachurus, in the southern
parts of its range, has workers producing the males as a strategy
to maximise total productivity thus enabling the queen to specialise in laying fertilised eggs.
    The social organisation of these four species will be compared
to that of other species of the genus for which data are available.

The reproductive strategy of the ponerine Ophthalmopone berthoudi: an insight into the evolution of ant eusociality

Christian P. Peeters
The University of the Witwatersrand

Ophthalmopone berthoudi is a ponerine species lacking a distinct reproductive caste. Colonies of a few hundred adults are
distributed into a small number of distinct, spatially separated
nests; frequent worker recruitment, ergatoid transfer and brood
transfer occurs between these. Colony foundation takes place
accidentally through the separation of one or more nests from the
complex; this is accompanied by the development of colony mate
recognition. All adult ants possess distinct ovaries with three
ovarioles each, as well as a spermatheca and accessory glands. A
variable number (up to 80) of inseminated ants occurs in each nest,
and all of them are reproductively active. Ovarian development is
confined to inseminated ants. The eggs are very large ($1,6 \times 0,6$ mm)
and laid singly. Ovaries never have more than two mature eggs
simultaneously, because of the limited capacity of the six ovarioles,
and the absence of trophallaxis between adults. Thus cooperative
breeding by many ergatoids appears necessary in this species. Males
are only active for a few weeks (January-February), and hence ants

of the appropriate age can only get fertilized during this period; this group then constitutes the reproductive force of the colony, all other ants are necessarily non-reproductives. This extreme polygyny would result in a lowering of the average genetic related-ness within the colony, and the data will be used to test theories accounting for the evolution of altruism by sterile workers.

Social and Physical Factors Affecting Agonism Among Paper Wasp Foundresses

Nancy Ross and George Gamboa*
The University of Kansas

Agonistic behavior between artificially-paired foundresses of Polistes metricus, prior to nest initiation, was investigated in the laboratory. Multivariate data analysis techniques were em-ployed to determine which of the following factors contribute most to individual variation in agonism: 1) "relatedness" of paired foundresses (were they formerly nestmates or non-nestmates?); 2) body size; 3) corpora allata and ovarian development. Results are compared to similar, published data on individuals in mature colonies.

*Present address: Oakland University, Rochester, Michigan

Sex Investment Ratios in Polistes exclamans

Joan E. Strassmann
Rice University

Sex investment ratios were measured in a Polistes exclamans population and found to be female-biased (1976:82% females; 1977: 64% females; 1979: 72% females) in 3 years and even in 1 year (1978: 50% females). Sex ratios are female-biased in years of high levels of bird predation on nests and low overall nest success. Males are the last to emerge from the nest so anything that results in an early end to the colony cycle decreases the numbers of males. Whether a female becomes a worker or a gyne probably depends on the situation when she emerges from the nest. Sex ratios on nests whose original queens were replaced by a mated worker did not differ from sex ratios on nests with original queens. Sex ratios may be generally female-biased because a fe-male can become either a worker or a gyne, so females are a more flexible asset to a nest than are males. This flexibility may be important for the nests given a highly variable end of the nesting season in central Texas.

Some aspects of the biology and social behavior of <u>Parischnogaster</u>
<u>nigricans serrei</u> (Hymenoptera, Stenogastrinae).

Stefano Turillazzi
Istituto di Zoologia, Università di Firenze, Italy

A colony of <u>P.n.s.</u> is founded by a single female that lays
5-7 eggs and then suffers an ovarian regression till the emergence
of the first daughter. Next usurpation by other females during the
pre-emergence period occurred on about 50% of the initial nests
observed (in rare cases the two females can remain together for a
while but without any sort of cooperation). The foundress spends
a lot of time on the nest defending it mainly from usurpers. The
first male emerges usually after 3-5 females, from an egg laid by
the foundress. After the emergence of the first wasps the foundress
begins to lay again and does not leave the nest any more. Young
females forage in flight on little insects or take them from spider-
webs; after fertilized they can remain on the nest and then replace
their mother or found new colonies. Nests can have as many as 45
cells but the number of individuals emerged is usually greater.
Males perform patrolling behavior with ritualized fights.

Juvenile Hormone Production in Polistine Wasps

Lisa Vawter and Joan E. Strassmann
Washington University in St. Louis and Rice University

Juvenile hormone production was studied in Polistine wasps
using a radioassay and it was compared with rank. High juvenile
hormone production was correlated with the aggressive behavior
associated with high rank, and subordinates produced no juvenile
hormone at all. Disturbing the nests caused the second-ranked
wasp to out-produce the queen, possibly indicating she was preparing
to usurp.

Diversity of Dominance Displays in <u>Polistes</u> and
its Possible Evolutionary Significance

Mary Jane West-Eberhard
Smithsonian Tropical Research Institute

Field observations of different <u>Polistes</u> species reveal
a striking diversity of ritualized agonistic behaviors. They include chewing over the body of another with the mandibles; an
antennae-upward faceoff without contact; mock stinging; lateral
bending of the abdomen; and a wagging rush -- a rapid run toward
another individual with head down and body rapidly moved from side
to side. A given display can be frequent in one species while rare
or absent in others, and none is performed by all species observed.
Specialization and diversity in such purely social characters
are unlikely to have an ecological explanation. Rather, they
probably occur due to the great importance of traits affecting
social status and, hence, reproductive success. A general theory of

social and sexual seclection predicts rapid evolution of such traits.
This possibility can be tested by examining patterns of subspecific
and interspecific variation in dominance behavior within the genus
Polistes.

On the evolution of the cell oviposition process
in stingless bees (Hym., Meliponinae)

R. Zucchi and S.F. Sakagami
São Paulo Univ., Ribeirão Preto, Brazil and Hokkaido Univ.,
Sapporo, Japan

Although uniformly brood-cell mass provisioners, the stingless
bees behavior in relation to the cell provisioning and oviposition
process (POP) is highly complex. Among the several studied species
the extreme diversity in the components of the POP is specially
remarkable in the following aspects: queen-worker interactions, cell
construction, queen patrolling, factors related to the starting
of the POP, food discharge, queen and worker oviposition, cell
operculation etc. (def. in Sakagami and Zucchi 1974, 1977). Visual-
ized as a whole the POP ranges from the UOP (unit oviposition process) till the IOP (integrated oviposition process) and the aim of
this communication is to promote an up to date discussion on the
origin and evolution of the POP.

CASTE AND ERGONOMICS
Organized by Bernd Heinrich

Introduction

Joan M. Herbers, University of Vermont

Fundamental to insect society organization are behavioral and morphological differentiation of individuals into distinct castes. Recent work has allowed us to understand caste phenomena both with respect to proximate, physiological mechanisms of differentiation (reviews in Schmidt 1974 and Brian 1979), and ultimate, evolutionary processes leading to highly organized colonies (Wilson 1971, Oster and Wilson 1978). Synthesis of these levels of explanation can now be attempted.

A hierarchy of caste differentiation is widespread among eusocial hymenopterans. All individuals are separable by sex, with striking differences between male and female behavior. Further, females can be distinguished reproductively: queens are fully fertile whereas workers are either completely sterile or incapable of copulation. Finally, workers themselves may be separable into subpopulations, each with specialized behavioral roles and, in some cases, distinct morphology. Caste differentiation has been investigated at each of these three levels in the hierarchy.

Sex determination in hymenoptera generally follows Dzierzon's rule; unfertilized eggs are hemizygous at sex-determining loci and become male whereas diploid eggs are usually heterozygous at these loci, thereby developing into females (Page and Metcalf 1982). Parenting of haploid eggs is variable among species. Queens produce all males in Formica polyctena (G. H. Schmidt's contribution) whereas workers can lay unfertilized eggs in Myrmica rubra (M. V. Brian and E. J. M. Evesham). The number of adult males reared depends on interactions among nestmates, as these and M. Breed's papers discuss.

Differentiation of females to become queens or workers depends on a variety of factors. Biochemical differences among eggs can produce an initial bias towards a caste (Schmidt), while worker behavior towards developing larvae ultimately determines the adult morphology of the larvae (Schmidt, Brian and Evesham). Again, interactions among nestmates are critical to the process.

Finally, differentiation of workers is dependent on larval nutrition; also, endocrine influences on worker development, notably juvenile hormone, have been implicated (M. Breed). Differences among workers with respect to reproductive capabilities are correlated with behavioral differences, showing the importance of interactions among nestmates.

Evolutionary implications of malleable female phenotypes among social insects were first suggested by Wilson (1968). Research on caste ergonomics links morphological and behavioral variation within a colony to fitness differences between colonies (Oster and Wilson 1978). Such analysis allows predictions about colony phenotypes expected under varying environmental conditions.

The most spectacular examples of caste specificity and its ergonomic payoffs are those ant species with polymorphic worker castes, such as leafcutter ants (E. O. Wilson). Behavior of these ant workers is strongly influenced by morphology (Wilson 1980a,b), such that natural selection has presumably molded the colony phenotype (i.e., proportions of the morphological worker castes) to produce those caste ratios which best meet the behavioral requirements of the nest. Similar reasoning can be applied to monomorphic species, which lack discrete worker subgroups (J. M. Herbers). Such species also exhibit behavioral differentiation among workers leading to a division of labor; hence arguments of ergonomic efficiency are equally appropriate in the absence of strong morphology-behavior associations.

Interactions among nestmates are critical determinants of the proximate mechanisms of caste differentiation (Breed, Brian and Evesham, Schmidt) and of ultimate production of the colony phenotype (Herbers, Wilson). The extent of worker physiological and behavioral autonomy depends on season, worker age, queen age, and overall colony composition (Schmidt 1974, Brian 1979). In some cases, the effects of a queen on her workers are dramatic (Brian and Evesham) whereas in others it is undetectable (Herbers). Moreover, worker-worker interactions may be mediated hormonally (Breed) or nutritionally (Schmidt). Further research on the roles of individuals in overall colony organization will provide useful details of these interactions.

Caste differentiation and caste ecology in social insects are exciting areas of current research because of the dual levels on which explanations can be based. Detailed understanding of developmental physiology can be utilized to suggest lines of evolutionary and ecological questioning which lend themselves to theoretical analysis. In turn, research motivated by evolutionary reasoning leads to further exploration of within-colony mechanisms of developmental control. If the past decade of research is any indication, we can look for a synthesis of these two fields of endeavor in the very near future.

REFERENCES

Brian, M. V., 1979.-Caste differentiation and division of labour. pp. 121-222. In: Hermann, H. R., ed. Social Insects, Vol. 1. Academic Press, New York, 437 p.

Oster, G. and E. O. Wilson, 1978.-Caste and Ecology in the Social Insects. Princeton University Press, Princeton, N. J.

Page, R. and R. Metcalf, 1982.-Multiple mating, sperm utilization, and social evolution. Amer. Natur. 119, 263-281.

Schmidt, G. H., 1974.-Sozialpolymorphismus bei Insekten. Wiss. Verlagsges., Stuttgart, 974 p.

Wilson, E. O., 1968.-The ergonomics of caste in the social insects. Amer. Natur. 102, 41-66.

Wilson, E. O., 1971.-The Insect Societies. Harvard University Press, Cambridge, 548 p.

Wilson, E. O., 1980.-Caste and division of labor in leaf-cutting ants (Hymenoptera:Formicidae:Atta). Behav. Ecol. Sociobiol. 7, 143-165.

The Role of Young Workers in *Myrmica* Colony Development

M. V. Brian and E.J.M. Evesham,
The Institute of Terrestrial Ecology,
Wareham, Dorset

This paper reports the present state of a long-term study of the development of a colony of Myrmica rubra, a polygyne, but not microgyne, an unspecialised ant of the subfamily Myrmicinae. Four ways in which young workers, whilst paler than average, and clearly recognisable, influenced the Myrmica system will be considered, (1) in male production, (2) seasonally as a preparation for hibernation, (3) in gyne production and (4) through queen/worker interaction.

RESULTS

1. Males

It is known that egg laying queens and clusters of their eggs, or clusters of the reproductive eggs laid by workers stimulate or release worker egg laying; this converts the eggs that they lay into genetically and micro-biologically sterile food packages (called 'trophic' eggs). It is reasonable, on the evidence available, to suppose that this is a social facilitation of egg laying. More smaller eggs are laid that lack the ooplasmic organisation and firm chorion of a normal reproductive egg. Contact between workers and new eggs, especially eggs as they are laid, appears to be necessary and probably involves a compound contact/chemical stimulus which, acting through the CNS and the endocrine organs (probably the corpora allata), changes the type of egg produced from large reproductive to small, trophic. This depends on worker/queen ratio and the general behaviour inside the nest. By general behaviour we have in mind the extent to which queens make contact with workers during their egg laying period.

Thus, male suppression in Myrmica depends largely on control of the type of egg that the workers lay and is successful in proportion to the probability and degree or duration of contact between egg laying queens and their attendant workers. Male larvae once they hatch and join the general larval population of the colony, survive with very little mortality. They are not subject to the control which, in say, Monormorium causes males to be destroyed in the presence of queens or fostered in their absence (Brian, Jones & Wardlaw 1981; Brian 1981).

Most male larvae are formed in late summer and overwinter as small, third instar, individuals. This must mean that they are laid as eggs in the late summer and, until recently, this has been rather a mystery. Smeeton (1981) has shown however that young workers up to 3 weeks of age lay reproductive eggs whatever the worker/queen ratio or queen status of the colony. Though only 1 or 2 eggs may be laid by each worker this adds up to a considerable number where the worker population is large and will, in fact, be a function of the number of new workers added to the colony earlier in the year. Between the third and fifth week of worker life they become sensitive to the queens' oviposition and start to lay trophic eggs instead of their earlier reproductive ones. Exactly at what time their sensitisation takes place is not yet known because there is certainly a lag of a few days, if not a week, between the perception of a queen laying and a change in their style of oogenesis. Thus, even in a well-queened colony, young workers can lay male producing eggs early in their life, in late summer, which mature after hibernation as larvae the following spring.

2. Dormancy

It is known that queens can stimulate regurgitatory feeding by workers so that larvae grow more quickly than they would otherwise do; in fact, the queens do this, not by demanding food, for they are able to do it even when freshly killed. These young workers in autumn and late summer are not responsive to the stimulatory effects of queens and, though they care and tend larvae satisfactorily and keep them alive, they absorb most of the food themselves and store it in their fat body. The larvae so tended grow slowly and become dormant at a definite developmental stage which depends on whether they are worker biassed or caste-labile (queen biassed). In the former case they stop growing at a smaller size than in the latter case. The effect of this is to stop any more metamorphosis and any more brood reaching the pupal stage so that the colony is ready for hibernation with third instar larvae and adults only (Brian 1975). Daylength and lower temperatures reinforce this social switch.

3. Gynes

Continuing our study of these young workers whilst they are recognisably pale in spring and have sharp, unworn mandibular teeth, wild nests were assessed for their ability to develop gynes out of a standard set of large caste-labile larvae, the same in all cases, randomly distributed and with the same fixed worker/queen ratio. Any experimental difference in these conditions would be due to a difference in quality of the worker population. Various characteristics of the worker population were measured: their age composition, based on colour and mandibular wear, their size, the worker/queen ratio in the wild nest sampled, their dry weight and a few other characteristics were all measured and after a correlation analysis with the gyne production it emerged that the young workers were still resistant to queen domination or control whereas the older ones were much more responsive. This means that the young workers reared more gynes out of the standard larvae. The proportion of pale workers with sharp teeth is positively correlated with mean dry weight (an expression of their stores) and so it is likely that the colonies with a fast growing, well fed worker force are the ones that mature most gynes; this is reasonable if production of sexuals is to be greatest when growth rate is maximal. As Smeeton (loc cit) found that worker stores were correlated with production of male eggs and, hence, male production through egg laying, the normal association of male and female sexuals appears to be arranged, through the production of many young workers under beneficial nutritive conditions in the previous year. It is assumed that the forager force of larger, older workers is adequate to collect food; in laboratory cultures very few, of course, were needed for this food collection function.

These correlation studies on wild colonies were confirmed experimentally by establishing population structures of young and old or small and large workers in different proportions and testing their ability to rear gynes out of gyne-labile larvae in the presence of queens at a fixed worker/queen ratio. The results obtained supported the observations of normal colonies.

4. Development

Though Carr (1962) showed that workers reared from the larval stage in a culture without queens responded less well to the presence of queens as adults, this was not analysed further until recently. Now we have taken worker pupae and allowed them to hatch with or without queens present after they had emerged, each culture was again divided equally in 2 halves, one with the queen and one without and they were allowed to incubate labile larvae. In this way queens were either present or absent at "birth" and present or absent (same queens) when the workers were given labile larvae to brood in a form of bio-assay. The whole procedure was duplicated using different colonies and the weights of labile larvae after culture are given in Table 1 below. Recall that the normal effect of queens on workers culturing labile larvae is to cause them to underfeed them by diluting their regurgitate and not to give them solid prey at all. Hence, looking at the Table it is clear that 4 cultures have given 318 mg without queens compared to 230 mg with queens, a substantial negative queen-effect. This difference of 88 mg can be divided into 2 parts according to whether the workers were born into a queen environment or a queenless environment. Those which were born into a queen environment gave 71 mg difference and those which were born into a queenless environment gave only 17 mg difference. Thus, earlier experience of queens has caused about 4 times as much response to queens during the bio-assay as no experience of queens and we are entitled to conclude that the effect of queens has been increased by worker experience at birth. This supports Carr's earlier result and adds that a substantial part of it takes place in the worker stage, not in the larval stage, though we have not excluded the possibility of an effect in the larval stage.

An interesting point about this is that queenless workers can feed larvae better if they have experienced queens at birth (169 mg compared to 249 mg). In other words they are responding better in the absence of queens because they were born into a queen society whilst the other set of workers are responding better in the presence of queens (by cutting larval growth from 132 mg to 98 mg) as a result of their early experience at birth. To sum up: workers born into a society with queens respond better later on, both positively

by giving more food (no queen) and negatively by giving less food
(queens present). One interpretation could be that they learn the
individuality of their own queen at birth and respond better than
they would to just any unknown queen. Another is that the queen
somehow trains them during a stage in life when they are especially
susceptible; she may, for example, teach them to regurgitate juices
to her and so help them to do so to larvae later on, when occasion
demands and is appropriate. Of course, besides training them to donate she must also train them to withold food in the circumstances
where larvae are caste-labile and she is in the same chamber. Whether she plays any part in training them for attacking labile larvae
is doubtful; since this is normally done by older workers who, perhaps, teach the younger ones.

In our next experiment workers were given a queen either as
they emerged from their pupal skin or after 1,2,3,4, or 5 weeks.
Immediately after the pre-treatment, each culture was divided into
2 equal parts, one with their own queen and one without any queen at
all. Larvae were provided to test worker ability to donate food.
At that time of year, late summer, larvae are all either worker-
biassed or else in diapause. None can grow into gynes until after
hibernation. Suppression of growth by larvae is not an aspect of
queen influence; on the contrary the queen stimulates growth. This
situation simulates the conditions that may be expected in a growing
colony in which each year the queens become more and more difficult
to find through a rising worker/queen ratio. How realistic the actual figures for the delay in finding the queen are is uncertain;
probably grossly exaggerated, except in very large colonies with
very few queens. The results, including 2 replicates, added together, are given as the weight of larvae in grams after a suitable
period of culture (Table 2). This time the drop in performance of
the queenless set in the bio-assay in relation to the days it took
them to 'find' the queen is astonishing and can be fitted significantly by a linear regression. The simplest interpretation is that the
total period with the queen has a lasting beneficial effect, at
least up to 5 weeks. An alternative possibility, however, is that
the earlier they encounter a queen the better, or at least that an
encounter at once or during the first week is necessary if the worker is to develop a full behaviour repertoire. Both, of course, may
contribute but there is certainly a lasting effect on the way workers rear larvae.

What of the influence of queens during the bio-assay? Overall,
queens in the bio-assay blotted out, erased or overcame the pre-
treatment differences so that there is no longer a significant regression. The set of cultures with queens in the bio-assay thus
benefitted from the immediate presence of queens if they had been
under-dosed before. Those that had a bigger dose prior to the
assay (3,4, or 5 weeks) did not benefit from their current session
with the queen at all. In fact, queens appear to have caused a
small depression in their performance which, if not due to chance,
could, we suggest, mean that the workers are beginning to produce
trophic eggs or (and less likely) that, after learning to regurgitate to larvae they have begun to learn to withhold food from gyne
potential larvae in the presence of queens.

DISCUSSION

So there is great scope for further experimental analysis. We need to know whether a given time with a queen gives different values at different ages, whether queens just invigorate workers and speed a learning process that occurs anyway or whether they make a special, individually distinctive contribution to worker activity and whether they first 'teach' response to larvae as a regurgitative function and then as an inhibitory one, using themselves as a model. Finally, we need to know whether queenless workers undergo a positive physiological change during their early development, (eg in their gonads), which is detrimental if not incompatible with their nursing function as shown by Smeeton (loc cit). Queens may thus have a very early influence on the developmental physiology and behaviour of a worker, by their presence encouraging it to nurse and lay trophic eggs and by their absence (or rarity) allowing it to neglect larvae and lay reproductive eggs.

Analogous situations in other wasps (sl) are well known and need no repetition (Brian 1980).

Table 1. Weight (mg) of larvae after culture.

Queens at birth	Queens during assay No	Queens during assay Yes	Total
No	80) 69) 149	86) 46) 132	281
Yes	92) 77) 169	57) 41) 98	267

Table 2. Weights (g) of larvae after culture.

Pretreatment		Queens during assay No	Queens during assay Yes	Queen effect
Weeks after 'birth' that a queen was added	0	1.37	1.33	- 0.04
	1	1.54	1.20	- 0.34
	2	1.15	0.98	- 0.17
	3	0.98	1.25	+ 0.27
	4	0.99	1.21	+ 0.22
	5	0.84	0.95	+ 0.11

References

Carr C.A.H., 1962. Further studies of the influence of the queen in ants of the genus Myrmica. Insectes Soc., 9, 197-211.

Smeeton, L., 1981. The source of males in Myrmica rubra L. (Hym. Formicidae). Insectes Soc., 28, 263-278.

Juvenile Hormone and Aggression in the Honey Bee

Michael D. Breed, The University of Colorado

Dominance hierarchies and associated agonistic behavior have been observed in a number of primitively eusocial species. Notable among observations of such behavior are those of Pardi (1948) on <u>Polistes</u> and Brothers and Michener (1974) on <u>Lasioglossum.</u> Dominance behavior in these species is correlated with reproductive status. Dissections of individuals indicate that dominance behavior and the development of ovaries is correlated (Pardi, 1948; Breed et al., 1978). Since individuals in primitively eusocial associations are all potential reproductives the presence of dominance behavior can be viewed as a consequence of reproductive competition.

Overt dominance interactions are rare or absent under normal conditions in highly eusocial species. While conflict over reproduction may exist between queens and workers (Trivers and Hare, 1976) workers do not possess full reproductive potential and usually do not mate; thus their offspring are all male. Queen dominance is thought, at least in some cases, to be established by the production of pheromones that inhibit worker reproduction (Wilson, 1971).

The purpose of this paper is to present data that elucidate the hormonal basis for the aggressive behavior in honey bee workers that is related to reproductive status. Such behavior was first reported by Sakagami (1954) in queenless hives. Velthuis (1976) corroborated the earlier result and presented data showing that worker bees displaying dominance behavior have elevated blood titers of vitellogenins. Recently Breed (1979) reported dominance interactions between queens and workers; similar observations were independently made by Weaver and Weaver (1980); the data presented here support the interpretation of certain aggressive acts by workers as being reproductive dominance.

MATERIALS AND METHODS

Worker dominance was measured in interactions with queens. Workers were recorded as either interacting aggressively with a queen (biting, stinging) or non-aggressively (feeding, licking, not interacting). To simplify recording of behavior workers were paired with a single queen. Since social interactions among

workers may affect their endocrine state, all workers tested were
maintained in social isolation until they were tested
behaviorally. Subsequent to behavioral testing the brain with
attached retrocerebral complex was dissected from the worker,
embedded in paraffin, sectioned, and stained. The volumes of the
corpora allata of each worker were measured using an ocular
micrometer. All workers tested for the experiments described in
this paper were five days old. Juvenile hormone treatments were
administered to the workers on the day of their emergence as
adults. JH III (Sigma) was administered in lul of safflower oil
vehicle.

RESULTS

1. Effects of juvenile hormone on worker corpora allata
    Treatment of workers on the day of their emergence with
juvenile hormone resulted in an inhibition of corpora allata
development. A clear dose-reponse relationship between the amount
of JH injected and the subsequent size of the CA is seen in figure
1.

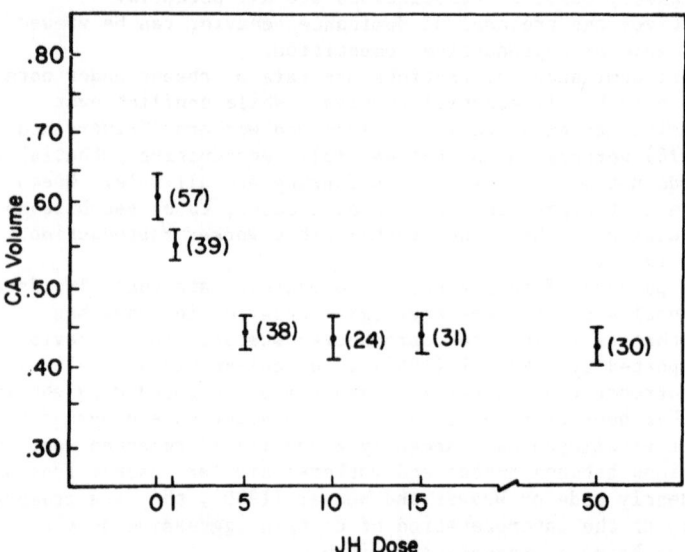

Figure 1. The relationship between juvenile hormone dosage and
worker corpora allata volume. Individuals were treated on the day
they emerged as adults and tested five days later.

While low doses slightly affect honey bee worker corpora allata
development doses of 5 ug or more have a highly significant
inhibitory effect on subsequent CA development.

2. Effects of social context on worker corpora allata

To test for the effects of social context on the endocrine system of worker honeybees workers were maintained in complete social isolation. Control groups were maintained containing 20 workers but no queen. Five day old individuals that had been maintained in social isolation had a mean corpora allata volume of .759 10^6u^3 (N=48). Bees from social groups had a mean corpora allata volume of .680 10^6u^3 (N=67). There is a statistically significant difference (t=2.31, P<.05) between these means. The mean volume of five day old worker bees from queenright hives was .707 10^6u^3 (N=8) and workers collected through the first six days of adult life had mean corpora allata volumes of .683 10^6u^3. Thus worker bees from the socially isolated context have larger corpora allata than bees of the same age from situations in which they interact socially with other workers.

3. The effects of juvenile hormone treatment on aggressive behavior

Juvenile hormone has a significant effect in inhibiting aggressive behavior towards queen honeybees expressed by five day old workers when administered to the worker on the day of emergence. This effect follows a dose response pattern so that maximal effect is achieved at a dose of 50ug (figure 2).

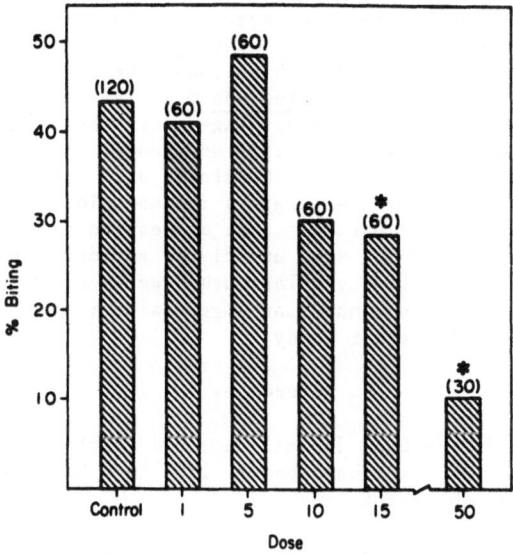

Figure 2. The relationship between juvenile hormone dosage and the probability that the worker will express aggressive behavior towards an unfamiliar queen.

Statistically significant differences in frequencies of attacks on queens were observed with doses of 15 and 50 ug juvenile hormone.
4. Effects of social isolation on aggressive behavior

Fifteen percent (N=20) of individuals from laboratory groups expressed aggression towards queens when tested at five days of age. Similar levels of aggression were obtained when five day old bees from queenright colonies were tested. The figure for laboratory groups is significantly different from the 49% (N=80) of five day old isolated bees that expressed aggression (x^2=7.48, P<.01).

DISCUSSION

Two separate sets of results are reported here; the first indicates that juvenile hormone treatment inhibits endocrine development in worker honeybees and that social isolation stimulates such development. The second set of results shows that juvenile hormone treatment inhibits the expression of aggressive behavior towards queens by workers while social isolation promotes aggressive behavior. In other words, factors that inhibit corpora allata development also inhibit aggressive behavior, while factors that stimulate corpora allata development also stimulate aggressive behavior.

It is clear, then, that there is a relationship between aggressive behavior expressed towards a queen and factors that affect the endocrine status of the worker honeybee. This is consistent with our knowledge of systems of dominance in other bees. For example, Lasioglossum zephyrum has a dominance hierarchy that is correlated with reproductive status (Brothers and Michener, 1974). As in Lasioglossum it is somewhat difficult in this case to distinguish between aggressive behavior expressed in a nest defense context and social dominance. In L. zephyrum the individual most active in nest defense also has ovaries that are larger than those of the foragers in the colony (Brothers and Michener, 1974). It is possible that aggression expressed in these two contexts has the same underlying endocrine basis; the question of the evolutionary relationship between aggression in the context of social dominance and aggression in the context of next defense merits further study.

References

Breed M.D., 1979. -- Does juvenile hormone regulate dominance and aggression in worker honeybees? 16th Int. Ethology Conference, Abstracts.

Breed M.D., Bell W.J., and Silverman J.M. 1978. -- Agonistic behavior, social interactions, and behavioral specialization in a primitively eusocial bee. Insectes Sociaux, 25, 351-364.

Brothers D.J., Michener C.D., 1974. — Interaction in colonies of
    primitively social bees. III. Ethometry of division of
    labor in Lasioglossum zephyrum. J. Comp. Physiol., 90,
    129-168.

Pardi L., 1948. -- Dominance order in Polistes wasps. Physiol.
    Zool., 21, 1-13.

Trivers R.L., Hare H., 1976. -- Haplodiploidy and the evolution
    of the social insects. Science, 191, 249-263.

Sakagami S.F., 1954. -- Occurrence of an aggressive behavior in
    queenless hives, with considerations on the social
    organization of honeybees. Insectes Sociaux, 1, 331-343.

Velthuis H.H.W., 1976. -- Egglaying, aggression, and dominance in
    bees. Proc. XV Int. Cong. Ent., 436-449.

Weaver E.C., Weaver N., 1980. -- Physical domination of workers
    by young queen honeybees. J. Kansas. Entomol. Soc., 53,
    752-762.

Wilson E.O., 1971. -- The insect societies. Harvard University
    Press: Cambridge, Massachusetts.

Queen Number and Colony Ergonomics in *Leptothorax longispinosus*

Joan M. Herbers, University of Vermont

To date, most studies of caste ergonomics in ants have focused on polymorphic species (Passera 1974, Herbers 1980, Wilson 1980a,b). In such groups the division of labor is correlated with morphological variation among individuals, such that inferences about a colony's efficiency can be made from knowledge of the morphological profiles or workers. For most species, however, workers are monomorphic; there are no distinct physical castes along which behavior can be organized. Even so, the question remains as to whether task specification in such species is biased by morphology.

This issue of morphologically-biased polyethism, interesting in itself, takes on added importance for understanding a phenomenon in Leptothorax ants. Several species in this genus are facultatively polygynous (Buschinger 1968) such that some colonies have single queens while others can have multiple queens. This phenotype of multiple queening, which violates assumptions of Hamiltonian kin selection (Hamilton 1964) is poorly understood (Holldobler and Wilson 1977). Several hypotheses for the evolution of polygyny in this genus are under investigation in my laboratory, but here I focus on ergonomic efficiency. Polygynous colonies, in which more than one queen contributes to the worker pool, are genotypically more diverse than monogynous colonies. Although worker morphology is heavily dependent on environmental influences in most hymenopterans (Schmidt 1974), increased genotypic variation among workers in polygynous colonies could result in a broader distribution of worker phenotypes. Then, if polyethism is organized along morphological lines, polygynous colonies might enjoy enhanced behavioral discretization (Wilson 1976), leading to higher efficiency and ultimately higher fitness.

MATERIALS AND METHODS

Four colonies of L. longispinosus Mayr were studied in summer 1981; these colonies differed primarily with respect to queen number. Colonies A and B were monogynous with 30 and 31 workers, respectively; Colony C was composed of five queens and 28 workers, while Colony D had four queens and 36 workers. Observations of workers were conducted by choosing individuals at random, measuring

Table 1. Ethogram variation in four colonies of L. longispinosus.
Colonies A and B were monogynous while C and D were polygynous. A
star (*) indicates behaviors observed casually but not quantified.

Behavior	Frequency observed in				
	Colony A (N=407)	Colony B (N=389)	Colony C (N=386)	Colony D (N=380)	Total (N=1562)
PERSONAL					
Rest	.263	.249	.207	.205	.232
Self-groom	.202	.234	.251	.295	.245
BROOD CARE					
Inspect egg	.015	.003	---	.003	.005
Groom egg	.005	---	---	---	.001
Carry egg	.007	---	---	.026	.008
Groom larva	.079	.057	.114	.076	.081
Carry larva	.027	---	.018	.032	.019
Inspect larva	.069	.069	.054	.103	.074
Feed larva solid	---	.018	---	---	.005
Regurgitate w/larva	.017	.026	.013	.021	.019
Inspect pupa	.054	.023	.003	.005	.022
Groom pupa	.027	.005	.003	.008	.011
Carry pupa	.003	---	.003	.003	.002
Assist larval ecdysis*					
Assist adult eclosion*					
SOCIAL INTERACTIONS					
Be groomed	---	---	.031	.010	.010
Regurgitate w/worker	.042	.064	.062	.058	.056
Antennal touch	.128	.190	.114	.121	.138
Antennate body	.005	.003	.005	---	.003
Allogroom worker	---	.008	.075	.008	.022
Fight worker	.003	---	.005	.003	.003
Antennate queen	.022	.003	.029	.008	.015
Allogroom queen	---	.003	---	---	.001
Regurgitate w/queen	.007	---	---	---	.002
Regurgitate w/male*					
Regurgitate with alate female*					
Fight male*					
Fight queen*					
COLONY MAINTENANCE					
Lick nest wall	.010	.013	.003	.005	.008
Handle nest material	.010	.010	.005	---	.006
Look out nest	---	.013	---	---	.003
Feed in nest	.005	.008	.003	.008	.005
Drink	---	.003	---	---	.001
Forage	.003	.003	.003	.003	.003
Carry live nestmate*					
Carry dead nestmate*					

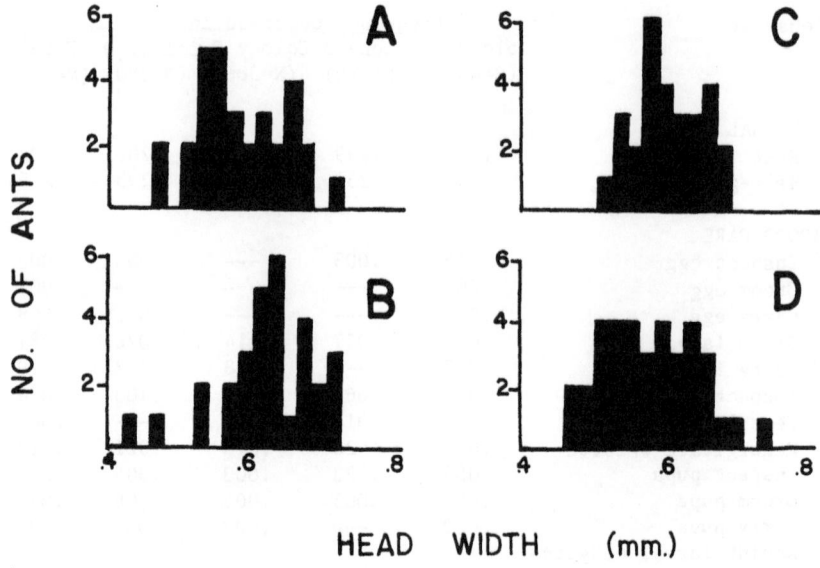

Figure 1. Worker size distributions in four colonies of L. longispinosus. Nests A and B were monogynous, whereas C and D had
multiple queens.

their head widths, and recording the sequence of behaviors over a 30
-minute time period. Head widths of all individuals were recorded
to estimate morphological variation, and were accurate to .01 mm.

RESULTS

The behavioral organization of workers is most simply represented by the colony ethogram. Table 1 shows the types of behavior
observed for workers in the four colonies. A total of 36 distinct
behaviors were recognized among the four worker populations. These
can be separated into four general categories. Some were personal,
fulfilling no social function. Of the social behaviors, some were
related to brood care, some involved interactions among adult nestmates, and a final category pertained to colony maintenance behavior.
There was considerable variation in the behavioral profiles of
colonies studied; not all behaviors were seen in all nests, and the
aggregate proportions of certain types of behaviors varied among
colonies. For example, Colonies B and C were observed more often
to exhibit brood care and less often social interactions, than were
A and D. This variation among colonies was, however, unrelated to
queen number. Even those acts associated with worker-queen interactions were no more common in polygynous than in monogynous
colonies. The profiles of single-queen colonies were no more

closely related to each other than to those of polygynous colonies, and vice-versa (2-way ANOVA, P >.05)

Figure 1 shows the morphological distributions of workers in the four nests. These ants are small, such that total body length of workers rarely exceeds 2 mm. The species is monomorphic, with the largest workers having head width about 1.8x that of the smallest ants. The distributions of Figure 1 differed neither with respect to their means nor their variances (LSD test and Fmax test, respectively; for both P >.05). Moreover, the distributions of monogynous colonies were not more similar to each other than to polygynous colonies, and vice-versa. Clearly, for this species, queen number had no measurable effect on the distribution of worker morphology.

DISCUSSION

The number of queens in a nest has consistent effects neither on worker morphology nor on overall colony behavior. In addition, studies of time budgets in these colonies showed there were no differences in the proportions of time engaged in particular behaviors. There is indeed a morphological bias in L. longispinosus workers such that large ants forage, whereas smaller ones stay in the nest to care for brood, maintain the nest, and so forth (Herbers, unpub. data). Nonetheless, this trait also is unaffected by queen number; the size distributions of workers engaged in particular tasks does not differ between monogynous and polygynous colonies.

Therefore, I can find no evidence to support the contention that ergonomic efficiency is the selective force maintaining multiple queen associations in this species. Not only are worker size distributions similar among colonies, but the division of labor is also organized along the same lines. In terms of gross behavioral characteristics, polygynous colonies are no different from monogynous colonies.

Moreover, the ergonomic efficiency hypothesis predicts that increased efficiency with queen number would result in a larger proportion of colony resources being available for production of virgin queens and males. Studies of alate production in this species showed that the average number of alates reared in a nest was independent of queen number; indeed, the actual investment in colony reproduction, measured as dry weight of males and females, declined with increasing queen number (Herbers, unpub. data). There was no observed fertility advantage in polygynous colonies; actually, the fertility per queen was lower in polygynous than in monogynous colonies.

These results are surprising in light of traditional views of social organization in hymenopteran populations. The lore of a queen maintaining control over her workers suggests that one should observe differences in behavioral profiles of those colonies which differ with respect to queen number. For this species, however, worker behavior appears to be quite independent of queen influence. This inference is further supported by field data which show an

astounding 28% of all nests lack queens. These results suggest
that, except for laying fertilized eggs, workers are behaviorally
autonomous. (In L. allardycei, workers even establish dominance
hierarchies within a nest (Cole 1981).) As they feed larvae, regurgitate with each other, and so on, workers appear relatively oblivi-
ous of the queen's presence. Beyond laying eggs, queens may have no
important role in structuring worker behavior; if queens do have
such a role, it is quite subtle.

Maintenance of polygynous colonies in nature remains a mystery;
clearly there is no ergonomic differential leading to higher fertility. To understand the phenomenon we must look elsewhere, to kinship associations among queens, population dynamics, and the role of
social parasites on colony survivorship. Tests of genetic and
ecological hypotheses for the evolution of multiple queening are
expected to further our understanding of social organization in
Leptothorax ants.

ACKNOWLEDGMENTS

I thank Michael Cunningham for his painstaking observations of
ant behavior. This work was supported by an Institutional Award
from the University of Vermont.

REFERENCES

Buschinger, A., 1968. - Monogynie und polygynie bei arten der
Gattung Leptothorax Mayr. Ins. Soc., 15, 217-226.

Cole, B. J., 1981. - Dominance hierarchies in Leptothorax ants.
Science, 212, 83-84.

Hamilton, W.D., 1964. - The genetical theory of social behavior. I,
II. J. theor. Biol.,7, 1-52.

Herbers, J.M., 1980. - On caste ratios in ant colonies: population
responses to changing environments. Evol.,34, 575-585.

Hölldobler,B., and E.O. Wilson, 1977. - The numbers of queens: an
important trait in ant evolution. Naturwiss.,64, 8-15.

Passera, L.,1974. - Differenciation des soldats chez la fourmi
Pheidole pallidula Nyl. (Formicidae:Myrmicinae). Ins. Soc.,21,
71-86.

Schmidt, G.,1974. - Sozialpolymorphismus bei Insekten. Wissenschaft-
liche Verlagsgesellschaft MbH: Stuttgart.

Wilson, E.O., 1976. - Behavioral discretization and the number of
castes in an ant species. Beh. Ecol. Sociobiol.,1, 141-154.

Wilson, E.O., 1980a,b. - Caste and division of labor in leaf-
cutter ants (Hymenoptera:Formicidae:Atta). I,II. Behav. Ecol.
Sociobiol. 7, 143-165.

Egg Dimorphism and Male Production
in *Formica polyctena* Foerster

Gerhard H. Schmidt,
Universität Hannover (FRG)

The structure of the society of F. polyctena can be
summarized as follows: In every nest there are numerous mated and
unmated queens and normally some hundred thousand workers.
According to Dzierzon's rule queens and workers develop from
fecundated eggs and males from unfecundated ones. All of the
three morphs grow from eggs produced by the queens. Workers of
F. polyctena are not fertile in the presence of queens and only
slightly fertile without queens. Therefore, the workers have no
role in the production of males (compare Schmidt, 1974).

MATERIALS AND METHODS

Formica ants were collected from the surroundings of
Würzburg and Hannover where F. polyctena mound nests are common
in the forests.

To determine fecundation the spermathecae of 2215 queens from
36 nests were prepared (Ehrhardt, 1970). Egg production and
growing experiments were carried out in the laboratory at $24°C$
(Winkler and Schmidt, 1982). For sex determination young larvae
were sectioned (Maidhoff, 1968). Paper chromatographic methods
were used to separate the flourescent compounds of the dimorphic
eggs and young larvae (Schmidt, 1972). Column-Sephadex G 10 and
thinlayer chromatography were used to separate RNA and protein
compounds, specific staining and UV absorption for quantitative
measurements (Winkler, 1979).

RESULTS AND DISCUSSION

1. Deposition of unfecundated eggs by the queens and growing of
   males

Gösswald and Bier (1957) observed that the first eggs laid by
queens in the early spring are normally unfecundated and develop
into males. But in all of the nests studied Ehrhardt (1970)
observed 19 to 34 per cent of the queens without living sperms
inside their spermathecae. Some of them had empty spermathecae,
others dead and more or less degenerated sperms which did not

depend on the age of the queens. After a normal copulation the number of sperm cells is sufficient for the queen to produce female castes. There is some evidence that this degeneration of the sperm begins in the males. Perhaps this is a normal process in polygynous _Formica_ species guaranteeing the production of males.

The young mated and unmated queens produce eggs immediately after tearing off their wings (Gösswald and Schmidt, 1960). They can be adopted by their own or any other society, but they cannot found their own family without the assistance of workers. Queens which are able to produce fecundated eggs lay the same number of winter-eggs as male-producing queens; the number of eggs laid is not dependent on the age of the queens.

Comparing the percentage of male-producing queens with the number of male pupae produced, Ehrhardt (1970) found 4 to 100 per cent for the latter. The sex relation depends on the fact that sexuals can repeatedly copulate inside the nest. Therefore, the sex present in small numbers will not leave the nest for a nuptial flight.

Depending on the percentage of sexuals leaving the nest we can differentiate between male producing societies with more than 50 per cent male pupae, where only males will leave the nest, females producing societies with less than 50 per cent male pupae, and ones that produce both sexes which release more than 50 per cent at first and later on less than 50 per cent males.

Males-determined eggs were found during the entire period of egg deposition. In June and July Ehrhardt (1970) found between 17 and 75 per cent of first instar larvae to be male. But all of these larvae die during the second instar and are used as food reserves (Maidhoff, 1968). Only during spring can males develop into adults; at this time females are also produced. However, in _Formica_ _pratensis_ Retz. it is possible to get males and females in late summer.

2. Division of labour between queens of polygynous _Formica_ societies

Gösswald and Bier (1954) demonstrated that the queens first produce winter-eggs, later on smaller ones, so-called summer-eggs. The former develop into sexuals, the latter into workers if they are fecundated. But the larvae of the female producing winter-eggs have to be determined to become queens, otherwise they develop into workers. Schmidt (1972) added that males also need determination in the same manner as the queens. For that determination, besides winter-eggs, physiologically young workers are necessary to produce post-pharyngeal secretions from their own fat body and to feed them to the first instar larvae. In later spring both of these necessities are absent, therefore no males can be produced; the male larvae die. This appears as an adaptation to queen growing, because in summer months the males cannot mate and the energy from the society would be lost for multiplication. In that connection there is a division of labor between the queens since some predominantly produce queens and later on, workers; others only produce males. This division of labor is determined only by the capacity for insemination. So far as known, there is no difference in the behavior of the mated and uncopulated queens after tearing off their wings.

3. Factors influencing the deposition of dimorphic eggs by the queen

From eggs laid by the F. polyctena queens, offspring can only be produced with assistance from the workers. High egg production by the queens is only possible in the presence of physiologically young workers which have stored large amounts of energetic reserves. Under laboratory conditions we obtained up to 2353 eggs from one queen with workers during a laying period of 18 weeks. Up to 200 of these were winter-eggs (Winkler and Schmidt, 1982). The highest number of eggs was laid in groups of 100-125 workers with one queen. In worker groups with two or more queens eggs were deposited in proportion to the queen-worker ratio.

Queens do not influence each other's egg production if an optimal number of workers is present. Too many workers have a disadvantageous effect, the same as if there were too few. Without workers the queens are unable to produce eggs or their number is very much reduced; the queens die after some days because they can not feed themselves. The storage capacity of the fat body of the queens is much reduced. That is understandable, as the queens have a diapause in summer and until December, and the storing process takes place during September-October.

This behavior may be responsible for the fact that the activation of the nest population in spring is done by the queens which have finished their diapause deep inside the nest. At that time it is very difficult to catch them. At the beginning of January they wait for the first sunbeams to reach the base of the mound; they enter the sunning period together with the non-diapausing workers and the storing workers are activated by the latter.

The longer cooling period of the queens is important for the subsequent deposition of dimorphic eggs. If queens were exposed to 4°C and compared with those kept at 24°C, the cooled queens produced eggs after activation at 24°C like winter-eggs. At 4°C egg deposition was stopped while it went on in queens continuously kept at 24°C as long as they were together with freshly activated workers, but all of the eggs were the size of summer eggs.

Our prior investigations demonstrate that the egg dimorphism of the queens is caused by a vernalisation of the queens which takes place during the early winter in the depths of the nest. The most sensitive oocytes are those three to five of each ovariole nearest to the oviduct. The number of deposited winter eggs is determined by the queen and partly by the nutritional potency of the workers. In the middle of the summer (July) the food reserves of the workers seem to be exhausted. Accordingly the queens stop their egg deposition and diapause begins.

246

4. Biochemical differences between the two differently predisposed egg types

Histochemically only the bigger winter-eggs demonstrate a basophilic oosom near the hind pole (Bier, 1953). That material belonged to the nurse cells from which it passed the fusoms and was deposited into the oocyte. Bier (1962) considered it as rRNA. Metabolic studies of both egg types demonstrated that pteridine metabolism goes on only in the winter-eggs after deposition. The synthesized compounds remain present until determination has taken place. These substances may not be very much involved in the embryonic processes (Schmidt, 1969).

During recent years biochemical studies were carried out by Winkler (1979). He studied the macromolecular substances of both the egg types to find differences which may be responsible for caste determination. Summarizing the results of the protein and RNA analyses, we can point out that the RNA content in winter eggs is higher than in summer eggs but there is no difference between the RNA concentration in mg/ml egg volume. The electrophoretic pattern of RNA of both of the egg types is very similar but nucleosides and other purin products are much higher in summer-eggs than in winter-eggs. During embryogenesis the RNA content increases more in the former than in the latter. Thus, at the beginning of embryogenesis there is an RNA difference of 40 per cent, after 6 days only 15 per cent between the two egg types. The synthesis of RNA is correlated with an increase of uric acid.

In correlation to the RNA fractions, the protein patterns demonstrated no real differences between the two egg types; no egg type specific proteins could be found.

The DNA content increases about 30 times in both of the egg types. Therefore, we can assume that there is the same number of body cells in both embryos. Then, the relations between DNA and RNA in the cells at the end of the embryogenesis must be different; it is significantly higher (15%) in those larvae developing from winter-eggs. Perhaps this fact has some importance in caste determination.

References

BIER K.-H., 1953. -Beziehungen zwischen Nahrzellgrösse und Ausbildung ribonukleinsäurehaltiger Strukturen in den Oocyten von Formica rufa rufo-pratensis minor Gössw. Verh. Dt. Zool. Ges. Freiburg 1952, 369-374.

BIER K.-H., 1962. _Autoradiographische Untersuchungen zur Dotterbildung. Naturw., 49, 332-333.

EHRHARDT H.-J., 1970. -Die Bedeutung von Königinnen mit steter arrhenotoker Parthenogenese für die Männchenerzeugung in den Staaten von Formica polyctena Först. (Hym., Form.). Diss., Nat. Fak. Univ. Würzburg, 106 p.

GÖSSWALD K., BIER K.-H., 1954. -Untersuchungen zur Kastendetermination in der Gattung Formica. 3. Die Kastendetermination von Formica rufa rufo-pratensis minor
Gössw. Insectes sociaux, 1, 229-246.

GÖSSWALD K., BIER K.-H., 1957. - Untersuchungen zur Kasten-
determination in der Gattung Formica. 5. Der Einfluss der
Temperatur auf die Eiablage und Geschlechtsbestimmung.
Insectes sociaux, 4, 335-348.

GÖSSWALD K., SCHMIDT G.H., 1960. - Untersuchungen zum Flügelabwurf
und Begattungsverhalten einiger Formica-Arten (Ins. Hym.)
im Hinblick auf ihre systematische Differenzierung.
Insectes sociaux, 7, 298-321.

MAIDHOFF A., 1968. - Morphologische und histologische zur
Kastendifferenzierung bei Formica polyctena Foerst. (Hym.,
Form.). Diss., Nat. Fak. Univ. Würzburg, 86 p.

SCHMIDT G.H., 1969. - Photoaktive Substanzen in verschieden
prädisponierten Eiern und Junglarven von Formica polyctena
Först. Proc. VI. Conqr. IUSSI Bern, 275-261.

SCHMIDT G.H., 1974. - Steuerung der Kastenbildung und
Geschlechtsregulation im Waldameisenstaat, 404-512, in
SCHMIDT G.H. (Ed.), Sozialpolymorphismus bei Insekten.
Probleme der Kastenbildung im Tierreich. Wiss. Verlagsges.
Stuttgart, 974 p.

WINKLER I., 1979. -Zur Bedeutung makromolekularer Zellbestandteile
für die Kastenprädisposition von Waldemeiseneiern (Formica
polyctena Foerst.). Diss., Math. - Nat. Fak. Univ. Hannover,
130 p.

WINKLER I., SCHMIDT G.H., 1982. -Beeinflussung des Eidimorphismus
der Königin von Formica polyctena Förster durch
Umweltfaktoren. Insectes sociaux, 29, in press.

Abstracts

Interaction of worker and queen genotypes on development,
size, and oviposition in Apis mellifera

Alan B. Bolten
University of Florida

Three demographic features expected to influence colony
reproductive rates in honey bees are worker bee development times,
queen maturation rates, and egg laying rates. Comparisons of
these features between Africanized and European populations in
South America were conducted under identical experimental conditions. Since Africanized populations are characterized by
smaller comb cell sizes, the effects of different cell sizes and
nurse bee gentoypes on worker bee development times were evaluated using a 2 x 4 experimental design. Africanized worker bees
emerged in 18.9 days from oviposition compared with 19.9 days
for European worker bees ($P< 0.01$). Development rates were independent of comb cell size and nurse bee genotype. In reciprocal
F_1 crosses, when comb cell size was held constant, worker bee
size was found to be determined by the maternal genotype. Queen
maturation from adult emergence to initiation of oviposition was
significantly more rapid in the European populations (7.7 vs.
9.2 days; $P< 0.01$). Numbers of eggs laid per 24 h for Africanized
and European queens were not significantly different and were
independent of worker bee genotypes.

Control of production of Melipona quadrifasciata diploid males by
worker bees (Hymenoptera, Apidae)[1]

Conceição Aparecida de Camargo
Dept. Parasitologia, Fac. Medicina - Ribeirão Preto-SP-Brasil

Melipona quadrifasciata produces 2n males and females at 1:1
ratio by inbreeding, the same way as Apis mellifera, but, unlike
Apis, does not seem to produce the cannibalism substance, since
it does not eliminate males, but rather the queens that produce
them. There are indications that the elimination of physogastric
queens occurs because of large numbers of males in the hive, which

248

lead the workers to identify the queen as old. From a genetic point of view, and according to Hamilton (1964), the workers "prefer" to care for their sisters, since in this way they protect 3/4 of their genotype, rather than for their daughters. This may partially explain the replacement of the queen rather than the elimination of diploid males, which also are 3/4 similar to workers.

[1]This research was supported by FAPESP and CNP_q.

On Dispelling the Myth of the "Lazy" Drone: Incubation
by Male Bumble Bees

Sydney A. Cameron
The University of Kansas

Pupal incubation behavior of males in laboratory-reared colonies of Bombus pennsylvanicus and B. griseocollis was studied. In light of the recent theory of Inclusive Fitness as an explanation of behavior of Hymenoptera the observation that males work for the colony takes on special interest. All drones sat over pupae and pumped their abdomens for the first few days after pupation. This was consistent with the usually observed incubation behavior of workers. In one nest composed solely of surviving drones, some continued incubating pupae for up to two weeks. Temperature measurements of pupae incubated by drones demonstrated that drones raised the temperature of the pupae as much as 3^0 C above their metabolic temperature. However, this increase was not as great as observed with workers, who in turn did not raise the pupal temperature as much as queens.

The Eggs of Founding Queens of the Imported Fire Ant

Daniel Cherix
Museum of Zoology, CH-1005 Lausanne, Switzerland

David J. C. Fletcher
Department of Entomology, University Georgia
Athens, GA 30602, U.S.A.

A detailed study was made of oviposition during the first six weeks of colony founding by queens of the imported fire ant (Solenopsis invicta Buren). The results elucidated some aspects of colony development as well as the problem of trophic eggs. Differences in egg sizes were not related to "trophic" or "embryonated" eggs. The first eggs laid were significantly larger than those laid later. No trophic eggs were found within the first 48 hours after mating, but about 40% were found after one week. The significance of the occurence of trophic eggs in the Formicidae is discussed.

Structure and Dynamics of the Caste Systems of Army and Leaf-Cutting
Ants: A Comparative Population Biology Study of Natural Colonies

Domiciano P. S. Dias
Departamento Regional de Pesquisas Ecológicas,
Fundação IBGE, Brasília, Brazil

The author is carrying out a long term project on the caste
systems of highly polymorphic Hymenoptera. The project involves
comparative studies on allometry, demography and division of labor
of natural colonies. The major objectives of the study have been
to determine the patterns of caste ratios and biomass allocation
and their relationship to division of labor and ecology of species
of different habitats (Amazonian tropical rain forest, Savanna
of Central Brazil, etc). The project began with a study of castes
in army ants of the genus Eciton, but attention has been concen-
trarted during the past ten years on leaf-cutting ants, of which 16
species have been investigated. The dynamics of the caste system
of a smaller number of species are under study at the present. More
recently the project has also involved studies on parental investment in both Atta and Acromyrmex.

Intra- and Interspecific Patterns of Allometry
in the Ant Genus Pheidole

Donald H. Feener Jr.
University of Iowa

The vast majority of species in the genus Pheidole possess completely dimorphic worker castes. Major and minor worker subcastes
can be distinguished by comparing the allometric relations of various body parts. Although in all these species major workers have
both larger bodies and disproportionately larger heads than do minor
workers, there is striking variation in the degree of morphological
disparity between the subcastes. This disparity can be quantified
as the linear distance in allometric space between the means of body
measurements for major and minor workers. A preliminary study on 20
North American species revealed that this distance between major and
minor workers in all allometric space defined by thorax length and
head width ranged from .24 for P.morrisi to .57 for P.militicida.
Interspecific variation in morphological disparity between subcastes
may be the evolutionary product of variation in the quality of environmental contingencies faced by the different species. Alternatively, the variation may not reflect adaptation per se, but instead
may be the consequence of rules governing relative growth rates
within the genus. This hypothesis is supported by the fact that head
size across species increases faster relative to thorax size in major
workers than it does in minor workers.

Pheromones and the Control of Queen
Number in Bees and Ants

David J. C. Fletcher
University of Georgia

Discrimination between nestmates and foreign conspecifics by
the members of a social insect colony is generally mediated by
olfactory stimuli. The odor cues are widely held to be
qualitative in nature and to be either environmentally acquired
or genetically determined. However, in the regulation of queen
number, workers sometimes kill gynes and pseydogynes that are
their own siblings raised in the same nest as themselves. In this
case, discrimination is unlikely to be based on qualitative odor
differences. The hypothesis is presented that quantitative
phermonal effects are involved and that these operate at the level
of both the colony and the individual. Experimental data that
support this hypothesis are also presented.

Constancy of Soldier Proportions in
Termite Colonies

Michael I. Haverty
Forest Service, U.S. Department of Agriculture
Berkeley, California

Adult soldiers are seldom, perhaps never, food producers; their
function is colony protection. The optimal proportion of soldiers
for a species is assumed to have evolved through the selection of a
mix of castes that will minimize the energy expended in producing
the maximum number of virgin males and females while providing
adequate defense of the colony, and is maintained by behavioral and
physiological mechanisms. There is little experimental or empirical
research to substantiate this assumption. Constancy of soldier proportions has been demonstrated experimentally with Coptotermes formosanus Shiraki, Reticulitermes flavipes (Kollar) and R. virginicus
(Banks). Counts of entire colonies and samples from colonies of
species in the Kalotermitidae, Rhinotermitidae and Termitidae support the concept of a constant proportion of soldiers. The correlation between soldier biomass and total colony biomass is more precise than between soldier numbers and total colony numbers. This
constant proportion of soldiers could be of practical value if the
proportion could be radically increased. Such an increase would
destroy the homeostasis of the colony, and has been shown to be
detrimental to colony survival.

Stochastic aspects of polyethism in the ant *Lasius niger*

A. Lenoir and H. Ataya

The University of Tours (FRA) and the University of Beyrouth (LEB)

Polyethism was studied in 5 young colonies comprising 225 ants individually marked. 24 behavioural parameters were noticed for each individual. Data have been treated by correspondence analysis (a form of principal component analysis). The first factor leads to distinguish, as known previously in numerous ants, nurses and foragers. The second factor is opposing among foragers carriers and explorers. We have in these young societies, reared in artificial nest with only one cell for the queen and the brood, only one caste of nurses. For outside tasks we observe simultaneously a great specialization (carriers and explorers) and a great plasticity for other foragers available for many tasks. The repartition for the levels of activity was studied for each task: it is always exponential, with a little number of hyperactive workers, the slope of curves is variable from very specialized tasks (for ex: egg nursing) to tasks involving the whole population (for ex: grooming). Nevertheless the repartition of the total activity is log-normal. This contradiction seems to indicate that individual potentialities are distributed according to a stochastic process with factors (perhaps experience) enhancing the expression of some groups of tasks.

Evolution of sociality in Apis mellifera. A Queen or Worker Strategy?

Robin F. A. Moritz

Institut für Bienenkunde der J.W. Goethe Universität Frankfurt/M
Federal Republic of Germany

In kinship selection theory of evolution of sociality in hymenoptera some problems rise when the coefficient of realtionship (WRIGHT 1922) is used. The relatedness between queen, workers and drones is obtained more easily by the coefficient of coancestry "s" defined by MALECOT (1948). "s" describes the probability of identical genes by descent in two individuals at a gametic level. The relationships between the members of a honey bee colony computed by introducing "s" into a pedigree model gives new aspects for evolution of sociality in kinship selection theory. Regarding single mating there is a probability of 3/16 for queen genes (P_q) and of 5/32 for worker genes (P_w) for transmission to the next generation. The queen's gain (Q) is calculated by $Q = P_q/P_w = (2z+4)/(z+4)$; (z=number of mating ♂♂ per ♀). Q amounts up to 1.67 in case of an eight drone mating as it happens under natural conditions. A queen strategy is no contradiction to the high frequency of sociality in hymenoptera in this model. The queen's profit in haplo-diploid compared to diplo-diploid populations is higher than the worker's. In diploidy P_q is equal to P_w whereas the above mentioned assymetry of relationship appears in haplo-diploidy. The missing of a haploid male worker caste as well as the extreme sex ratio in honey bee colonies are also signs for a successful queen strategy in Apis mellifera.

Comparative study of social carrying in Pseudomyrmex and Ectatomma
(Hymenoptera: Formicidae: Pseudomyrmecinae and Ponerinae).

William L. Overal
Museu Paraense Emílio Goeldi, Belém, Pará, Brazil

Social carrying of adults in the "parasol" position is common
to the behavioral repertoires of Pseudomyreme termitarius (Fr. Smith)
and Ectatomma quadridens (Fabr.), two soil-nesting ants abundant in
clearings and pastures near Belém, Brazil. This behavior occurs
sporadically in undisturbed colonies of both species, although with
lesser frequency in E. quadridens, and involves the inward transport
of conspecifics from other nests. At one site with a high nest
density (> 5 nests m^{-2}) 3 to 12 instances of social carrying were
observed per day among 19 nests of P. termitarius, suggesting an
important role in recruitment. This behavior is a constant feature
of nest-moving in both species, and could be elicited by uncovering
the brood chambers of nests. Colonies were repeatedly forced to
emigrate (as many as five times for a single colony) and transporter
and transportee ants marked. Social carrying by P. termitarius was
done by less than 15 percent of the workers, and more than 95 percent of the social carrying was done by transporter ants which
assumed the same role during other emigrations. There was much less
consistancy in the composition of the transportee class. In contrast, E. quadridens has no recognizable elite of transporter ants:
even after 3 forced migrations within 8 days, colonies exhibited
new (unmarked) transporter ants.

Pollen Ball Weight and Sex Determination in Halictinae

Cécile Plateaux-Quénu
Laboratoire d'Evolution, Paris

The first brood of Evylaeus calceatus (Scop.) includes workers
and a few males generally laid first, on the smaller pollen balls.
   We produce a situation whereby numerous males are naturally
produced in the spring brood. If a foundress, which has just laid
her first set of eggs, is experimentally deprived of them by the
removal of the earth filled frame which houses them, she lays another cluster of eggs in the same place, when provided with a
second earth filled frame; she then lays another cluster of eggs
when a third earth filled frame replaces the second used one.
Such a foundress, deprived of her first set of eggs, will not
as a result, directly give big males and future foundresses but
a new spring brood composed of fewer workers and more males. The
increase in the number of males will be even greater in the third
laying of the foundress.
   After having estimated the percentage of males in 3 successive layings produced by identified foundresses, we are looking for
the influence, on the sex-ratio of an experimental adding of
pollen in each cell of the three successive spring nests of
identified foundresses.

Caste Partitioned Survivorship and Route Fidelity of
Leaf-Cutting Ant Workers

Sanford D. Porter and Michael A. Bowers
Florida State University and University of Arizona

The survivorship and route fidelity of Atta colombica
workers were studied on Barro Colorado Island, Panama. Foragers
and refuse workers from several colonies were sprayed with in-
visible fluorescent ink as they travelled on their trails. Capture
samples were taken periodically to monitor changes in the frequency
and dispersal of marked ants.

Results are as follows: 1) Refuse workers and foragers were
essentially independent populations with only a minor exchange
of workers between them. 2) Route fidelity of marked foragers decreased with worker size; approximately 60% of maxima, 30% of media
and 20% of minima workers changed trails within 24 h of marking.
3) The frequency of marked ants in forager and refuse worker
capture samples declined to extinction in about 3 wk. Further analysis revealed that forager longevity is proportional to body
size. Minima workers survived about 2 wk while media workers
survived about 3 wk. 4) Survivorship for foragers and refuse
workers was approximately equal for workers of equivalent body size.

Colony Food Reserves and Intranidal Ecology:
Models for Melipona and Trigona

David W. Roubik
Smithsonian Tropical Research Institute

Intranidal ecology of meliponine bees was studied in lowland
forest of Panama and French Guiana. Long-term intensive studies of
Melipona in observation hives were combined with extensive colony
dissections of Trigona. Significant trends and correlations in
food harvest and storage, brood and gyne production, adult populations and worker life spans were identified. Small, resource
poor colonies of Melipona fulva and M. favosa were conservative in
brood production and worker foraging expenditure. Worker life
spans varied inversely with resource abundance, and brood production
tracked colony pollen stores, regardless of pollen abundance. 75
nests of Trigona cupira, dissected at 4 month intervals, showed
negligible seasonal shifts in adult and brood populations, gyne
production and pollen stores. Stored honey was significantly more
abundant in the dry season, and prepupae were absent in many
nests during the late wet season. Both studies suggest that
meliponines adjust to and probably anticipate seasonal changes in
resource abundance. Conservatism by Melipona would lead to colony
death during extended resource scarcity. Colonies with abundant
stored food, however, may respond as did T. cupira by maintaining
steady growth even during periods when food outside the nest is
relatively scarce. Such periods coincide with the late wet season
in the seasonal tropics.

Spatial Efficiency and the Evolution of Temporal Polyethism
in Honeybee Colonies

Thomas D. Seeley
Yale University

The adaptive significance of the temporal polyethism schedule
of honeybee workers was studied by determining the patterns of
task performance probability in time and space for a variety of
tasks (clean cell, tend queen, ventilate, receive nectar, forage,
etc.) Workers of a given age work in just one nest region (central
brood-nest, peripheral food storage area, or outside the nest),
but can perform all the various tasks occurring in their nest
region. The temporal pattern of labor changes with increasing age
mirrors the spatial pattern of work site changes in the life of a
worker bee.

Patterns in time reflect patterns in space.

Division of labor and social food flow
in the stingless bee Melipona favosa F.

Marinus J. Sommeijer
State University of Utrecht

Division of labor was studied in colonies of Melipona favosa,
housed in observation hives. Different worker tasks related to
reproduction appeared to be performed simultaneously by the same
individuals. Workers that discharged larval food into a particular
brood cell had always been among the principal constructors of that
cell. Division of labor is age dependent. Cell building,
provisioning and sealing is mostly performed by workers of 8 - 12
days. Worker ovipositions are most frequent between 9 and 18 days.
Except for the offering by returning nectar foragers
spontaneous food offerings do not occur. Trophallactic interactions
in the brood nest are always initiated by soliciting. A distinct
food flow from non-discharging towards discharging workers could
be observed: workers participating in the previous or the next
discharging sequence demonstrate a pronounced trophallactic
advantage over more distant dischargers and non-dischargers. Also
the pollen is collected from the storage pots by other workers
than the recent dischargers.

Factors Influencing the Production of Sexuals
in the Fire Ant, Solenopsis invicta

Edward Vargo
The University of Georgia

Colonies of the fire ant, Solenopsis invicta Buren, were
founded in the laboratory and their development was closely
monitored. They were fed on a high protein diet. Some colonies
produced both male and female reproductives at an early stage of
their development, but production of sexuals was not correlated
with colony size. To further investigate the factors associated
with the production of sexual brood, large laboratory colonies
were divided into several parts of known but different composition.
Their subsequent change in composition over time was then measured
and correlated with the production of sexuals. The most important
factors were found to be associated with the queen.

Forager Polymorphism in Atta texana Buckley

Deborah Waller
Zoology Department, University of Texas, Austin 78712

The leaf-cutting ant Atta texana Buckley (Formicidae; Attini)
forages on a predictable sequence of host plants throughout the
year in a central Texas habitat. The ants harvest leaves and
occasionally seeds from trees, shrubs and herbs for use as
substrate for their symbiotic fungus.
Different-sized foragers harvest different host species, with
larger foragers cutting tougher leaves. Small and large ants
carrying different host plants often utilize the same trail and
nest entrance.
Tree foragers are larger than ground foragers, even when both
use the same nest entrance and forage on the same plant species.
This pattern is the reverse of that observed for A.sexdens in
Paraguay (Fowler and Robinson, 1979) in which small tree foragers
clip petioles and return down trees unladen. A. texana foragers
cut and carry burdens down trunks. Thus burden retrieval may be
a major determinant of large tree forager size. Unladen outgoing
ants on tree trails are larger than unladen outgoing ants on
corresponding ground trails.

Soldier determination in <u>Pheidole</u> <u>bicarinata</u>:
Inhibition by soldiers

Diana E. Wheeler
Duke University

Topical application of methoprene to final instar worker
larvae can induce soldier development. Soldier induction takes
place if hormone levels are above a soldier-determining threshold
during a critical period of JH-sensitivity. The threshold itself
can change as a function of which worker caste rears the larvae.
When acetone treated larvae are reared by minor workers, about 5%
become soldiers. By contrast, no soldiers are produced in acetone
treated groups reared by soldiers. A concentration of 1 ul/ml
methoprene applied during the critical period induces most larvae
to become soldiers, if they are reared by minor workers. However,
only a few larvae respond to the treatment if soldiers are used
as nurses. Treatment with a concentration of 5 ul/ml of methoprene
almost completely compensates for this inhibition. Supression is
seen only in larvae that were treated near the beginning of the
instar. Growth patterns of larvae reared by both worker castes
were monitored and were not significantly different. Therefore
differences in soldier production cannot be attributed to the
ineptitude of soldiers as nurses.

PREDATION, SOCIAL PARASITISM, AND DEFENSE
Organized by Robert W. Matthews

Introduction

Robert W. Matthews,
The University of Georgia

When planning this symposium, the decision made was to permit a few speakers to develop specific topics in detail. Time constraints, rather than a lack of potential participants, dictated that some areas within the broad scope of this symposium would receive less emphasis than they properly deserved. The result is a series of definitive papers focusing on slavery in ants (Buschinger and Winter; Alloway), ant predation (Franks; Chadab-Crepet and Rettenmeyer), and wasp and bee colony defensive strategies (Jeanne; Seeley).

The complete omission of the termites is regrettable and perhaps reflects the organizer's bias toward the Hymenoptera. However, recent reviews do give a good flavor of research progress on termite defensive behavior (see Deligne et al., 1981; Prestwich, 1979, 1982). A recent report of yet another unique method of rapid repair of nest damage in the termite Prohamitermes mirabilis of Malaysia by Tho (1981), involves the use of preformed pellets which are stored in galleries until needed at which time they are apparently rolled into place, thereby forming an instant barrier blocking off penetrated galleries. Defensive mechanisms in the social Hymenoptera have been thoroughly reviewed by Hermann and Blum (1981)

Social parasitism in the vespine wasps and bumblebees is another area in which there is much work in progress, certain to yield new insights at all levels of analysis. For example, the genetic relationships between social parasites and their hosts is explored by Pamilio et al. (1982); Matthews (1982) examines social parasitism in Vespula. The earliest stage of social parasitism, that of facultative intra-specific usurpation of nests, is a virtually universal behavioral attribute of nesting wasps and bees. Its significance has only recently begun to be appreciated and interpreted from the standpoint of adaptive benefits accruing to the concerned individuals. Brockmann et al.'s (1979) elegant analysis using evolutionarily stable strategy theory applied to nesting in the digger wasp Sphex, points the way to what will surely become a fruitful means of viewing decision-making behavior in social insect species. As they have said "Time is a currency which an animal spends. Decisions are the moments at which the down payments are made."

References

BROCKMANN, H. J., GRAFEN, A, DAWKINS, R., 1979. - Evolutionarily stable nesting strategy in a digger wasp. J. Theor. Biol., 77, 473-496.

DELIGNE, J., QUENNEDEY, A., BLUM, M. S., 1981. - The enemies and defense mechanisms of termites., In, Social Insects, vol II. (H.R. Hermann, ed.) Academic Press, New York, pp. 3-76.

HERMANN, H. R., BLUM, M. S., 1981. - Defensive mechanisms in the social Hymenoptera., In, Social Insects, vol. II. (H.R. Hermann, ed.) Academic Press, New York, pp. 78-97.

MATTHEWS, R. W., 1982. - Social parasitism in yellowjackets., In, Social Insects in the Tropics. (P.Jaisson, ed.) Proc. I.U.S.S.I. Symp. (Cocoyoc) Mexico. (in press).

PAMILIO, P., PEKKARINEN, A., VARVIO-AHO, S.-L., 1981. - Phylogenetic relationships and the origin of social parasitism in Vespidae and in Bombus and Psithyrus as revealed by enzyme genes., In, Biosystematics of Social Insects. (P.E. Howse and J.-L. Clement, eds.) Systematics Association Special Volume No. 19, Academic Press, New York, pp. 37-48.

PRESTWICH, G. D., 1979. - Chemical defense by termite soldiers. J. Chem. Ecol., 5, 459-480.

PRESTWICH, G. D., 1982. - Evolution of chemical defenses in termites. In, Social Insects in the Tropics. (P.Jaisson, ed.) Proc. I.U.S. S.I. Symp. (Cocoyoc) Mexico. (in press).

THO, Y. P., 1981. - A unique defense strategy in the termite Prohamitermes mirabilis (Haviland) of peninsular Malaysia. Biotropica, 13, 236-238.

How the Slave-Making Ant
Harpagoxenus americanus (Emery)
Affects the Pupa-Acceptance Behavior
of Its Slaves

Thomas M. Alloway,
Erindale College, University of Toronto

Slavery in ants is a form of social parsitism in which slave-making parasite species exploit the labor of workers of other species. The slave-making parasite <u>Harpagoxenus americanus</u> (Emery) enslaves workers of three very closely related species (<u>Leptothorax ambiguus</u> Emery, <u>L. curvispinosus</u> Mayr, and <u>L. longispinosus</u> Roger). The frequent presence of slaves of more than one species in <u>H. americanus</u> nests (Alloway 1979; Headley 1943; Wesson 1939) indicates that enslaved workers of one host species often accept and rear brood of another host species. However, in these three host species, the rearing of brood from other colonies is not a completely unique behavior seen only in slave-maker nests. Territorial battles between unenslaved conspecific and allospecic colonies within this species triad sometimes result in one colony's capturing and rearing pupae from another colony. Nevertheless, unenslaved colonies usually eat many of the pupae which they capture (Alloway 1980). Hence, it seemed likely that <u>H. americanus</u> somehow enhances the pupa-acceptance behavior of its slaves. This paper reports the results of an experiment designed to test this hypothesis.

MATERIALS AND METHODS

The observations reported here were made during the summers of 1978, 1979, and 1980. Fresh colonies of <u>H. americanus</u> and its hosts were collected in the spring of each year in southern Ontario and in Illinois, and this material was kept throughout the summer under conditions described by Alloway (1979, 1980).

The pupa-acceptance behavior of <u>L. ambiguus</u>, <u>L. curvispinosus</u>, and <u>L. longispinosus</u> workers was studied under three conditions of worker "servitude":

a. "free worker groups", consisting of all the workers from unenslaved host-species nests.
b. "supervised slave groups", consisting of the entire worker force (both slave-makers and slaves) from <u>H. americanus</u> nests which had contained only one species of slaves when collected.
c. "unsupervised slave groups", consisting of the slave workers

only from H. americanus nests which had contained only one species of slaves when collected. The slave-makers were removed from these groups 2 weeks before the introduction of test pupae.
These coditiions made it possible to determine whether enslavement affected the pupa-acceptance behavior of host-species workers and, if so, the extent to which the effect depended upon the physical presence of H. americanus workers.

Altogether, the experimental design involved the manipulation of three independent variables. The first variable was the three kinds of "servitude" outlined above. The second variable was the species of host workers in recipient groups (L. ambiguus, L. curvispinosus, and L. longispinosus). The third variable was the species of pupae offered to recipient groups (L. ambiguus, L. curvispinosus, and L. longispinosus). These variables were factorially combined in a 3 X 3 X 3 design to produce 27 experimental conditions, each of which was replicated 6 times (yielding a total of 162 experimental observations). Two replicates of each condition were run in each of the 3 years.

The experimental pupae were taken from surplus colonies not otherwise used in the experiment. To exclude possible effects of queens and brood belonging to the colonies from which the recipient worker groups were derived, the worker groups were isolated in freshly cleaned artificial nests 2 weeks before the introduction of test pupae. Each recipient group received one worker pupa for each host-species worker in the group. The pupae were presented in piles located about 2 cm outside the entrances to the group nests. The dependent measure was the proportion of the test pupae which had matured or which remained intact inside each group's nest 5 days after presentation.

RESULTS

Every group carried all the pupae into its nest within a few hours. Inside the nests, the pupae were either eaten or reared. Intact colonies have been observed sometimes to attack workers eclosing from alien brood (Wilson 1975). However, so far as I could determine, no eclosing workers were attacked during this experiment.

The main body of data was analyzed using an analysis of variance. Duncan's range test was employed to make certain non-orthogonal comparisons (Edwards, 1960).

On average, "free workers" accepted 64.87% of the host-species pupae presented to them; "unsupervised slaves" accepted 70.17%; and "supervised slaves" accepted 75.52%. The main effect of the "servitude" variable was significant ($F=11.22$, $df=2/135$, $p<.001$), as was the difference between each of the three conditions of "servitude" ($.01<p.05$ by Duncan's range test). Thus, "supervised slaves" accepted the largest proportion of pupae, "free workers" accepted the smallest, and "unsupervised slaves" were intermediate.

Pupa acceptance also depended upon the species of the accepting host-species workers ($F=77.43$, $df=2/135$, $p<.001$) and the species of the pupae offered (F=27.18, df=2/135, p<.001). On average, L. longispinosus workers accepted 85.81% of the pupae offered, L. ambiguus workers accepted 65.83%, and L. curvispinosus workers accepted 58.91%. The overall proportions of pupae accepted were 78.93% for L. ambiguus pupae, 69.19% for L. longispinosus pupae, and 62.44% for L. curvispinosus pupae. A very large worker-species by pupa-species interaction (F=106.71, df=4/135, p<.001) arose primarily because host workers of all three species laboring under all three conditions of "servitude" accepted a larger proportion of conspecific than allospecific pupae.

Behavioral observation revealed that, in the "supervised-slave groups", the slave-makers had little direct contact with the test pupae or the eclosing workers. The slaves carried most of the pupae into the nest, either ate or tended them, and assisted the eclosion of young workers.

DISCUSSION

The enhanced pupa acceptance of "supervised", as compared to "unsupervised", slaves was probably mediated by a pheromone which was either applied to the pupae by the slave-makers or transmitted trophallactically by the slave-makers to their slaves. The small amount of direct contact which the slave-makers had with the pupae makes the latter possibility seem the more likely. However, whatever the nature of the mechanism, it is clearly adaptive for slave-makers to enhance the pupa-acceptance behavior of their slaves; and it is scarcely surprising that the effect is maximal in the presence of slave-maker workers.

What is somewhat surprising is the fact that the behavior of "unsupervised slaves" differed from that of "free workers". The enhanced pupa-acceptance of "unsupervised slaves" indicates that some experiential effect of slavery changes the behavioral ontogeny of slaves. Various authors (e.g. Jaisson 1975; Wilson 1955) have speculated that enslavement might somehow alter slave behavior. However, this study is the first demonstration of any relatively long-term behavioral effect of slavery which clearly does not depend upon the continual presence of slave-maker adults or brood. Further research is needed to discover which aspects of the enslavement experience produce the effect. However, since the behavioral difference is a relatively long-lasting and depends upon a difference in experience, it seems likely that some kind of learning is involved (Alloway 1972).

Most of the other findings can be explained rather straightforwardly. The largest effect was the tendency for host workers of all species and in all conditions of "servitude" to accept a larger proportion of conspecific than allospecific pupae. This effect is what one would expect on either one of two bases. Jaisson (1975) and others have reported that ant workers learn to recognize and care for whatever species of pupa they find in their environment during the first days of their adult lives, a phenomenon which has been regarded as an example of "imprinting". Since all except possibly the youngest of the host-species workers in all the conditions of "servitude" employed in the present study would have been exposed to conspecific pupae, this kind of imprinting might well account for the observed preference for conspecific pupae. However, imprinting to pupae has not yet been demonstrated in Leptothoracine ants. Thus, the observed preference could also be a reflection of an unlearned species-recognition process.

The main effects of worker species and pupa species on pupa acceptance are somewhat more problematic. It is not clear why one host species should on average accept more alien pupae than any other or why another host species' pupae should on average be the most acceptable. However, all nests containing Leptothorax ambiguus and L. longispinosus used in this study were collected in southern Ontario, a region where L. curvispinosus is extremely rare or absent. Furthermore, all the nests containing L. curvispinosus were collected in a part of southern Illinois where L. ambiguus and L. longispinosus are quite rare or absent. These practices which expedited the collection of sufficient numbers of hard-to-find slave-maker colonies containing the desired species of slaves may have influenced the findings. Had the material used in the study been collected in a region where all three host species are abundant, the observed main effects of worker species and pupa species on pupa acceptance might have been different.

REFERENCES

ALLOWAY, T.M., 1972. - Learning and memory in insects. Annual Rev. Entomol., 17, 43-56.

ALLOWAY, T.M., 1979. - Raiding behaviour of two species of slave-making ants, Harpagoxenus americanus (Emery) and Leptothorax duloticus Wesson (Hymenoptera: Formicidae). Anim. Behav., 27, 202-210.

ALLOWAY, T.M., 1980. - The origins of slavery in Leptothoracine ants (Hymenoptera: Formicidae). Am. Nat., 115, 247-261.

BUSCHINGER, A., ALLOWAY, T.M., 1977. - Population structure and polymorphism in the slave-making ant Harpagoxenus americanus (Emery) (Hymenoptera: Formicidae). Psyche, 83, 233-242.

EDWARDS, A.L., 1960. - Experimental Design in Psychological Research. Holt, Rinhart, and Winston, 398 p.

HEADLEY, A.E., 1943. - Population studies of two species of ants, Leptothorax longispinosus Roger and Leptothorax curvispinosus Mayr. Ecol., 38, 912-914.

JAISSON, P., 1975. - L'impregnation dans l'ontogenese des comportments de soins aux cocons chex la jeune fourmi rousse (Formica polyctena Forst.). Behav., 52, 1-37.

WESSON, L.G., Jr., 1939. - Contribution to the natural history of _Harpagoxenus americanus_ (Hymenoptera: Formicidae). _Trans. Am. Entomol. Soc._, 65, 97-122.

WILSON, E.O., 1955. - Division of labor in a nest of the slave-making ant _Formica wheeleri_ Creighton. _Psyche_, 62, 130-133.

WILSON, E.O., 1975. - _Leptothorax duloticus_ and the origins of slavery in ants. _Evolution_, 29, 108-119.

ACKNOWLEGDEMENT

This research was supported by Grant AO302 from the Natural Sciences and Engineering Research Council of Canada. I would like to thank Mrs. Alla Kazakov for her assistance in collecting the data and Mr. James Beckwith for his help in preparing the manuscript.

Evolutionary Trends in the Parasitic Ant Genus *Epimyrma*

Alfred Buschinger and Ursula Winter,
Technische Hochschule Darmstadt

The evolutionary pathways from independent to socially parasitic life in ants have been discussed by various myrmecologists.
Among the social' parasites, two or even three evolutionary lines
through slavery, temporary parasitism, and xenobiosis are supposed
to end up in permanent parasitism or inquilinism, following the
schema depicted by Wilson (1971).

During the past few years we have studied several species of
the myrmicine genus Epimyrma, which represent a number of steps in
the evolution of social parasitism though they do not fit well to
the above-cited schema.

About a dozen of Epimyrma species have been described mainly
from southern Europe, and from North Africa. They all are parasites
of Leptothorax species belonging to the subgenera Myrafant and
Temnothorax. Colony foundation has been observed in several species.
The parasitic queen eliminates the host queen by throttling her to
death. She is accepted by the host colony workers. There exist
some species of Epimyrma with considerable numbers of workers,
others have few or lack them completely. All Epimyrma species are
obligatorily monogynous.

Gösswald (1930), Wilson (1971), and others put forth the hypothesis that in the genus Epimyrma a degeneration of the worker caste
occurred, beginning with the numerous but functionless workers of
E. goesswaldi, and ending with E. ravouxi, a workerless inquiline
which coexists with the host colony queens.

With our investigations we are now able to rectify this
picture. Epimyrma is a slavemaker genus in which evolution can be
observed towards a kind of permanent parasitism which, however,
differs profoundly from inquilinism.

MATERIALS AND METHODS

We collected Epimyrma colonies preferably in or near to the
type localities of E. goesswaldi Menozzi, E. ravouxi (André), E.
vandeli Santschi, and E. stumperi Kutter. E. kraussei Emery was
found in northern Italy (Buschinger, in press; Buschinger and Winter,
in prep.; Winter and Buschinger, in prep.). Laboratory cultures
were maintained following Buschinger (1974). Winter (1979) describes

arena experiments for the observation of slave raiding behavior. For
population studies the colonies were censused after collecting in
the field, and the production of sexuals and workers was counted
both in field and laboratory colonies.

RESULTS

In this paper we present a summary of our recent work, which
has been or will be published elsewhere in detail.

1. E. goesswaldi Menozzi is a junior synonym of E. ravouxi
(André) (Buschinger, in press). Colonies from the type localities
of both exhibit identical structure, with up to 70 Epimyrma workers.
Queens of both populations kill the host queens by throttling them.
No significant differences could be found in a morphological comparison of material from both type localities and with the E. ravouxi
type series. The main host species is Leptothorax (Myrafant)
unifasciatus.

2. E. ravouxi is a true slavemaker (Winter, 1979). Raids upon
neighbouring L. unifasciatus nests are initiated by scouts which
recruit small groups of Epimyrma workers on a short-lived pheromone
trail to the target nest. They kill the defending host workers by
stinging them. All brood stages are carried back by the Epimyrma
workers to their nest. Host workers can develop from these broods
up to 2 years later. Reproduction of slaves in E. ravouxi nests by
thelytokous parthenogenesis (Gösswald, 1933; Wilson, 1971) thus
becomes unlikely. The life span of E. ravouxi colonies is estimated
to reach more than 10 years.

3. E. stumperi Kutter, living in the High Alps with
Leptothorax (Myrafant) tuberum, has not yet been observed during
slave raiding. However, the numbers of Epimyrma workers in the
nests would enable them to conduct raids. We also found host worker
pupae in colonies which already produced Epimyrma sexuals. These
pupae supposedly were newly gathered by raiding.

4. E. kraussei Emery, living together with L. (Temnothorax)
recedens, exhibits a population structure which differs profoundly
from that of E. stumperi or E. ravouxi (Buschinger and Winter, in
prep., Table 1). About one third of the E. kraussei colonies are
workerless, even those already producing sexuals. The remaining
colonies have extremely low numbers of Epimyrma workers. The
population data suggest that raiding in this species occurs rarely,
if at all, in the field. In laboratory experiments, however, we
released slave raids of two colonies with the exceptionally high
numbers of 7 and 9 Epimyrma workers. The organisation and success
of these raids were essentially the same as in E. ravouxi.

Further important differences between E. kraussei and E.
ravouxi refer to their sexual production and mating habits. E.
ravouxi sexuals are produced in much higher numbers (Table 2), and
they mate in a swarm outside the nest. E. kraussei sexuals, on the
contrary, mate inside the mother nest, the females shed their wings
and remain there until spring. Then they search on foot for host
nests in the immediate vicinity.

5. E. vandeli Santschi was collected in its type locality in
southern France. No workers were found in 11 colonies, as in the 6

Table 1. -- Population structure of Epimyrma ravouxi and Epimyrma kraussei. (In parentheses: Mean numbers of slaves per colony).

	n col.	% without E. workers	% with 1-5 E. workers	% with 6-10 E. Workers	% with > 10 E. workers
E. rav-ouxi	56	7.1 (150.7)	17.9 (37.3)	10.7 (79.0)	64.3 (164.4)
E. kraus-sei	82	46.3 (32.5)	40.2 (28.2)	13.4 (35.2)	-- --

described by Santschi (1927). However, in laboratory culture, we obtained 1 or 2 workers in a few nests. In addition, the sexuals mate inside the nest as in E. kraussei, and mating was observed between sexuals of the two species when experimentally put together in one nest. Thus, the two species might be subspecies, or even synonymous.

Table 2. -- Sexual production of Epimyrma ravouxi and Epimyrma kraussei. (In parentheses: Production per colony.)

	n col.	Males produced	Females produced	Sex ratio male/female
E. rav-ouxi	37	1915 (51.8)	1257 (34.0)	1.52
E. kraus-sei	41	183 (4.5)	762 (18.6)	0.24

A further argument in favour of this opinion is the fact that both have the same host, L. (Temnothorax) recedens.

6. E. cf. corsica (Emery): Only one female of E. corsica has ever been found, and the host species is unknown. However, guided by the papers of W. Faber, we collected numerous colonies of a species in Yugoslavia, which lives together with Leptothorax exilis, and which exhibits some morphological characters of E. corsica. This species is truly workerless. Among 25 colonies we found not one with an Epimyrma worker. Its sexual biology resembles that of E. kraussei.

7. The remaining Epimyrma species have not yet been investigated in detail. We can only refer to a paper of Cagniant (1968) suggesting that E. algeriana queens do not kill the host colony queens but invade already orphaned colonies.

DISCUSSION

Apparently the original type of social parasitism in Epimyrma is dulosis in the way which is characteristic of E. ravouxi and probably E. stumperi. However, we observe a trend toward reduction of worker numbers, and thus giving up slave raiding despite the behavioral patterns still present (E. kraussei). The worker caste finally disappears completely (E. cf. corsica). A second, parallel trend leads from normal swarming (E. ravouxi) to intranidal mating and adelphogamy, to hibernation of young queens in the mother nest, and to very short-lived colonies which die out together with the host worker stock after 2 or 3 years (E. kraussei, E. vandeli, E. cf. corsica).

This evolution is rendered possible only the the fact that the Epimyrma queens manage to be accepted by the adult host workers when invading their colonies. In other slavemaker genera, e.g., Harpagoxenus or Chalepoxenus, the parasitic queen has to eliminate all adults, queens, and workers, during colony foundation, and she gathers only their broods from which the first few slaves emerge.

The final stage of this evolution in Epimyrma is a workerless parasite that kills the host queen, or, if the observations of Cagniant (1968) prove true, a parasite who invades already orphaned host colonies. This differs from inquilinism in that inquilines associate with the host queens.

REFERENCES

Buschinger A., 1974. -- Experiment und Beobachtungen zur Gründung und Entwicklung neuer Sozietäten der sklavenhaltenden Ameise Harpagoxenus sublaevis (Nyl.). Insectes Sociaux, 21, 381-406.

Buschinger A. (in press). -- Epimyrma goesswaldi Menozzi 1931 = Epimyrma ravouxi (André 1896) - Morphologischer und biologischer Nachweis der Synonymie (Hym., Formicidae). Zool. Anz.

Buschinger A., Winter U. (in prep.). -- Population studies of the dulotic ant, Epimyrma ravouxi (André), and the degenerate slavemaker, E. kraussei Emery (Hym., Formicidae).

Cagniant H., 1968. -- Description d'Epimyrma algeriana (nov. sp.) (Hyménoptères Formicidae, Myrmicinae), fourmi parasite. Représentation des trois castes. Quelques observations biologiques, écologiques et éthologiques. Insectes Sociaux, 15, 157-170.

Gösswald K., 1930. -- Die Biologie einer neuen Epimyrmart aus dem mittleren Maingebiet. Z. wiss. Zool., 136, 464-484.

Gösswald K., 1933. -- Weitere Untersuchungen über die Biologie von Epimyrma gösswaldi Men. und Bemerkungen über andere parasitische Ameisen. Z. wiss. Zool., 144, 262-288.

Santschi F., 1927. -- Notes myrmécologiques. Sur quelques nouvelles fourmis de France. Bull. Soc. Ent. Fr., 32, 126-128.

Wilson E. O., 1971. -- The Insect Societies. Harvard Univ. Press, Cambridge (Mass.).

Winter U., 1979. -- Untersuchungen zum Raubzugverhalten der dulotischen Ameise Harpagoxenus sublaevis (Nyl.). Insectes Sociaux, 26, 123-135.

Winter U., 1979. -- Epimyrma goesswaldi Menozzi eine sklavenhaltende Ameise. Die Naturwissenschaften, 66, 581.

Winter U., Buschinger A. (in prep.). -- The reproductive biology of a slavemaker ant, Epimyrma ravouxi, and a degenerate slavemaker, E. kraussei.

Comparative Behavior of Social Wasps When Attacked by Army Ants or Other Predators and Parasites

Ruth Chadab-Crepet and
Carl W. Rettenmeyer,
The University of Connecticut

Neotropical social wasps (Vespidae: Polistinae) are frequently attacked by species of large army ants (Formicidae: Ecitoninae) and are the second most common prey for army ants, which eat primarily other non-ecitonine ants. Some wasps attempt to defend their colony against approaching ants, but such attempts only succeed against weak raids. We believe that army ants can successfully raid all species of Neotropical Polistinae except species living in association with some dolichoderine ants. Although the attacked colonies lose all their brood, most adult wasps survive and quickly reestablish a nest. The importance and frequency of army-ant attacks are reflected in a diverse array of alarm and defensive behaviors by the wasps. Those alarm-defense behaviors often are qualitatively or quantitatively different from the behaviors shown when the wasps are attacked or disturbed by other invertebrates or vertebrates.

METHODS AND FIELD RESEARCH SITES

Research was conducted at two sites: Barro Colorado Island, Panama (Smithsonian Tropical Research Institute), and Limoncocha, Ecuador (00° 24' S, 76° 36' W), a center of the Summer Institute of Linguistics. Barro Colorado Is. has a wet, lowland, tropical forest with a distinct dry season from January to May, and most of our observations were made during January-March, 1976, and April-May, 1977. Limoncocha is near the western border of the Amazon Basin rain forest, and there is no predictable, distinct dry season. The forest has been greatly disturbed, and there are large clearings at the study site. Data were collected from June-August, 1972 and 1973; May-July, 1975; and June-July, 1977.

Raids on wasp nests were observed by following the raiding columns of the army ants. By the time we located the wasp nests, however, the important initial events of the raid had already occurred. To observe the first contacts between wasps and ants, we manipulated the raid columns of the ants by shifting vegetation or by transferring ants via leaves or branches to vegetation immediately adjacent to a wasp nest. We also carried wasp nests on their substrate vegetation to the site of army-ant raids or collected army ants in a plastic bag and transferred some of those ants to vegetation near nests. The last technique made it possible to control the number of attacking ants so that the wasp nest was not destroyed or abandoned and could be used again for further tests.

Additional details on methods and study sites can be found in Chadab (1979).

RESULTS

1. How army ants find and attack wasp nests

The army ants, Labidus praedator, Nomamyrmex esenbecki, and all
species of Eciton, raid social wasps, but E. burchelli, E. hamatum,
E. lucanoides, and E. rapax probably account for most of the mortality caused by invertebrates. The first two species of Eciton commonly raid to the tops of canopy trees, and all four species regularly raid from the forest floor to several meters above the ground.
Although the ants raid in a consistent direction and follow topo-
graphical features, their raiding behavior otherwise appears to be
random. The ants are blind and show no ability to detect a wasp nest
at a distance. Even ants running over the envelope of a wasp nest
may not become excited and sometimes do not appear to recognize the
nest as a source of food. Consequently, it may be adaptive for wasp
colonies to keep their adults within the nest. Not only does this
make the nest less noticeable to a vertebrate, but it would prevent
provocation of individual army ants along weak raid fronts. Nests
which have populations of 100-300 adults but no wasps or only one or
two on the outside of the nest include Parachartergus fraternus,
Para. fulgidipennis, Polybia velutina, Protopolybia exigua, and
Synoeca chalibea. Compared to contact with nest carton material,
contact with adult wasps is much more important for stimulating army
ants to recruit additional workers and to invade the nest.

Once a wasp nest is perceived as a source of food, recruitment of
additional workers is very fast, hundreds can be attracted in a few
minutes (Chadab and Rettenmeyer, 1975). The ants run all over the
nest, and occasionally so many ants run off the vegetation that ropes
of living ants connect leaves to objects 20-40 cm below. The ants
often plug the nest entrance with their bodies, effectively trapping
any wasps which do not evacuate the nest. The ants remove all the
adults, larvae, and pupae, leaving behind only the eggs. Some of the
prey is taken out through the entrance, but the ants usually cut
holes in the envelope facilitating removal of the brood via shorter
pathways. These small holes up to 2 cm wide with ragged edges are
characteristic of nests raided by army ants, distinguishing such
nests from those destroyed by birds or invaded by other ants. Once
the nest has been invaded by the ants, the removal of the brood may
take more than ten hours depending on the size of the wasp colony.

2. Alarm responses of wasps to army ants

Wasps exhibit some alarm responses to an approaching army ant
that are identical to the responses shown toward any ant or parasitic
wasp that approaches or runs on the nest envelope. The odor, visual
image of running ants, and vibration of ants on the nest or substrate
leaf all stimulate alarm and evacuation behavior (Chadab, 1981). The
most common responses are raised wings, head and thorax elevated,
front legs extended, and the wings flipped one to several times. We
have seen such behavior by Apoica pallens, Mischocyttarus spp., Parachartergus fulgidipennis, Polistes spp., Polybia catillifex, P. emaciata, P. occidentalis, P. rejecta, P. velutina, Protopolybia exigua,
Pseudopolybia vespiceps, Synoeca chalibea, and S. septentrionalis.

In Charterginus fulvus, Protopolybia acutiscutis, and

Protopolybia exigua, group fanning accompanied by audible buzzing
occurs among all or most of the wasps on the outer surface of the
envelope (and probably by many wasps within the nest). One to
several wasps begin the fanning soon joined by others. The wasps fan
and buzz intermittently and synchronously, tilting their bodies
upward at a 70° angle as they fan. When their heads are elevated,
the tips of their gasters press against the envelope and the wings
fan — transmitting vibrations to the nest envelope and to the substrate leaves. This behavior serves as an intracolony alarm, alert-
ing the wasps within the nest and stimulating more wasps to exit.
Odor tests performed on Pr. exigua showed that of several ant odors
presented to the wasps including Camponotus, Paraponera, and Ectatomma, the most effective in eliciting this behavior was that of Eciton (Chadab, 1981). The greater response to Eciton odor demonstrates
specificity in the cues that elicit alarm behavior. A strong rap on
the nest substrate or surface of the carton will also cause a similar
alarm, while gently shaking the nest causes the wasps to retreat
inside. Although synchronization of the behavior increases the
amplitude, it is still so faint that the sound would not seem to be
effective at repelling vertebrates. Most people, however, will back
away from such a nest covered with bobbing wasps.

Angiopolybia pallens has a different alarm response: the wasps
rush out of the entrance at the end of the long downward spout, then
run upward until most of the envelope is covered, turn to face downward, and rap on the nest with their mandibles. Simultaneously with
the mandible rapping, the wings are rapidly flipped and the tip of
the gaster is also rapped against the nest. The behavior may begin
with one to a few individual wasps but is quickly synchronized so
that all the wasps move in unison to produce a distinct buzz or rat-
tle. Large colonies, in which over 200 wasps participate in this
behavior, produce a sound that can be clearly heard for distances of
at least eight meters. This alarm response, which is elicited when
the nest vegetation is disturbed, frightens most people and presumably protects the wasps against vertebrates. The response can also be
elicited by a single army ant running on the carton and momentarily
contacting a wasp. A greater response was obtained to a single Eciton than with the larger ants, Ectatomma or Camponotus.

3. Defensive behavior of wasps against army ants

The intermediate workers of Eciton do most of the attacking, and
wasps such as Protopolybia and Leipomeles, which are considerably
smaller than those workers, do not actively fight the ants. Other
wasps attempt to bite and sting the ants. We never observed any
injuries from stings to either the wasps or ants. Ants grasped by
the wasp's mandibles usually just rolled off the nest. If other army
ants were in the vicinity of a wasp locked in combat with an army
ant, the wasp was almost invariably captured quickly. The large
wasps, Apoica pallens and Polybia velutina, grasped approaching ants
behind the head, flew, and dropped the ants one to several meters
away. If the ant column is a weak one, this defense might save a
colony. If the ants are dropped or fall beneath a nest, however,
that could further stimulate army ants to run up adjacent vegetation
and increase the probability that the wasp nest will be invaded.

Once recruitment of ants has begun, fighting or ant removal by the wasps is futile. Regardless of their superior size and sting, no wasp species can cope with the overwhelming numbers of army ants.

Another type of defense is directed fanning, in which a wasp turns toward an approaching insect, raises its body, and fans its wings. This behavior repelled Azteca workers and small flies approaching nests of Metapolybia and Occipitalia. It is unclear whether the invader is repelled primarily by an air current or by the visual image of the fanning wasp. Both those wasp genera also peck at the substrate in front of the approaching insect. We refer to this as directed pecking to distinguish the behavior from other common types of pecking on the nest that are not oriented toward an intruder. Although the wasps do not actually hit the invader in directed pecking, the vibration caused may help repel the approaching insect. In the case of O. sulcata, which lives on trees occupied by Azteca, the pecking removes bits of nest carton and moss and bark adjacent to the nest. That could also remove ant trail chemicals so that the Azteca are less likely to get close to the wasp nest.

Some wasps plug the nest entrance with their heads, for example, Polybia velutina, Synoeca chalibea, and S. septentrionalis. This behavior successfully protected one nest of S. septentrionalis from an Eciton raid, but the ants (same colony?) returned the following day and penetrated the wasps' defenses (Skutch 1971). We staged a similar raid against the same species of wasp in Panama and for approximately 30 minutes the wasps' heads kept the ants out. The army ants persisted, however, and were able to pull the wasps out by their antennae. Once the entrance was breached, the wasps had no further defense.

We predict that wasps living in clay nests, such as Polybia emaciata, or wasps with dense felt-like carton, such as Chartergus chartarius, might be able to save their colonies from weak raids by plugging the nest entrance. However, one colony of P. emaciata we tested did not effectively plug its entrance.

The only defense which we found to be consistently effective against army ants is for the wasps to live in trees occupied by certain species of dolichoderine ants in the genus Azteca. These tiny ants mob any ant attempting to walk on their tree. The first advancing army ants are each pinned down by 50-100 Azteca eliminating the possibility that the ants will recruit additional followers. Wasps which nest in Azteca trees thereby receive protection from army ants.

DISCUSSION

Since social wasps do not have any effective direct defense against army ants, the wasps' most important alternative is rapid evacuation. If they exit the nest and fly as soon as army ants arrive, the adults can escape. Fighting is nonadaptive because adult wasps never win, and fighting attracts more ants. Various alarm behaviors of the wasps stimulate exit from the nest before any ants have entered it. A prime example of this is Protopolybia exigua which becomes alarmed by the odor of army ants enabling that species to evacuate before the ants have plugged its tiny entrance.

The nest envelope has long been considered a defense against parasites and predators (Richards and Richards, 1951), but it is a hindrance for rapid nest evacuation. Some nest entrances, such as that of Protopolybia exigua, will allow only a single wasp to pass, and even the widest entrances are bottlenecks for rapid evacuation. Early alarm of the colony is important in order to get the maximum number of wasps outside the nest.

In social wasps the queens tend to stay inside nests when colonies are disturbed. Such behavior would promote the death of queens when army ants attack. The development of polygyny in Polistinae helps ensure that at least one queen will survive (Richards and Richards, 1951). Moreover, multiple queens can supply many eggs to replace the brood lost during the raid. Since queens must be able to fly, the development of a physogastric queen with poor flight ability would also be impossible.

The frequency of army-ant raids is greatest in the vegetation up to two meters above the forest floor. Thus, selection should favor nests at higher elevations, but the difficulty of finding such nests has made it impossible to test that hypothesis. Higher nests may be subject to greater bird, bat, or monkey predation as well as greater wind damage.

Although we found no evidence that chemical repellents applied by the wasps to their nests are effective against army ants, it is clear that the wasps have evolved a wide variety of alarm-defense behaviors which are somewhat specific towards army ants.

References

CHADAB R., 1979. - Army-ant predation on social wasps. Ph. D. dissertation: University of Connecticut, Storrs, Connecticut. 260 p.

CHADAB R., 1981. - Early warning cues for social wasps attacked by army ants. Psyche, 86, 115-123.

CHADAB R., RETTENMEYER, C.W., 1975. Mass recruitment by army ants. Science, 188, 1125-1125.

RICHARDS O.W., RICHARDS M.J., 1951. Observations on the social wasps of South America. Trans. Ent. Soc. London, 102, 1-170, 4 pls.

SKUTCH A.F., 1971. A naturalist in Costa Rica. University of Florida Press, Gainesville. 378 p.

Social Insects in the Aftermath of Swarm Raids of the Army Ant *Eciton burchelli*

Nigel R. Franks, The University of Leeds

The composition of the leaf litter ant fauna of Barro Colorado Island, Panama is determined in part by competition and in part by army ant predation. A series of observations and experiments suggest that for the ant prey of Eciton burchelli both competition and the probability of army ant predation are lowest in areas just raided by these dorylines. For these reasons, queens of the ant species most heavily preyed upon by E.burchelli are expected to found new colonies in greater abundance in recently raided areas. Before describing an experiment to test this hypothesis, the ecological factors which suggested this investigation will be reviewed.

Eciton burchelli colonies live in Neotropical rain forest where they stage massive swarm raids. On average, each colony's daily raid covers an area 105m long by 6m wide (Willis,1967; Franks, 1980, 1981). These army ants hunt a wide variety of leaf litter invertebrates but they prey mostly on other ant species (Rettenmeyer, 1963; Franks, 1980, 1982).

Eciton burchelli lowers substantially the abundance of its ant prey. Colonies of prey species are more than twice as abundant in areas of the forest where E.burchelli does not occur, than in areas frequented by the army ants (Franks, 1980). This has been demonstrated by comparing the forest floor ant faunas of Barro Colorado Island (BCI) and Orchid Island. The latter is an 18 hectare neighbour of BCI from which Eciton burchelli had probably been excluded since 1914 when the islands were formed during the creation of the Panama Canal.

To test the hypothesis that the differences in the ant faunas between these islands was due to the earlier absence of these army ants, an E.burchelli colony was transferred from BCI to Orchid Island The diet of the introduced colony was significantly different from the diets of BCI colonies. On BCI 39%, and on Orchid Island 67%,of all prey items of E.burchelli were social insect larvae. This suggests that E.burchelli is significantly lowering the density of its prey on Barro Colorado Island, and that it is having a major impact on the structure of the leaf litter ant community.

The Orchid Island experiment has further emphasized that social insects are the most important component of the diet of E.burchelli.

This is entirely consistent with our accumulating knowledge of the diets of the Dorylinae, in general (Rettenmeyer, 1963; Schneirla, 1971; Gotwald, 1974; Mirenda et al.,1980; Rettenmeyer et al., 1980). Competition for nesting sites and foraging space limits the abundance of the ant prey of E.burchelli, such competition is most intense in the absence of these predators (Franks, 1980). Before the army ants were introduced to Orchid Island, forest floor ant nests were mapped in 100m^2 quadrats placed at random on BCI and Orchid Island. In all quadrats, on both island, mapped nests were significantly overdispersed; however, in all quadrats on Orchid Island, nests were more perfectly overdispersed than in any of the quadrats on BCI (Franks, 1980). Overdispersion may be attributed to workers of established colonies killing queens that attempt to found nests near others. The greater tendency to hexagonal spacing on Orchid Island was probably due to the larger size of established ant colonies on that island. These colonies were larger than those on BCI because they had never been cropped by army ants. Bigger forest floor ant colonies can patrol larger areas and kill colony-founding queens at greater distances - thereby maintaining a more perfect overdispersion of nests. Detailed analyses of maps from BCI (Levings and Franks, 1982) further suggest that predation on colony founding queens by established colonies occurs preferentially within genera.

E.burchelli raids generally either kill or crop to smaller size established colonies of their ant prey. For this reason incipient colonies of prey species should experience less intraspecific predation and competition in recently raided areas. Prey queens should be more successful at founding colonies in these areas. Furthermore, queens of non-prey species should not be more success-ful at nesting in old army ant raid paths than in non-raid areas, because established colonies of non-prey species will be equally abundant in both types of area.

There is another reason why prey queens should be more abundant in areas just raided by E.burchelli. These army ants have foraging and movement patterns that lower the probability that they will forage over their own earlier paths. E.burchelli colonies seem also to avoid areas just raided by conspecifics (Franks,1980; and in prep). By nesting in recently raied areas prey queens will be lowering the probability that they will be consumed by Eciton burchelli in the near future.

To examine the relative abundance of incipient colonies of prey and non-prey species in areas recently raided by Eciton burchelli and in other areas of the forest floor, potential nest sites were placed in the two kinds of area.

MATERIALS AND METHODS

The experiment to determine the abundance of incipient ant colonies was conducted in 1979 on Barro Colorado Island, Panama (Croat, 1978). Sections of bamboo were used as potential nest sites to sample incipient colonies of large formicines and ponerines that commonly nest in rotten logs. Each bamboo section was a hollow cylinder about 35 cm long and about 4cm in diameter. The ends of each

cylinder were sealed by the partitions at the bamboo's nodes. A
small hole about 0.5cm in diameter was made in one of these partitions in each piece of bamboo for a nest entrance.

Four raid systems of 4 different nomadic (schneirla, 1971) E.
burchelli colonies were mapped by laying a 100m tape along each
raid's principal trail. This trail is a column of ants that
continuosly limks the army ants' nest and swarm. The principal
trail generally passes down the center of the area covered by the
raiding swarm (Schneirla, 1940). One day after each raid a transect
of 30 bamboo nests, set at 2m intervals, was layed along the
measuring tape in each of the four raided areas. To sample areas
that were unlikely to have been recently raided identical transects
were placed 20m to one side and parallel to each of the four raid
transects. The pieces of bamboo were individually numbered and
flagged with red ribbon to facilitate their recovery. All transects
were layed out on the forest floor within the first two weeks of
May 1979. This month is at the transition of the dry to wet seasons
in Panama. After the first four months of the wet season had
elapsed, the bamboo nests were returned to the laboratory. Each
section was split open and its contents placed in a vial of alcohol
for later examination.

The numbers of incipient colonies belonging to the subfamilies
Formicinae and Ponerinae occuring along the various transects are
presented in Table 1. These "incipient colonies" all had a queen
with a retinue of unusually small, nanitic workers and a small
population of brood.

The incipient colonies have been categorized as prey and non-
prey species (Table, 1). This division was based on extensive
collections of the prey of E.burchelli (Franks,1980). Corpses of
adult workers of all "prey" species have been collected from the
principal raid trails of E.burchelli colonies on BCI. Dead workers
of "non-prey" species have never been found in these raid trails.
Moreover, swarms have been seen to pass harmlessly over the nests of
these non-prey species.

RESULTS

Prey species occurring in the transects were Camponotus
abdominalis, Camponotus landolti (Formicinae), and Odontomachus
bauri (Ponerinae). Non-prey species found in the bamboo nests were
all ponerines, Pachycondyla impressa, Pachycondyla apicalis and
Odontomachus chelifer.

Table 1. -- Numbers of incipient colonies in nest transects.

Transect Pair	Recently Raided Areas		Not-Recently Raided Areas	
	Prey	Non-prey	Prey	Non-prey
A	1	1	0	2
B	2	0	0	1
C	2	1	1	0
D	1	0	1	0
Data Transform x	1.403	0.966	0.966	1.055
$\sqrt{(x + 0.5)}$ S.D.	0.206	0.299	0.299	0.427

The abundance of incipient prey colonies is significantly different in the two types of area (one-tailed t-test, t = 2.407 p < 0.05). The abundance of incipient non-prey colonies is insignificantly different in the two types of area (t = 0.341).

DISCUSSION

Incipient colonies of ants preyed upon by Eciton burchelli occur in greater abundance in the old raid paths of these army ants than in not-recently raided areas. Young colonies of non-prey species are equally abundant in both types of area. In general, incipient colonies of formicines and ponerines appear to be rare: only 13 were found in all 240 potential nest sites. Without the provision of suitable nest sites, incipient colonies may be even more rare. Nevertheless, if incipient colonies are only slightly more abundant in raided areas the impact of Eciton burchelli on the population structure of its prey is likely to be significant. The experiment described here sampled only 240m of old raid systems. The E.burchelli population on BCI produces over 140 kilometers of raid systems every month (Franks,1980,1981).

There are two hypotheses which might explain the greater abundance of new prey colonies in old raid paths: a) Queens attempt to found colonies in equal abundance everywhere -- but they are more successful in raided areas due to the paucity of established con-specifics in these areas, Or, b) Queens actively seek old raid paths to avoid competition from established colonies and to lower the probability that they will be discovered by these army ants in the near future. Sampling the abundance of newly de-alate queens in recently raided and not-recently raided areas could discriminate between these hypotheses.

In the short term, prey in old raid paths will be safe from E.burchelli raids because these army ants have foraging patterns which lower the probability that they will forage over their own recent raid path. In addition, E.burchelli may also avoid areas, which have been recently foraged and bear the trail pheromones of conspecific army ants (Franks, 1980; and in prep.). These trail pheromones are persistent (Schneirla, 1971) and may serve as cues to these "temporary safe sites" for colony founding queens.

The potential prey of army ants employ diverse and sometimes elaborate escape techniques (Chadab, 1979; Rettenmeyer et al., 1980). The data reported here raise the possibility that the queens of the ant species most heavily preyed upon by Eciton burchelli may also be able to elude their predators by founding their colonies in the aftermath of army ant raids.

Acknowledgments.
I wish to thank C.M.Philips for assistance in the field, S.C.Levings for suggesting the use of bamboo nests, W.L.Brown Jr. for taxonomic help, and S.H.Bartz for reviewing this manuscript. This work was supported by a Natural Environment Research Council (U.K.) Studenship. Facilities were generously provided by the Smithsonian Tropical Research Institute.

References

CHADAB R., 1979. - Early warning cues for social wasps attacked by army ants. Psyche, Camb., 85,115-123.

CROAT T.B., 1978. - Flora of Barro Colorado Island. Stanford Univ. Press, viii+943p.

FRANKS N.R., 1980 - The evolutionary ecology of the army ant Eciton burchelli on Barro Colorado Island, Panama. PhD thesis: University of Leeds, U.K.

FRANKS N.R., 1981. - A new method for censusing animal populations: the number of Eciton burchelli army ant colonies on Barro Colorado Island, Panama.,Oecologia (Berl.)(in press).

FRANKS N.R., 1982. - Ecology and population regulation in army ants, Eciton burchelli. In, The Ecology of a Neotropical Forest, Seasonal Rythms and Longer-Term Changes. (E.G.Leigh, Jr., A.S.Rand, and D.Windsor. eds.), Smithsonian Institution Press, Washington D.C. (in press).

FRANKS N.R., (in prep.). - Spatial patterns in army ant foraging and migration: Eciton burchelli on Barro Colorado Island, Panama.

GOTWALD W.H.Jr., 1974. - Predatory behavior and food preferences of driver ants in selected African habitats., Ann. Ent. Soc. Am., 67, 877-886.

LEVINGS S.C., FRANKS N.R., 1982. - Patterns of nest dispersion in a tropical ground ant community., Ecology (in press).

MIRENDA J.T., EAKINS D.G., GRAVELLE K., TOPOFF H., 1980. - Predatory behavior and prey selection by army ants in a desert-grassland habitat. Behav. Ecol. Sociobiol.,7,119-127

RETTENMEYER C.W., 1963. - Behavioral studies of army ants. Univ. Kansas Sci. Bull., 44, 281-465.

RETTENMEYER C.W., CHADAB-CREPET R., NAUMANN M.G., MORALES L., 1980. - Comparative foraging by neotropical army ants., In, Social Insects in the Tropics.(P.Jaisson, ed.) Proc. I.U.S.S.I. Symp. (Cocoyoc) Mexico. (in press).

SCHNEIRLA T.C., 1940. - Further studies of the army-ant behavior pattern. Mass organization in the swarm raiders. J.Comp.Psychol. , 29, 401 -450.

SCHNEIRLA T.C., 1971. - Army ants: a study of social organization. (H.R. Topoff, ed.) Freeman, San Francisco. xx + 249 p.

WILLIS E.O., 1967. - The behavior of bicolored antbirds. Univ.Calif. Publ. Zool.,79,1-127.

Predation, Defense, and Colony Size and Cycle in the Social Wasps

Robert L. Jeanne,
The University of Wisconsin

Colony size among social wasp species varies within a range of at least four orders of magnitude. Intrinsic factors -- limits to the physiological and behavioral. capabilities of the insects -- probably partly determine upper and lower limits to a species' colony size. Although it is common to speak of "species-typical" colony size, mature colony size may vary considerably from place to place within the range of the species. The most conspicuous examples among the social wasps are summarized in the observation that for some species colonies are larger at the edges of the tropics than near the equator (Richards & Richards, 1951). It is likely that these differences represent relatively labile adaptations to local conditions made within the limits of more fundamental demographic parameters. Here I examine the ways in which colony size and cycle might respond to external influences. I will deal mainly with the polistine wasps.

EXTRINSIC FACTORS AFFECTING COLONY SIZE AND CYCLE

Extrinsic factors fall into three main groups: natural enemies, seasonal patterns, and competition. I will be concerned primarily with natural enemies that destroy the brood, since these are believed to have played a major role in the evolution of the social wasps. Natural enemies fall into several categories according to the nature of the pressure they exert on social wasps colonies.

1. Scouting and recruiting ants
In most ant species foragers search for food singly, and upon finding a source recruit nestmates to it, typically via a chemical trail. In the wet tropics an undefended wasp larva can expect to fall prey to these ants in less than half a day (Jeanne, 1979). For this reason probably all species of Polistinae are obliged to allocate a fraction of their total energy budget to defense against scouting and recruiting ants. The independent-founding polistines, whose nests are single, open combs suspended from a tough petiole, appear to rely on a combination of visual detection and removal of scout ants and the maintenance of an ant repellent secretion on the nest petiole (Jeanne, 1970; Post & Jeanne, 1981).

The swarm-founding polistines build nests of one or several combs enclosed in an envelope, with only a narrow entrance to the outside. There is little evidence that these species maintain ant-repelling chemicals on the nest. Most of these species seem to rely on the ability of the numerous workers to detect and physically remove approaching scout ants.

2. Mass foraging ants

These are more serious enemies of social wasps than are the scouting and recruiting species because their raids, involving large numbers of individuals, are much more difficult to defend against. Despite the intensity of army ant attacks, however, some wasps appear to be able to protect their nests, for example by blocking the nest entrance, or by nesting with ants that can repulse army ant raids (Chadab, 1979). The fact that army ant raids occur relatively infrequently -- Chadab's (1979) estimate is once every 200-460 days in rain forest habitats in Central and South America -- opens the option for social wasps not even to attempt defense, but to adapt their life cycles to tolerate the occasional raid. Most of the polistine species seem to have taken this route. When attacked, the adult wasps quickly abandon the nest and brood and renest elsewhere.

3. Vertebrates

The sting is undoubtedly effective against at least some potential vertebrate predators, notwithstanding the numerous accounts in the literature of vertebrates -- particularly certain birds and bats -- attacking social wasp colonies, even large ones, with seeming impunity (Jeanne, 1975; Windsor, 1976).

4. Parasitoids

Most parasitoids attack individual larvae and pupae and cause a reduction in colony output, but rarely destroy the entire brood. Some species reinfest their natal nests, a practice that can lead to an increase in the percentage of brood destroyed as the colony ages (Strassmann, 1981).

FUNDAMENTAL FEATURES OF COLONY SIZE AND CYCLE

Data available for several species of social wasp suggest that colony growth is exponential, but that the growth rate, r, measured as the fractional increase in adult population per unit time, declines as the colony ages (Michener, 1964; Jeanne, 1972; Archer, 1980). The values of the intrinsic parameters that determine r (including oviposition rate, worker life span, and the brood rearing efficiency of workers) are probably rather conservative demographic features that evolve slowly. The same is probably true of d, the egg-to-adult development time. In contrast, such colony characteristics as the size of the founding group and the length of the cycle (in d-units) probably respond more readily to local environmental conditions facing the population. It is the response of these more labile parameters to the external pressures discussed above that I now examine.

ADAPTATION OF SIZE AND CYCLE

If, as evidence suggests, r declines from colony initiation onward, then in the absence of any external constraints, the colony should reproduce with the first generation of offspring. That is, the species would revert to solitary behavior (Michener, 1964). Under these conditions, no amount of predation pressure on the population will shift the optimal reproduction time to $d > 1$, as long as the risk of failure is constant throughout the cycle. If, however, the risk of failure is greater during the pre-emergence period than later, then the optimal switching time will come later in the cycle, precisely when depending on the shape of the intrinsic growth curve and on the mortality rate in the pre-emergence period relative to subsequent stages.

For social wasps the pre-emergence stage is indeed the most vulnerable to failure. This is especially evident among independent-founding species, where pre-emergence failure rates as high as 80% have been documented (Litte, 1981). Predators against which active defense is partially effective (scouting and recruiting ants, vertebrates) are likely to be more successful in their attacks on the pre-emergence stage, when the colony is still small. Yet some vertebrates seem equally capable of attacking large, mature colonies, and in some cases seem to favor such colonies (Jeanne, pers. obs.), perhaps because of the greater yield of food. Army ants probably are equally effective at raiding post- and pre-emergence colonies, and therefore may have little lengthening effect on colony cycle, except to the extent that raids on pre-emergence colonies mean that it is more likely that the swarm will die of old age before it can replace itself. Parasitoids have a greater impact on post-emergence colonies, and should therefore exert a shortening effect on the colony cycle.

The cycle-lengthening effect of high pre-emergence mortality can be partially countered if colony mortality increases rapidly in later stages of the colony cycle. The most clearcut examples of this are the extremely truncated colony cycles of Polistes colonies at the northern limit of their ranges. These colonies are obliged by the short season to switch to reproduction after only 2-3 weeks of worker production (Yamane & Kawamichi, 1975). It is unlikely that any kind of predation is specific enough and intense enough on older colonies to have a significant effect in this direction, but certain kinds of parasitization may be so.

The heightened aggressiveness and greater numbers of defending workers should make larger colonies of a species more able to defend against vertebrate attack, and this should select for larger colony size. For the independent-founding species, the route to larger colony size is through a longer cycle, but it is difficult to see how vertebrate predation could select for larger colony size via this route, because the colony must still pass through the early stages of small size. Moreover, Strassmann (1981) found that colony size made no difference in the risk of bird predation on Polistes exclamans colonies in Texas, and Litte (1977) found the same to be true for Mischocyttarus mexicanus in Florida. In the swarm-founding species, on the other hand, vertebrate predation could favor the

evolution of larger colonies via larger founding swarms.
Unfortunately, we have no evidence for these wasps that vertebrate
predators selectively prey on smaller colonies of a given species.

It has been suggested that high rates of army ant predation on
swarm-founding polistines will select for small colonies (Forsyth,
1978). Forsyth offered this as a reason why colonies of Polybia
occidentalis and Metapolybia azteca are much smaller in rain forest,
where Eciton is abundant, than in seasonal dry forest, where it is
rare. But the selective mechanism is not clear. Conceivably,
smaller colonies could result through the effect of shortening the
colony cycle, as described above, but is seems more likely that the
smaller colonies of the rain forest are at least in part the result
of correspondingly smaller swarms. Increasing rates of army ant
predation could depress swarm size if small swarms had higher
reproductivity per female than large ones, but there is no evidence
for this. The decrease in reproductive output per female attributed
by Michener (1964) to increasing colony size is in fact just as
likely to have been due to the effects of colony age, a variable
which Michener was unable to control in the data available to him.
Forsyth's (1978) data for Metapolybia azteca showed the number of
cells constructed in the pre-emergence period to be linearly related
to swarm size: about 5 cells were constructed per worker,
regardless of the size of the swarm. If this relation holds
throughout the colony cycle, then swarm size can have no effect on
reproductivity, and would not be expected to respond to changes in
rates of army ant predation.

CONCLUSION

Although it is likely that natural enemies influence colony
cycle and size in wasps, perhaps in the ways suggested above, the
extent and even the directions of the effects will remain largely
matters of speculation until we gain a better understanding of the
dynamics of the colony cycle and obtain reliable values for colony
demographic parameters. It seems important, for example, to know
more reliably how r varies during the development of a colony, and
to know the causes of the variation. Experimental manipulation may
ultimately be the most effective way to dissect apart the dynamics
of colony growth. For quantifying the effects of natural enemies on
colony size and cycle, comparing intraspecific populations in
habitats chosen so as to vary levels of predation, parasitization,
and seasonality independently of one another may be the most
fruitful approach.

References

Archer, M.E., 1980. -- Population dynamics. In Social Wasps: Their
    Biology and Control. R. Edwards. Rentokil, East Grinstead,
    W. Sussex, England, pp. 172-207.

Chadab, R., 1979. -- Army-ant predation on social wasps. Ph.D.
    thesis: University of Connecticut, 260 pp.

Forsyth, A.B., 1978. -- Studies on the behavioral ecology of poly-
    gynous social wasps. Ph.D. thesis: Harvard University, 226
    pp.

Jeanne, R.L., 1970. -- Chemical defense of brood by a social
    wasp. Science, 168, 1465-1466.

Jeanne, R.L., 1972. -- Social biology of the Neotropical wasp
    Mischocyttarus drewseni. Bull. Mus. Comp. Zool., 144, 63-150.

Jeanne, R.L., 1975. -- The adaptiveness of social wasp nest
    architecture. Quart. Rev. Biol., 50, 267-287.

Jeanne, R.L., 1979. -- A latitudinal gradient in rates of ant
    predation. Ecology, 60, 1211-1224.

Litte, M., 1977. -- Behavioral ecology of the social wasp, Mischo-
    cyttarus mexicanus. Behav. Ecol. Sociobiol., 2, 229-246.

Litte, M., 1981. -- Social biology of the polistine wasp Mischo-
    cyttarus labiatus: survival in a Colombian rain forest.
    Smiths. Contr. Zool., 327, 1-27.

Michener, C.D., 1964. -- Reproductive efficiency in relation to
    colony size in hymenopterous societies. Ins. Soc., 11, 317-
    342.

Post, D.C., Jeanne, R.L., 1981. -- Colony defense against ants by
    Polistes fuscatus (Hymenoptera: Vespidae) in Wisconsin. J.
    Kansas Ent. Soc., 54, 599-615.

Richards, O.W., Richards, M.J., 1951. -- Observations on the social
    wasps of South America (Hymenoptera Vespidae). Trans. R. Ent.
    Soc. London, 102, 1-170.

Strassmann, J., 1981. -- Parasitoids, predators, and group size in
    the paper wasp, Polistes exclamans. Ecology, 62, 1225-1233.

Windsor, D.M., 1979. -- Birds as predators on the brood of Polybia
    wasps (Hymenoptera: Vespidae: Polistinae) in a Costa Rican
    deciduous forest. Biotropica, 8, 111-116.

Yamane, So., Kawamichi, T., 1975. -- Bionomic comparison of
    Polistes biglumis (Hymenoptera: Vespidae) at two different
    localities in Hokkaido, northern Japan, with reference to its
    probable adaptation to cold climate. Kontyu, Tokyo, 43, 214-
    232.

Colony Defense Strategies of Honeybees in Thailand

Thomas D. Seeley, Yale University

Students of the social insects have recently begun examining the ecological forces shaping the properties of insect societies. A powerful tool for this type of study is the comparative approach: differences in colony design among closely-related species are related to differences in ecology. I studied the role of predation in shaping colony design among the honeybees (Apis spp.) of tropical Asia by comparing their methods of colony defense (see Seeley et al. (1982) for a complete report).

MATERIALS AND METHODS

The three species of honeybees in tropical Asia (Apis florea, A. cerana, A. dorsata) were studied for 7 months in and around the Khao Yai National Park, Thailand. Nests were collected to measure features of nest architecture and certain colony properties. The behaviors of predators attacking colonies and of bees defending them were studied by observing colonies without interference using binoculars and telescopes.

RESULTS

Each species focusses its defenses upon different stages in the predation sequence of detection-approach-consumption (see Table 1). This radiation in defense strategies apparently reflects each species' preadaptation by worker size (small, medium-sized, or large) and nest site (cavity or tree branch) to a different pattern of colony defense.

Wasps, birds, and primates probably have difficulty finding the small, dispersed colonies of Apis florea, whose nests are built low on the branches of dense, shrubby vegetation. Once found, however, they are easily approached and overpowered because their low, exposed nests are accessible and their small workers inflict relatively painless stings. When overwhelmed, the bees quickly abandon their nest; later they return to salvage wax. Ants find A. florea nests easily and at least one species (Oecophylla smaragdina) easily kills these small bees. However, sticky bands of resin encircling the nests' slender substrate branches prevent ants from invading A. florea nests.

285

Table 1. -- Summary and comparison of the colony defense strategies
of the honeybee species in Thailand in relation to the predation
sequence of detection-approach-consumption.

	A. florea	A. cerana	A. dorsata
Detection	Difficult, except for ants	Moderate	Easy
Approach	Easy, except for ants	Difficult	Difficult, except for good climbers and fliers
Consumption	Easy	Moderate	Difficult

Cavity-nesting colonies of Apis cerana are conspicuous with
their medium-sized bees streaming in and out of low, clearly visible
entrance holes in caves and hollow trees. However, gaining access
to A. cerana nests is difficult. Large predators cannot pass
through the small entrance opening and small predators are overpowered by entrance guards. But if a large predator can breach a
nest cavity's walls, it faces an only moderately powerful stinging
defense. Apis cerana colonies are relatively small and their workers
are not fiercely aggressive.

Predators easily find the large, sometimes aggregated colonies
of Apis dorsata, whose nests hang in the crowns of the tallest
forest trees. But only skilled fliers and climbers can reach these
lofty nests. Those which do face massive stinging attacks from the
large colonies of these relatively giant, ferocious bees.

Nests of both open-nesting species, Apis florea and A. dorsata,
are protected by a 3-6 layer thick curtain of bees over the comb.
Apis cerana colonies lack these curtains but are protected by their
nest cavities' walls. A curtain of inactive guards requires a large
labor force. The high worker-to-brood ratio in A. florea relative
to A. cerana colonies suggests that the age polyethism schedules of
the open- and cavity-nesting species are tuned differently to
generate the appropriate proportions of guard bees.

DISCUSSION

The colony defense system of each honeybee species consists of
numerous interwoven lines of adaptation, including nest site; nest
architecture; colony population; labor allocation to defense; age
polyethism schedule; colony mobility; and worker morphology,
physiology, and behavior. Predation has been a pervasive and
powerful force in the evolution of these tropical bee societies.

Predation appears to have also extensively shaped the polybiine
wasp and stingless bee societies of tropical America. In these two
groups as well, a wide array of colony properties, including nest
site, nest architecture, colony population size, mode of colony

reproduction, colony dispersion, and worker defense behaviors can be seen as integrated, adaptive complexes for colony defense (Evans and West Eberhard 1970, Michener 1974, Jeanne 1975, 1978, Forsyth 1978). Thus it seems likely that for the bee and wasp societies throughout the tropics, predation has been a major agent of natural selection. The shared characteristic of bee and wasp societies of rearing large amounts of nutritious brood (and, for bees, of storing honey) in fixed nests evidently underlies the evolution by both groups of elaborate colony defenses in the predator-rich tropics.

References

Evans, H.E., West Eberhard, M.J., 1970. -- The Wasps. Univ. of Michigan Press, 265 p.

Forsyth, A., 1978. -- Studies on the behavioral ecology of polygynous social wasps. Ph.D. thesis: Harvard University, Cambridge, Massachusetts.

Jeanne, R.L., 1975. -- The adaptiveness of social wasp nest architecture. Q. Rev. Biol., 50, 267-287.

Jeanne, R.L., 1978. -- Intraspecific nesting associations in the neotropical social wasp Polybia rejecta (Hymenoptera: Vespidae). Biotropica, 10, 234-245.

Michener, C.D., 1974. -- The Social Behavior of the Bees. A Comparative Study. Harvard Univ. Press, 404 p.

Seeley, T.D., Seeley, R.H., Akratanakul, P., 1982. -- Colony defense strategies of the honeybees in Thailand. Ecol. Monog. In press.

Abstracts

Distribution of the bee louse on honey bees

Dewey M. Caron & I. Barton Smith, Jr.
University of Delaware & MD Dept. of Agriculture

The bee louse Braula coeca Nitzsch is a wingless Dipteran
commensilate in honey bee, Apis mellifera L. colonies. It has
been reported on every continent and in 13 of the United States.
In a survey in Maryland, 18% of colonies in 28% of apiaries examined
harbored lice, In infested apiaries, one-half of the bee colonies
examined had Braula.
On worker honey bees, only 1.4% contained more than one lice
while 62% of queens had lice. One queen individual hosted 29 lice.
Drones also had lice in proportion to their colony numbers. When
offered choices, lice adults preferred hive workers over older
worker bees and mated queens over virgin queens. Queens were preferred over workers or drones. Efforts to infest honey bee nucs
with lice were not entirely successful for unknown reasons.

Nest structure and soldier defense: an integrated
strategy in Termites

Jean Deligne and Jacques M. Pasteels
The University of Brussels (Belgium)

The nest structure and the soldier behavior are two main defense
mechanisms of Termites. They can vary in different species but appear to be precisely adapted to one another as shown in this study.
In several species of Microcerotermes, Cephalotermes,
Procubitermes, Cubitermes and Noditermes the nests are composed of
many little chambers connected by small openings. These openings
are few and of constant and minimal size: they allow only one termite
to pass at a time, fitting exactly the maximum diametrical dimensions
of the workers abdomen and of the soldiers head. This allows the
workers to wall up quickly a few small openings to isolate a damaged
or invaded part of the nest. Besides, although the soldiers are
quite scarce (<3%), they can suffice to block efficiently some
strategic openings with their sclerified head, armed with strong
mandibles.. It is a static warfare.

288

In several species of Nasutitermes and Trinervitermes the nests
are composed of labyrinthine irregular spaces not clearly divided
into chambers. The openings between spaces are numerous, they vary
greatly in size and always allow 2 or more termites to pass at a
time. The long-legged nasute soldiers are numerous (>10%). They
can quickly move to an invaded part of the nest and repell an enemy
with the efficient spray of their frontal fluid. It is a movement
warfare.

Why Don't Psithyrus ashtoni Females Fight With Their Bumblebee Host Queens?

Richard M. Fischer
University of Toronto

A study of 100 attempted invasions by females of the social
parasite Psithyrus ashtoni Cr in laboratory nests of its hosts
Bombus affinis Cr and B. terricola Kby has shown that host queen and
parasite never fight. Invading Psithyrus are however attacked by
host workers, and invasion success is limited to small colony sizes.
Once established inside host nests parasites allow growth to continue until optimum worker numbers have been reared. Psithyrus
females cannot dominate workers in the absence of the host queen.
Fewer parasite reproductives are reared if the queen dies or is
removed. Notwithstanding large variations in the effects of parasitization on colony reproductive success, in most cases Psithyrus
females allow host sexuals to be produced. These findings are discussed in the context of a 'compromise' reached between parasite and
host.

Aggression and predation in wood ants

Abraham Mabelis
Research Institute for Nature Management

The relationship between aggression and predation was studied in
wood ants (Formica polyctena), in the field as well as under
laboratory conditions. The aggression between wood ant nests
appeared to be maximum in early spring when many wood ant wars
broke out. The wars ended at the time when they prey density
increased strongly. Since the casualties were taken as food
to the warring nests, the hypothesis was formulated that warfare
between wood ant nests occurs only in periods when the demand for
prey is greater than the supply. Experimental evidence supporting
this hypothesis was provided by experiments performed in the
laboratory: populations given a prey-poor diet became more
aggressive after a few weeks than did sister populations which
received a prey-rich diet.
Protein-rich food in early spring is mainly for the benefit
of the queens and the sexual larvae. The main function of warfare
may be the acceleration of the wedding flight of the queens,
because this gives a better chance for the survival of their genes.

Tool Use by an Assassin Bug in Capturing Termites

Elizabeth A. McMahan
The University of North Carolina

A Neotropical reduviid, Salyavata variegata, captures Nasutitermes workers by enticing them out of nest openings with the spent carcasses of freshly killed termites. The carcass serves not only as bait but probably also as a buffer against worker mandibles and soldier toxins. Nymphal bugs cover themselves with carton crumbs, a camouflaging procedure that presumably not only foils visually active predators such as lizards and birds, but also conceals the bugs against olfactory and tactile detection by their blind termite prey. Opportunity for using the bait-and-capture technique occurs during spontaneous nest expansion by termites and following accidental breaching of the nest. In laboratory choice tests all instars of S. variegata appeared to prefer large workers over small workers, and the latter over soldiers.

Adoption of Broods of the Ant Leptothorax nylanderi
by Workers of Myrmica rubra

Luc Plateaux
Laboratoire d'Evolution, Paris

During 33 experiments, 2714 larvae of Leptothorax (and also some pupae, prepupae and eggs) were given to breed by 518 workers of Myrmica generally without queen, and sometimes with queens. These breedings produced 177 pupae which gave 131 adults of Leptothorax. Mortality of the broods happened mainly at the beginning of the experiments and at the start of the pupation. Of all the experiments, 4 were negative, 7 gave growing larvae, one ended in pupae without emergency and 21 produced adults. The results of these experiments were classed according to the presence or absence of Myrmica queens, the presence or absence of broods of Myrmica, the age of the breeders, the cyclic condition of the breeders, and the production of queens and workers (and intercastes). Some observations were made concerning the longevity of Leptothorax and the behaviour of both species.

Biosynthesis and Detoxication of
Termite-Produced Contact Poisons

Glenn D. Prestwich, Stephen G. Spanton,
Mark Schneider, and Charles Bleecher
Department of Chemistry, State University of New York
Stony Brook, NY 11794

Termite soldiers of Prorhinotermes simplex biosynthesize an unusual fifteen-carbon nitroalkene defense secretion in the frontal
gland. Injection of soldiers with radiolabeled precursors and isolation of the nitroalkene by HPLC suggest that the biosynthesis involves a modified sphingolipid pathway. Thus, myristate (C_{14}) con-
denses with serine to give an intermediate which suffers N-oxidation
dehydration, and decarboxylation to give the C_{15} nitroalkene. Pal-
mitate is poorly incorporated, ruling out an α-amination/decarboxy-
lation pathway.

P. simplex workers detoxify the nitroalkene via a rapid sub-
strate-specific reduction requiring NAD(P)H and an alkene reductase.
The nitroalkene is then catabolized to acetate for recycling. The
fate of the nitrogen of the defense secretion and the substrate specificity of the reductase have been examined. Workers of the higher
rhinotermitid Schedorhinotermes lamanianus also possess a detoxify-
ing alkene reductase, which in this case reduces vinyl ketones to
ethyl ketones. The details of this enzymic conversion have also
been examined.

Divergent strategies of colony usurpation among two obligate
vespine parasites (Hymenoptera:Vespinae)

Hal C. Reed and Roger D. Akre
Washington State University

Investigations into the invasion and colony behavior of two
obligate social parasites (=inquilines), Dolichovespula arctica
(Rohwer) and Vespula austriaca (Panzer), revealed differing usurpation methods and modes of maintaining colony cohesion. V.austriaca
primarily practices forceful invasion tactics, like some temporary
social parasites [e.g. Vespula squamosa (Drury)], while D. arctica
passively invades a host colony; an invasion likely mediated by
chemicals. D. arctica usually invades embryo colonies and coexists
with the host queen prior to the emergence of the host workers.
Thus, D. arctica does not immediately kill the host queen, but allows
her to live and oviposit even after worker emergence. However, no
coexistence period exists in V. austriaca parasitism, as this
inquiline invades a host colony after worker emergence and immediately kills the queen. Although both species exhibited similar

behaviors during occupation of their respective colonies, $\underline{V}$. austriaca aggressively dominated the host workers early during occupation. In contrast, $\underline{D}$. arctica usually did not physically dominate the hosts; however, aggression was sometimes exhibited in the later stages of occupation.

The Importance of Venoms in Eusocial Hymenoptera

Justin O. Schmidt and Stephen L. Buchmann
Carl Hayden Bee Research Center, Tucson, Arizona

The frequent threat of macropredators attacking the nest and its inhabitants is a great hazard encountered mainly by eusocial, but not by solitary Hymenoptera. This threat is almost unique to eusocial insects because they represent a rich food bonanza for predators. Venoms can be useful against predators in two ways: they can be toxic, or debilitating to the predator; or they can induce excruciating pain. We suggest that selection pressure has acted to make both venom properties important in eusocial Hymenoptera. Solitary aculeate Hymenoptera do not defend nests, do not participate in mass stinging, and do not represent a large food reward for a predator; thus mainly venom painfulness is important for them (with only one sting available, lethally injuring an enormously larger predator is remote).

To test these hypotheses, the lethalities to mice of the venoms of 2 solitary bee, 2 solitary wasp, and 30 eusocial hymenopterous species were determined. Of the solitary Aculeata, only Xylocopa virginica venom was lethal in quantities (22 mg/kg) under 50 mg/kg; whereas the venoms of only 2 of the 30 eusocial species required quantities greater than 50 mg/kg and all but 4 species required quantities of 15 mg/kg or less. Subjective analyses of painfulness of the stings is also discussed. Thus, both toxicity and painfulness of venoms appear important to eusocial Hymenoptera.

Raiding Behavior of the Slave-Making Ant, Polyergus breviceps

Howard Topoff and Brent LaMon
Hunter College of C.U.N.Y.

The ecology and orientation of Polyergus breviceps was studied in southeastern Arizona, where the slave species is Formica gnava. Raids began in late afternoon, following the peak in diurnal temperature. Most raids were preceded by a period of high colony arousal, in which workers circle rapidly near the nest entrance. On days when Polyergus penetrated large nests of F. gnava, up to 2,800 individuals (mostly pupae) were captured. On the day after each raid, a pile of empty cocoons appeared outside of the Polyergus nest. This suggests that most Formica pupae are eaten. Most raids are organized by a single scout. When we placed a glass jar over a marked scout during a raid, the column stopped advancing and the workers began circling. Tests show that scouts use optical orientation to locate <u>Formica</u> colonies, to return to their own nest, and to lead the raid column to the target. Workers in the raiding column deposit a chemical trail en route to the <u>Formica</u> nest. After the raid, <u>Polyergus</u> workers use both optical orientation and trail-following behavior to return to their own nest.

COMMUNICATION
Organized by Rolf Boch

Introduction

Rolf Boch, Ottawa Research Station

The literature on the subject of communication in social insects is very large; yet it is difficult to decide what is communication and what is not. One could consider all activity in a colony as communication or the product of communication. How is communication defined?

Wilson in Sociobiology gave one carefully worded definition as follows:

"Action on the part of one organism (or cell) that alters the probability pattern of behavior in another organism (or cell) in an adaptive fashion."

For the pragmatic purpose of analysing and understanding a communication process, we may consider 4 aspects separately:

1. The sending of a signal by one individual;
2. The observable response by another;
3. The probability of message transmission;
4. The adaptive significance.

Signals

The signals consist of visual, auditory or chemical stimuli, or any combination of modes. The signals can range from simple on-off releasers to highly complex and graded "messages". The honeybee dance "language" is an example of the latter.

During the evolution of man, visual and sound signals became the predominant modes of communication, and seeing and hearing our most significant senses. In social insects, on the other hand, odor, touch and sound, rather than vision, are the primary modes. Thus the human observer is handicapped when studying the signals of social insects. First, it is difficult to observe the insects without disturbing them in their dark nests. Second, sometimes we assume that they only use signals that we can readily perceive. Because our senses can scarcely detect them, we have until recently neglected the chemical signals (pheromones) which are perhaps the most powerful and versatile means of communication in the social insects. GLC, HPLC, MS, NMR and other microanalytical methods are currently in use to determine the compositions of the communicative secretions that social insects produce.

Response

Generally, each signal evokes a specific behavioral or physiological response from the receiver animal. But if there is no overt
reaction, we cannot determine in most cases whether or not the
signal has been picked up. We are just beginning to develop the
necessary technical expertise in order to track the reception of
signals, especially odors by neuro-physiological means.

Probability

Observations of the behavior of animals produce data that allow us
to correlate signals with different responses, or vice-versa. We
may construct computerized models to explain and predict possible
outcomes under changing conditions and in various contexts. This
may involve tests with artificial signals, e.g. synthetic
pheromones.

Fitness

For a certain communication process to be retained in the
repertoire of a species, it must have adaptive value. The benefit
is assessed in terms of increased inclusive fitness.

The Adaptive Value of Probabilistic Behavior During Food Recruitment in Ants: Experimental and Theoretical Approaches

Jacques M. Pasteels, Jean-Claude
Verhaeghe, and Jean-Louis Deneubourg,
The University of Brussels

Communicative efficiency can be evaluated in terms of the velocity at which information is transmitted (e.g. recruitment rate), or in terms of the accuracy of communication (e.g. dispersion around the target).

Although the recruitment rates are usually considered when comparing food recruitment methods in ants (e.g. Chadab and Rettenmeyer, 1975), accuracy has been nearly completely neglected. It is well known, however, that animal behavior is probabilistic, in particular, ant communication (Wilson, 1962), and that accordingly, only a fraction of recruited ants will actually reach the food source, "lost" ants exploring the environment.

We will present here experimental evidence showing that the amount of "noise" during food recruitment communication varies considerably between species, and that "lost" ants, exploring the environment, can discover new food sources and initiate new recruitment with possible advantages for the society.

It is almost a truism to consider that when only one food source is present, the most efficient recruitment is the most accurate. However, we will demonstrate with the help of a mathematical model, that a certain amount of noise is needed in order to minimize food collection time when a multiple source situation is encountered, and when the sources show some degree of aggregation.

1. Comparison of recruitment accuracy in Tetramorium impurum and Tapinoma erraticum.

These two species are sympatric. T. impurum performs group recruitment : the recruiter leads a small group of recruits. However, some ants are able to follow the trail alone. T. erraticum uses mass recruitment: the recruitment trail suffices to guide the recruits which travel independently of the recruiter.

METHODS

Both species were exposed to the same experimental conditions. Food recruitment was initiated by offering 1M sucrose solution at 10 cm from the nest entrance after 4 days of starvation. The behavior of recruiters and recruits were videorecorded. The accuracy of

297

recruitment was measured by the proportion of ants recruited by a
first recruit which actually reached the food sources, and also by
the mean length of single recruitment trail actually followed by
recruiters travelling alone.

RESULTS

Clearly, a recruitment trail is a far better orientation cue
for Tapinoma than for Tetramorium. In addition, mass recruitment in
Tapinoma is much more accurate, even when group leading is considered together with solitary trail following in Tetramorium (table 1,
Verhaeghe et al. 1980).

Table 1. - Comparison between recruitment accuracy in two ant
species.

	Tetramorium impurum		Tapinoma erraticum
length (%) of single recruitment trails actually followed	17 (40)		67.7 (47)
% of recruits reaching the food source	alone	8.9 (45)	73.6 (216)
	in group	60 (10)	-
	total	18.2 (55)	73.6 (216)

Numbers in parentheses are those of ants actually observed.

2.   The value of "lost" ants in a two source situation.
      This was experimentally investigated by Parro (1981) for
T. impurum.

METHODS

Two polyvinylchloride (5 X 20 X 30 cm) containers were connected by a bridge. The nests were placed in one container, the food
in the other. Recruitment towards sucrose solutions was initiated
after 8 days of starvation. Food sources were 11 cm distant from the
bridge and 14 cm from each other. They were deposited either simultaneously, or with a 1 hour delay between them. Traffic was videorecorded on the bridge. The spatial distribution of ants was determined by photographs taken every 5 min during the whole experiment
(up to 5 h). The ants were familiar with the experimental device,
but the substrate (paper sheet) was replaced after each experiment.

RESULTS

Recruitment appeared to be only slightly accurate. Only 20 to
40% of recruits participated in the collection of food. Most of the
ants lost the recruitment trail and explored the foraging area. Such
very high values for "lost" ants could be partly induced by the experimental conditions, in which this territorial species recruited on
an unmarked substrate.
      "Lost" ants quickly discovered the second source, even when

recruitment towards the first source was already well established. When the 2 sources were of equal quality, the first one discovered was exploited maximally until it was exhausted. However, the second source was colonized by the ants, and its exploitation intensified, when the first became exhausted, so that no interruption or discontinuity in food-collection rate occured.

When the second source was more concentrated than the first, the ants shifted their collecting efforts towards the most rewarding source, the latter, without however, completely abandoning the first source. The exploitation ratio of the 2 sources is function of the sugar concentration ratio of the sources. (more details in Parro 1981).

3. Optimal noise in a multiple sources situation.

Ants losing the trail are able to discover new food sources, but they do not exploit the already known food sources. What is the best balance between noise, allowing new discoveries, and accuracy, allowing immediate exploitation ? This was investigated with the help of a mathematical model in an still very idealistic situation.

MODEL

All food sources are identical, equidistant from the nest, and regularly distributed. Recruitment is initiated towards the median source, and it is assumed that recruited ants distribute themselves normally around the source. Ants hitting a source are able to initiate recruitment towards it.

Recruitment is described by a logistic equation generalized for several sources :

$$\frac{dX_i}{dt} = (a_{ii} X_i + \sum a_{ji} X_j) (N-X_1-X_2-\ldots X_k) - bX_i$$

and 'when a source is exhausted by

$$\frac{dX_i}{dt} = - bX_i$$

in which $X_i,\ldots X_k$ are the number of recruited ants towards the source specified by the index $1,\ldots k$, N the total number of possible recruits, and b the inverse of mean time of staying near the food source and coming back to the nest.

a_{ii} is the recruitment rate for the ants recruited to i and effectively reaching it, and a_{ji} is the recruitment rate for the ants recruited to j and erroneously going to the source i. They were calculated by :

$$a_{ji} = a \int_{r_{i-d}}^{e_{i+d}} \exp\left[-\theta^2 (r_j-r)^2\right] dr/S$$

in which S is a normalisation factor, 2d the dimension of the food sources, r_i-r_j the distance between sources i and j; a is the total recruitment rate of the society (divided by N), and θ is the inverse of the standard deviation of the Gaussian distribution : the greater θ, the more accurate the recruitment.

SIMULATIONS

During the present simulations, a, b, N and d were maintained constant. The performances of recruitments for various values of θ, total food quantity available (Q), quantity of food collected by the ants (q), and number of sources (k), were evaluated by the time needed for food collection.

These simulations demonstrated that :

1) There is one value of θ, θopt, which minimizes the time of food collection in a multisource situation.

2) For the same number of food sources, θopt increases when Q increases. This implies that species living in poor environment, where food is parcelled, should be more stochastic in their recruitment.

3) When q is close to Q, the curves relating θ and time of collection show a second minimum for a lesser value of θ than θopt and a higher value of time of collection, which suggests that species could be evolutionarily "trapped" at a suboptimal level of efficiency (fig. 1).

4) The relationship between θ and time of collection depends in a complex way on the proportion of food collected to the total available (q/Q). Roughly, two different situations could occur, which could correspond to different strategies for ants. When q represents a higher proportion of Q, species should be more stochastic in their recruitment. Besides ants collecting most (above 50%) of the discovered food should possess a rather fixed θ (depending on Q, see 2.), and be able to maintain possession of the source for long period. For ants which rely on the rapidity of exploitation, but with collected food representing less than 50% of total discovered, θ could be far less critical unless the species avoids being too deterministic. This suggests that in rich environment, when most of the time only a small fraction of discovered food can be collected, recruitment strategies could be more varied depending on other biological and ecological factors that those considered here (fig.1).

5) If species could be characterized by a certain value of θ, one species will be more competitive than another under the same conditions, but less than the same in others, according to the level of food parcelling.

More details about the model and simulations will be given elsewhere (Deneubourg, Verhaeghe and Pasteels, in preparation).

DISCUSSION

Several parameters, both internal and external to the society, can affect the efficiency of food recruitment. Experimental and theoretical evidence presented here indicate that one of them is the amount of noise introduced during communication. Some level of noise during communication can be advantageous for the society by increasing the probability of discoveries, allowing the society to focus its collecting activies on the most rewarding resources, and also promotes the colonization of resources which will be fully exploited later.

The level of noise can be optimally tuned to parameters like food quantity and parcelling. Thus, species cannot be simplistically ranked along a linear evolutionary scale towards an idealistic and deterministic system of communication.

References

Chadab R., Rettenmeyer C., 1975. - Mass recruitment by army ants. Science, 188, 1124-1125.

Parro M., 1981. - Valeur adaptative du comportement probabiliste lors du recrutement alimentaire chez Tetramorium impurum (Hym. Formicidae). Mémoire de licence, Université libre de Bruxelles, 103 p.

Verhaeghe J.-C., Champagne Ph., Pasteels J.M., 1980. - Le recrutement alimentaire chez Tapinoma erraticum (Hym. Form.). Biol.-Ecol. méditerranéenne, 7, 167-168.

Wilson E.O., 1962. - Chemical communication among workers of the fire ant Solenopsis saevissima (Fr. Smith). 2. An information analysis of the odour trail. Anim. Behav., 10, 148-158.

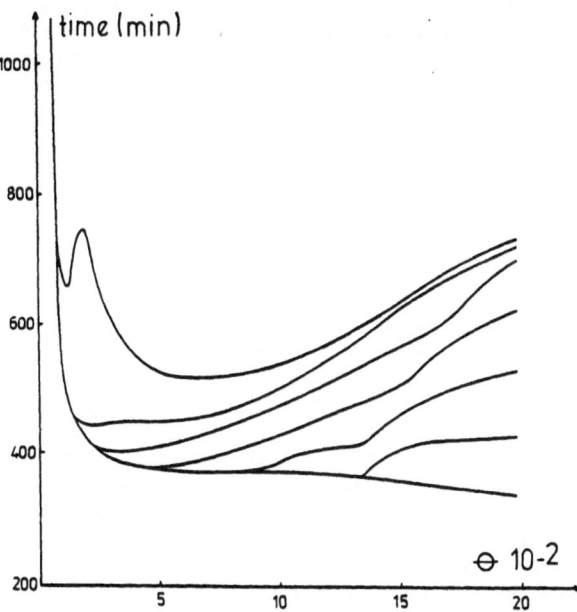

Fig. 1 - Exploitation time in function of θ when the amount of collected food (q) represents various fractions of the total available (Q). q/Q equals respectively 1, 0.9, 0.8, 0.6, 0.4, 0.2, 0.1 for curves 1 to 7. Q is maintained constant.
q = 1 ml (each ant collected 0.26 µl), N = 650, $\alpha = 10^{-3}$ min^{-1}, K = 9, r_1-r_9 = 100 cm, 2d = 1.4 cm.

Antennal Communication and Food Exchange in the Domestic Bee *Apis mellifica* L.

Hubert Montagner and Gérard Galliot,
Université de Franche-Comté Besançon

If entomologists often speak of antennal play in their studies of the communication systems of social Hymenoptera either among members of the same colony and the same species (Hölldobler and Wilson, 1978;Cammaerts-Tricot,1981;etc...)or with myrmecophiles (Hölldobler, 1971) the dynamics and the function of this behavior have not yet been extensively studied.It is most probably Wallis (1961) who,when describing the antennal play in *Formica fusca* and *Formica sanguinea*, was the first to suggest that this could be an important communication element among social insects.Free (1956) underlined the role of stimulation given by the antennae in establishing contact between worker bees.Montagner (1967) showed that ritualized antennal movements play an important role in food solicitation and reinforcement or refusal of regurgitations in *Paravespula germanica* and *Paravespula vulgaris*,and also in the domestic bee (Montagner and Pain,1971).

Over the last years we have tried to pinpoint more accurately the dynamics ,the patterning and the function of the antennal play of *Apis mellifica* L.(Galliot et al.,1979,1982).We will discuss some of the results that we have obtained.

MATERIALS AND METHODS

Our research was carried out on black bees from the Besançon region in France.In all the experiments the workers were marked immediately after imaginal hatching and followed for variable periods going up to 25 days either in the hive or in experimental cages (Galliot,1978).The quantitative study of the food exchanges was done in the cages which had 3 compartments : the central one which received 10 day old workers which fed on honey labelled with ^{198}Au.The side compartments received either 20 workers whose age was known or 20 males.All the bees were then put together once the radioactive food had been withdrawn and the perforated partitions between the cage compartments taken away.We measured the radioactivity of each bee 30 minutes after all the three groups of bees had been put together (for futher details see Galliot,1978 and Galliot et al.1979,1982). The quantitative study of the duration of the antennal contacts between bees was done in the cages under exactly the same conditions as above (the temperature was maintained at 32°C.).The duration of each contact was obtained by closing an electric circuit at the be-

ginning and the end.This gave a double electric signal on an actogr-
aph with an operating speed of 252 cms per hour.Each bee was observed for at least 10 minutes.Studies of the structure,the frequency
and the duration of each antennal movement were able to be made at
any moment with Ektachrome 16 mm 200 ASA film.Each film was then analysed frame by frame either through a hand projector or after development onto paper according to the method already used (Montagner,
1967;Montagner and Pain,1971).

RESULTS

1.Quantitative study of the frequency and duration of antennal movements during contact between workers

a.The worker in the solicitation position (fig.1.1.)

Figure 1.1. shows that the beginning of the contact is characterized by a relatively rapid succession of forward and backward movements of the both antennae (phases 1 to 5).However the movements
become slower and slower when the contact begins to come to an end
(from phase 13).Just before the end the active antenna (the right
one) applies a stronger pressure on the mouth parts of the bee that
is offering.This was shown by an accentuated curving of the antenna
(phases 20 to 23).This antennal patterning is probably related to
the decrease or the stopping of the regurgitations of the offering
bee.The dynamics of the antennal movements were similar in the 50
contacts that were studied.

b.The worker in the offering position

Figure 1.2. shows that the movements of antennae of the offering bee are more rapid than those of the soliciting bee (the duration is lower than that in figure 1.1.).Usually they do not come into
contact with the soliciting bee.It is only at the end of regurgita-
tion that one of the antennae and sometimes two have a more or less
regular succession of slow forward movements and fast return movements at the mouth parts of the soliciting bee after the ending of each
antennal contact (phases 10 and 24).The dynamics of the antennal movements were similar in the 50 contacts that were studied and which
were distinct from the above paragraph because our method did not allow us to quantify at the same time the duration of the whole movements of the soliciting bee and the offering bee.

2.Quantitative study of the food obtained by the soliciting worker of varied age

The study of the food transfer of workers of 10 days to workers
from 12-24 hours,4,10 and 25 days grouped into series of 20 show that
1.the percentage of radioactive food obtained by the soliciting bees
varied considerably from one experimental group to another (from 12.9
to 15.3 % with 12-24 hour old bees;from 16.8 to 27.7 % with 4 day old
bees;from 29.8 to 35.8 % with 10 day old bees;from 30.6 to 39 % with
25 day old bees);2.the percentage is even higher when the bee is older (the averages are respectively 14,22,33 and 35 % for workers of
12-24 hours,4,10 and 25 days).

3.Quantitative study of the duration of antennal contacts between workers and males

We have found that the proportion of brief contacts (duration
of less than 0.5 second) between workers is 38 % whereas that betwe-

304

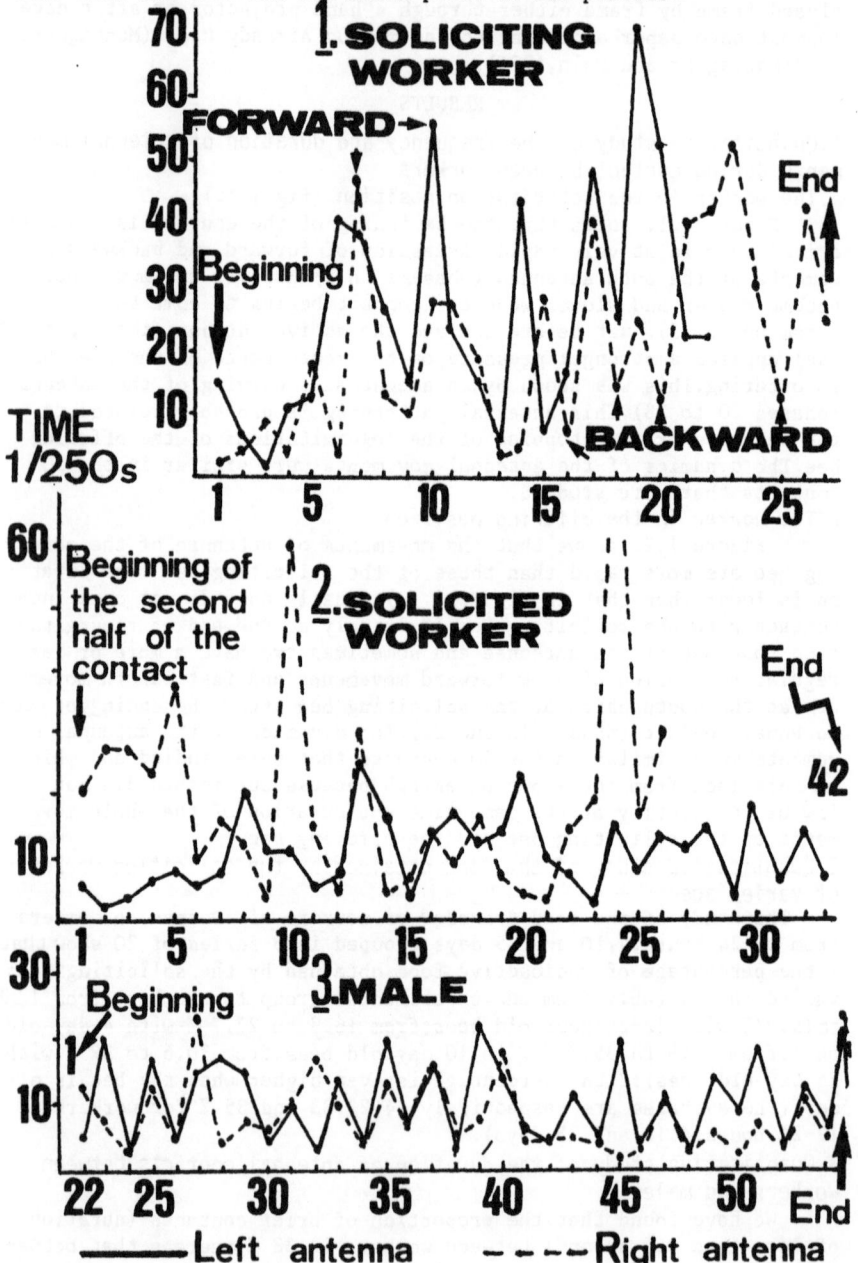

Figure 1 The duration of the successive forward and backward movements (or phases) of both antennae of 1.a soliciting worker,2.a solicited worker and 3.a soliciting male in a trophallaxis with a worker. It is during the forward phase that the antenna is coming into contact with the head and the mouth parts of the partner.

en workers and males is 73 %.The contacts from 0.5 to 5 seconds are more frequent between workers than between workers and males.However 12.5 % of the contacts between workers and males last more than 5 seconds whereas this type of contact is only 5.5. % between workers. Observation shows that the relatively high proportion of these long lasting contacts with the males is due to the male who frequently takes advantage of a situation and takes food when two workers are exchanging food.Contrary to the worker,the soliciting male uses almost all the lenght of his antenna during trophallaxis.The movements of his antennae are more sustained and more rapid than that of the workers.The speed of the movements and the antennal pressure on the offering bee practically do not vary as the trophallactic contact proceeds (fig.1.3. :the evolution of the 50 contacts that were studied was similar).The males do not modulate the movements of their antennae in relation to the behaviour of the offering worker.

4.Quantitative study of the food transfer between workers and males

We found that the males which had complete antennae received 7% of the total quantity of the food that was taken during one hour by the workers whereas the workers put in the same conditions received from 16.8% to 39% at least when they were older than 2 days (Galliot et al.,1982).In our experimental conditions the soliciting males thus received much less food than in the experiments of Roger,Pain and Douault (1975) where each male isolated with a worker received on average 47% of the content of the crop of the worker.Thus,if a male can obtain from an isolated worker as much food as another worker it is not the same for a male within a group where the solicitations of the male are less frequent than those of the worker and where the refusals to regurgitate by the workers are more frequent when they are solicited by a male (Montagner and Pain,1971).

DISCUSSION

All the studies show that the ritualized antennal movements of the worker in the soliciting position play an essential role in the establishing of the contact with another worker and the releasing and the reinforcing of regurgitations by the solicited bee.In a recent work Korst and Velthuis (1982) have confirmed the importance of antennal solicitations in establishing contact between two workers.The structure of the soliciting antennal movements does not seem to vary very much in Hymenoptera that were studied.However,they woul appear to be differences from one group of Hymenoptera to another in the dynamics of the interactions between workers and in particular at the beginning of the breaking off of contact:1.in the domestic bee the movements of the antennae of the soliciting worker and her partner appear to be significantly modified at the same moment and just before the end of contact : there is an increase in the duration and the pressure of the active antenna of the soliciting worker;at the same time there is longer forward and fast return movements of the antennae of the offering worker in contact with her partner. The temporal relationship between these patterns is such (the regurgitations of the offering bee have then stopped or are very weak) that it is probable that the antennal forward movements of the offering worker form a signal for the breaking off of contact.The same type of events has also been observed in *Paravespula*;2.These pre-breaking off rituals do not appear to exist in *Camponotus vagus* (Bonavita-Cougourdan,1981) and *Myrmica* (Lenoir,1979).It can,thus,be thought,as Bonavita-Cougourdan (1981) suggests,that the absence of modulations in the antennal res- ponses between workers during contacts has perhaps facilitated the co -lonisation of societies of many species of ants by a relatively lar- ge number of myrmecophile species.However the existence of rituali- zed antennal patterns in the domestic bee and in *Paravespula* Wasps has perhaps enabled them to "recognize"more easely the presence of foreign species and to reject these species.This is also what the non-ritualized soliciting behaviour of the males of the bee and of the wasp and the responses of workers to this type of behaviour would suggest.Finally,it is possible that the ritualized antennal movements of the worker bee could have even more specific and diversified com- munication functions : 1.the frequency and the conspicuousness of the offering behaviour decreases as for the 4th-5th day which is the op- timum for food exchanges;2.the efficiency of the soliciting behaviour increases with age;3.within the same age group the workers differ gr- eatly one from the other.It can,thus,be thought that the trophallac- tic contact has many other functions than just food transfer.This hy- pothesis is even more plausable in that if the exchange of food can last several minutes,the majority of contacts are not accompanied by food transfer.Indeed we found that in groups of 20 workers 49% of the contacts lasted less than 1 second (this appear to be incompatible with food transfer)and only 5.5% of the contacts lasted more than 5 seconds.

REFERENCES

Bonavita-Cougourdan A.,1981 - Analyse des communications antennaires chez la fourmi *Camponotus vagus* Scop..Meeting of the French Section of I.U.S.S.I.,Toulouse 1981,in press.

Cammaerts-Tricot M.C.,1981- Systèmes d'approvisionnement chez *Myrmica scabrinodis.Ins.Soc.,28*,n°4,in press.

Free J.B.,1956 - A study of the stimuli which release the food beggi- ng and offering responses of worker honeybees.*British Journal of Animal Behaviour,5,*94-101.

Galliot G.,1978 - Etude qualitative et quantitative des transferts de nourrirure chez *Apis mellifica L.*.Thesis , Univ.of Besançon.

Galliot G.,Azoeuf P., 1979 -Etude quantitative des transferts de nour -riture entre ouvrières d'âge connu chez l'abeille domestique *Apis mellifica L.,Ins.Soc.,26,*39-49.

Galliot G.,Montagner H.,Azoeuf P.,1982-Etude quantitative des trans- ferts de nourriture entre ouvrières et mâles chez l'abeille domestique *Apis mellifica L.,Ins.Soc.,29,*in press.

Hölldobler B.,1971-Communication between ants and their guests.*Scient. Amer.,224,*86-93.

Hölldobler B.,Wilson E.O.,1978-The multiple recruitment systems of the African weaver ant *Oecophylla longinoda* (Latreille). *Behav.Ecol.Sociobiol.,3,*19-60.

Korst P.J.A.M.,Velthuis H.H.W.,1982-The nature of trophallaxis in honeybees,*Ins.Soc.,29,*in press.

Lenoir A.,1979-Le comportement alimentaire et la division du travail chez la fourmi *Lasius niger,Bull.Biol.Fr.Belg.,123,*79-314.

For Montagner H.;Montagner H,Pain J.;Roger B.et al.:see Galliot et al.

Behavior Genetics of
Honey Bee Alarm Communication

Anita M. Collins,
USDA-ARS, Baton Rouge, Louisiana

Communication of alarm by <u>Apis</u> <u>mellifera</u> involves the release of alarm pheromones associated with the sting and accompanying structures. The Committee on the African Honey Bee (Michener, 1972) suggested that the increased aggressiveness of the Africanized bee might be due to an enhanced responsiveness to pheromones, or the release of greater quantities of pheromone.

In order to investigate these possibilities, a measurement procedure for responsiveness to pheromones was developed and used to evaluate this character in several populations. The patterns of inheritance for the measured aspects of alarm response were also examined, as genetic manipulation has been suggested as a feasible way to cope with the Africanized bee.

MATERIALS AND METHODS[*]

A bioassay procedure to evaluate the response of caged honey bee workers was developed for use under controlled laboratory conditions (Collins & Rothenbuhler, 1978). Frames of emerging worker brood from individual queens were caged separately and held in a 35°C incubator for 24 h. The newly emerged workers were transferred in single-source groups of 30 bees to glass-fronted wooden cages (Kulinčević & Rothenbuhler, 1973) and arranged on shelves in a 35°C walk-in incubator for testing. After at least 24 h for acclimation and aging, testing was initiated using a double-blind identification system. Each group/cage of bees was tested several times per day for three days using a single pheromone, or a mix of pheromones, diluted in paraffin oil 1:9 v/v. At least 1 h elapsed between successive test sessions in the incubator.

The test was initiated by removing the paper liner from the glass front and counting the no. of bees on the front, top, sides, and floor of the cage (initial activity level - IAL). The pheromone was then presented to the bees by placing .03 ml on a slice of cork and holding it under the wire cage floor for 1 m. The time at which the bees began to flicker their wings and increase their locomotion was noted (speed of response - SR), and the intensity of that response (IR) was quantified. Occasionally the bees did not respond at all (none), otherwise a response was barely discernible (weak),

[*]
In cooperation with Louisiana Agricultural Experiment Station.

clearly given by a few of the bees (medium), simultaneously given by
most of the bees (strong), or explosive in appearance (very strong).
The time of cessation of the response was noted and the duration (DR)
calculated. For analysis, SR and DR were adjusted for IAL using a
least-squares analysis of covariance. IR was analyzed by chi-square.

RESULTS

1. Variation in inbred European lines

In 1973 four inbred lines of bees available at The Ohio State
University were chosen for bioassay on the basis of their behavior
in the field (Collins & Rothenbuhler, 1978). Two, Van Scoy and
Susceptible, were relatively non-defensive; Brown-Caucasian (Br-Cau)
and Resistant were defensive. Isopentyl acetate (IPA), (the only
sting alarm pheromone identified at that time, Boch et al., 1962) was
used as a stimulus.

Br-Cau and Resistant were not significantly different in SR or
DR, but the Br-Cau had more intense responses (Table 1). The
Susceptible bees were slower to react, but had similar DR. The Van
Scoy bees were slowest, reacted for shorter periods of time, and had
the greatest percentage of none responses.

Table 1. -- Response to IPA by bees from four inbred lines. Means
(± std. dev.) followed by different letters are significantly
different.

	Line			
	Van Scoy	Susceptible	Resistant	Br-Cau
SR(s)	30.8 ± 23.5a	22.0 ± 21.3b	15.9 ± 19.9c	16.0 ± 20.6c
	F = 6.05	d.f. = 3, 357		P < 0.01
DR(s)	21.1 ± 24.1g	43.2 ± 21.7h	43.3 ± 24.3h	37.0 ± 20.1h
	F = 5.57	d.f. = 3, 357		P < 0.01
IR(% obs.)				
none	25	18	12	12
weak	15	26	25	9
medium	25	35	27	16
strong	32	21	35	25
very strong	3	0	1	38
	χ^2 = 295.67	d.f. = 16		P < 0.01

2. Mode of inheritance in inbred European lines

The Br-Cau line and another nondefensive line, YD, were used as
parents for F_1 and backcross matings following the scheme of
Rothenbuhler (1960) (Collins, 1979). Worker offspring of inbred
queens inseminated by a single drone were assayed using the cage
test procedure (Table 2). DR was not measured. A genetic analysis
of the data indicated that a more responsive phenotype was partially

dominant to a less responsive phenotype.

Table 2. -- Response to IPA by workers from YD, Br-Cau, F_1, and back-cross matings. All means significantly different but those *.

	YD	Br-Cau	F_1	Backcross to: YD	Br-Cau
SR(s)	12.6	4.8*	6.3	8.7	5.2*

IR(% obs.)				In comparison to the parental patterns, the colonies tested were:	
none	8	2	1		
weak	34	3	6	7 ≃ Br-Cau	10 ≃ Br-Cau
medium	45	18	25	12 inter-	5 inter-
strong	11	50	49	mediate	mediate
very strong	2	27	19	7 ≃ YD	

Using the segregation pattern in the backcrosses, it was estimated that 2 or 3 genes controlled each of the behavioral components.

3. Variation in free-mated European stocks.

Compounds derived from the honey bee sting were identified by gas chromatography and mass spectrometry (Blum et al., 1978, in prep). The stimulation of caged worker bees was used to bioassay synthetic forms of these compounds and many did show alarm pheromone activity (Collins & Blum, in press, in prep). Among the several colonies used as sources for worker bees, differences in response were observed. Some colonies were significantly slower to respond to a particular pheromone. However, response to one compound did not always indicate the level of response to another. Table 3 shows an example of 3 colonies, with #9 slowest to react to isopentyl acetate and benzyl acetate, but fastest to respond to isopentyl alcohol.

Table 3. -- Significant colony differences in speed of response (SR) (s) to alarm pheromone by caged worker bees.

Pheromone	Colony number 7	8	9
isopentyl acetate	5.08	4.20	7.54*
isopentyl alcohol	5.63	7.58*	4.93
benzyl acetate	5.65	5.56	11.17*

* Significantly different from means in column and row, P < 0.05.

4. Population differences between European and Africanized bees.

Ten field colonies each of European (multiple-drone insemina-tions) and Africanized (free-mated) stock from a single apiary in Venezuela were used as sources for worker bees. Bees were emerged, caged, and tested as described. The pheromone used was a mixture of butyl acetate (6.74%), isopentyl acetate (13.47%), isopentyl alcohol (6.74%), 2-heptanol (0.01%), hexyl acetate (6.74%), 2-heptyl acetate (0.05%), 2-nonanol (42.66%), benzyl acetate (0.01%), octanol (13.47%), and 2-nonyl acetate (10.11%) diluted 1:9 v/v in paraffin oil. The two populations were similar in how quickly they responded, but the

Africanized workers reacted more strongly and for longer periods of time (Table 4).

Table 4. -- Response to an alarm pheromone mixture by worker bees of European or Africanized genotype (means ± std. error).

	European	Africanized		
SR(s)	3.4 ± .3	4.1 ± .3		
	F = 2.07	d.f. = 1, 17	P = 0.17	
DR(s)	73.5 ± 3.3	100.8 ± 3.6		
	F = 32.03	d.f. = 1, 17	P < 0.01	
IR(% obs.)				
none	2	0		
weak	11	6		
medium	55	37		
strong	30	48		
very strong	2	9		
	χ^2 = 34.65	d.f. = 4	P < 0.01	

5. Estimates of heritability

Drones from each of six European colonies (Louisiana, USA) and six Africanized colonies (Monagas, Venezuela) were used to singly inseminate three queens each from three inbred lines. Workers from each of the matings were tested for response to IPA. Data for IAL and SR were analyzed by a mixed model least-squares analysis of variance and the variance components calculated were used to estimate heritability, a genetic parameter indicating the ease of modification of a character by selection (Collins et al., in prep). Duration was not measured due to time limitations, and the intensity observations were in inappropriate form for such analysis. Estimates were made for the European population alone, the Africanized alone, and the combined population. The values in Table 5 indicate that SR could be readily altered by selection based on the lab test results, but not IAL. However, the high correlations between the two must be considered. A selection program reducing the response to alarm pheromones, which occurs early in the defensive behavior sequence (Collins et al., 1980) could be used to reduce the defensiveness shown by Africanized bees. However, if a bee results that shows reduced activity levels as well, other economically important traits such as honey production could be adversely affected.

Table 5. -- Estimates of heritability and phenotypic and genetic correlations for two behaviors of caged worker bees.

Population	Heritability		Phenotypic correlation
	IAL	SR	
European	.05	1.28	-.57
Africanized	.12	.31	-.26
Combined	.04	.83	-.49

DISCUSSION

If genetic selection is to be a viable way to combat the defensiveness of the Africanized bee, there must be existing variation in the behavior and we must have an adequate system to measure that behavior. The assay procedure described does in fact show that considerable variation in aspects of alarm communication exists in populations of the honey bee.

The genetic analyses indicate that the parameters measured are probably inherited in a quantitative manner, i.e. several genes are involved in the control of each aspect, and that at least some of them are sufficiently heritable that a selection program might significantly alter the behavior.

REFERENCES

BLUM, M. S., FALES, H. M., TUCKER, K. W., COLLINS, A. M., 1978. - Chemistry of the sting apparatus of the worker honeybee. J. Apic. Res., 17(4), 218-221.

BLUM, M. S., FALES, H. M., TUCKER, K. W., COLLINS, A. M., (in prep). - More compounds identified from the honeybee sting apparatus.

BOCH, R., SHEARER, D. A., STONE, B. C., 1962. - Identification of iso-amyl acetate as an active component in the sting pheromone of the honey bee. Nature, 195, 1018-1020.

COLLINS, A. M., 1979. - Genetics of the response of the honeybee to an alarm chemical, isopentyl acetate. J. Apic. Res., 18(4), 285-291.

COLLINS, A. M., BLUM, M. S., (in press). - Bioassay of compounds derived from the honey bee sting. J. Chem. Ecol.

COLLINS, A. M., BLUM, M. S., (in prep). - Bioassay of additional compounds derived from the honey bee sting.

COLLINS, A. M., RINDERER, T. E., HARBO, J. R., BROWN, M. A., (in prep). - Estimates of heritabilities and correlations for several characters of the honey bee, Apis mellifera.

COLLINS, A. M., RINDERER, T. E., TUCKER, K. W., SYLVESTER, H. A., LACKETT, J. J., 1980. - A model of honeybee defensive behaviour. J. Apic. Res., 19(4), 224-231.

COLLINS, A. M., ROTHENBUHLER, W. C., 1978. - Laboratory test of the response to an alarm chemical, isopentyl acetate, by Apis mellifera. Ann. Entomol. Soc. Amer., 71(6), 906-909.

KULINČEVIĆ, J. M., ROTHENBUHLER, W. C., 1973. - Laboratory and field measurements of hoarding behaviour in the honeybee. J. Apic. Res., 12(3), 179-182.

MICHENER, C. D. (chairman), 1972. - Report of the committee on the African honey bee. Nat'l Tech. Inf. Ser., Springfield, VA.

ROTHENBUHLER, W. C., 1960. - A technique for studying genetics of colony behavior in honey bees. Amer. Bee J., 100(5), 176, 198.

Chemical Communication in Ants: New Exocrine Glands and Their Behavioral Function

Bert Hölldobler, Harvard University

Chemical signals play a central role in the complex communication sytem of ant societies. During the last 20 years a number of exocrine glands were identified as anatomical sources for a diversity of pheromones that mediate sexual and social behavior in ants (for reviews see Wilson 1971, Blum 1977, Hölldobler 1978).

In recent years,however, several previously unknown exocrine glandular structures have been discovered in ants. Studies of their functions have led to a partial revision of our understanding of the mechanisms and evolution of chemical communication in ants.

RESULTS AND DISCUSSION

1. Tergal glands

In a detailed anatomical study of Myrmica rubra Janet (1898) described a pair of clusters of a few glandular cells, located under the 6th abdominal tergite. Each cell sends a duct through the intersegmental membrane between the 6th and 7th abdominal tergites. This glandular structure was then ignored for almost 80 years. We rediscovered it in Novomessor cockerelli and N. albisetosus(Hölldobler et al. 1976), and subsequently Kugler (1978), and Hölldobler and Engel (1978) found it to be present in many genera of the subfamily Myrmicinae. Since the gland opens at the 7th tergite (pygidium) it is now called pygidial gland (Kugler 1978). Only in four myrmicine species has the function of the pygidial gland been analyzed. In Novomessor cockerelli and N. albisetosus the strongly smelling secretion of the pygidial gland releases a "panic alarm" response in workers, apparently specifically designed against army ant predation. Kugler (1979) demonstrated that in Pheidole biconstricta the pygidial glands produce an alarm-defense secretion, and recently we found that the dacetine ant Orectognathus versicolor lays alarm recruitment trails with the secretion of the pygidial gland (Hölldobler 1982). The recent anatomical investigations revealed that the pygidial gland is much more common in ants than previously assumed. It was found in the Myrmeciinae, Ponerinae, Dorylinae, Pseudomyrmecinae, Myrmicinae, Nothomyrmeciinae, Aneuretinae and Dolichoderinae (Hölldobler and Engel 1978).

We agree with Kugler (1978) that it is very suggestive that the "anal glands" of the Dolichoderinae and Aneuretinae are homologous to the pygidial glands of other ant subfamilies. In fact the term "anal gland" is misleading, because the glands do not exist from the anal opening of the gaster, as is sometimes inferred, but between the 6th and 7th abdominal terga. Thus, the "anal gland" of the Dolichoderinae and Aneuretinae should be called pygidial glands. The function of the pygidial gland secretions in the Dolichoderinae and Aneuretinae is very similar to that found in the Myrmicinae. In dolichoderine species they were found to repel intruders and alarm nestmates (for review see Blum and Hermann 1978); and in Aneuretus simoni the sole living representative of the Aneuretinae, they release aggressive alarm in nestmates, but have no repugnatorial effect (Traniello and Jayasuriya 1982).

Generally the Aneuretinae are considered phylogenetically very close to the Dolichoderinae: in fact it is assumed that they are ancestral to the Dolichoderinae. The recent morphological and functional study of the sternal gland and pygidial glands in Aneuretus simoni (Traniello and Jayasuriya 1982) lend further support to this hypothesis. On the other hand, Brown, (discussed in Kugler 1978) hypothesizes that the Aneuretinae "might just be closer to the Myrmicinae than has been thought". Again, the anatomical and functional analysis of the pygidial glands in both subfamilies do not contradict this speculation. Thus the three subfamilies Aneuretinae, Dolichoderinae and Myrmicinae might indeed be phylogenetically much closer than previously thought. In fact, based on some striking similarities in the chemistry of the secretions of the mandibular glands in several myrmicine and dolichoderine species, Blum and Hermann (1978) concluded: "from an exocrinological standpoint, the Dolichoderinae currently have far more in common with the Myrmicinae than any other formicid subfamily." Since Taylor (1978) considers the Nothomyrmeciinae ancestral to the Aneuretinae, it is noteworthy that the pygidial gland secretions of Nothomyrmecia macrops elicits also an aggressive alarm response in nestmates and in addition have a slight repellent effect on some other ant species, which occur sympatrically with Nothomyrmecia (Hölldobler and Taylor in prep.). Thus in the Nothomyrmeciinae, Aneuretinae, Dolichoderinae and Myrmicinae the secretions of the pygidial gland appears to function primarily as alarm pheromones and/or as defensive substances.

The pygidial gland is also very common in the Ponerinae where its structural features vary greatly in different species (Hölldobler and Engel 1978; Jessen et al. 1979). In those species where its behavioral function has already been investigated, its secretions invariably elicit recruitment response or sex attraction. In several ponerine species that recruit by tandem running technique we recently found that the tandem running pheromone derives from the pygidial gland (Hölldobler and Traniello 1980, Hölldobler unpublished data). Furthermore Maschwitz and Schönegge (1977) demonstrated that the pygidial gland secretions of the termite raiding Leptogenys chinensis serve as a recruitment pheromone in conjuction with poison gland substance. We obtained similar results with several Australian Leptogenys species (Hölldobler unpublished data) and with the primitive ponerine species Cerapachys (turneri group) and Sphincto-

myrmex steinheili, both of which conduct group raids on ants
(Hölldobler 1982). Similarily in the termite raiding ponerine ant
Pachycondyla laevigata the recruitment trail pheromone originates
from the pygidial gland and not, as previously assumed, from the
hindgut (Hölldobler and Traniello 1980). A quite different function
of the pygidial gland has been discovered in the ponerine ant
Rhytidoponera metallica. Here the wingless virgin females attract
males by the release of a pheromone from the pygidial gland (Hölldobler and Haskins 1977). Since Rhytidoponera workers also have well-
developed pygidial glands and are attracted to its secretions, it is
possible that they might also function in Rhytidoponera as a recruitment signal.

Very little has yet been learned about the function of the tergal
glands in the Dorylinae, although some pilot experiments suggest that
the glandular secretions might also serve as a short lasting recruitment pheromone (Hölldobler and Engel 1978). We know nothing about the
behavioral function of the tergal glands in the Myrmeciinae and
Pseudomyrmecinae.

Since we found the pygidial gland widespread in the subfamilies
Myrmeciinae and Ponerinae and since also the most primitive ant,
Nothomyrmecia macrops has a well developed pygidial gland, we hypothesized that this gland might be a primitive monophylogenetic trait
in ants generally. The only subfamily where this gland was not found
is the Formicinae.

2. Sternal glands
A variety of intersegmental sternal glands have recently been
described in ants (Hölldobler and Engel 1978; Jessen et al. 1979),
but only in a few species have the behavioral functions of the
glandular secretions been analyzed. Like the Dolichoderinae, also
the Aneuretinae have a well developed sternal gland located beneath
the 7th sternite (Miradoli et al. 1957), and as in the dolichoderine
ants, this gland produces a mass recruitment pheromone in Aneuretus
simoni (Traniello and Jayasuriya 1982). Traniello and Jayasuriya
hypothesize that this sternal gland probably arose de novo in
Aneuretinae and that this structure was present initially in the
ancestral stock that gave rise to the dolichoderines. Also many
myrmicine species possess paired clusters of glandular cells beneath
the 7th sternite (Hölldobler and Engel 1978), however their very
different morphology does not suggest that they are homologous with
the sternal glands of the Dolichoderinae and Aneuretinae. Nothing is
yet known about the function of these myrmicine sternal glands.

The greatest variety of sternal glands have been found in the
Ponerinae (Hölldobler and Engel 1978; Jessen et al. 1979), and only
in a few cases have their behavioral function been studied. In Paltothyreus tarsatus we found well developed sternal glands between the
7th and 6th, 6th and 5th, and 5th and 4th sternites, but no reservoir.
Instead, the duct openings are associated with filament-like pro-
trusions of the intersegmental membrane. We were able to demonstrate
that foragers of Paltothyreus lay orientation and recruitment trails
with these sternal gland secretions (Hölldobler in prep.).
Onychomyrmex, a genus of the primitive ponerine tribe Amblyoponini,
possesses a large single sternal gland between the 5th and 6th

abdominal sternites. Secretions of this gland serve as a powerful trail and recruitment pheromone during predatory raids and colony emigrations (Hölldobler et al. 1982). We also investigated several other species of the tribe Amblyoponini (<u>Amblyopone australis</u>, <u>A.</u> <u>longidens</u>, <u>A. reclinata</u>, <u>A. pallipes</u>, <u>Mystrium camillae</u>, <u>Myopopone</u> <u>castanea</u>); in none of these did we detect the sternal gland. Thus it appears that this major pheromone producing sternal gland is unique to the genus <u>Onychomyrmex</u>.

The sternal glands recently found in some species of the subfamiliy Formicinae are very different from the intersegmental glands described above. In <u>Oecophylla</u> <u>longinoda</u> and <u>O.</u> <u>smaragdina</u> we discovered a sternal gland beneath the 7th sternite (Hölldobler and Wilson 1976, 1978; Hölldobler unpublished data). This structure consists of an array of single glandular cells that send short channels into cuticular cups on the outer surface of the sternite. The secretions of this gland function as a short range recruitment signal. In none of the other formicine species investigated did we find this type of sternal gland. But in several <u>Camponotus</u> species we detected a different kind of sternal gland that I called the "cloacal gland" (Hölldobler 1982). This gland consists of a paired cluster of glandular cells located at the base of the 7th abdominal sternite. Each cluster is associated with a major duct, comprised an invagination of the ventral membraneous wall of the cloacal chamber. The channels of the glandular cells of each cluster open in dense bundles into these two major ducts. Experimental evidence obtained with <u>Camponotus ephippium</u> suggests that the secretions of this cloacal gland serves as a group recruitment pheromone.

3.   Other abdominal glands

Several other glandular structures, associated with the sting apparatus, have recently been described (Hölldobler and Engel 1978, Jessen et al. 1979, Bazire-Benazet and Zylberberg 1979), but only in one case has the function of one of these glands been examined. In several species of the myrmicine genus <u>Atta</u>, a single cluster of glandular cells, the secretory channels of which open close to the sting base, appears to secret a pheromone which apparently is used by the ants as a home range marker. Whether or not it functions also as a true colony specific territorial pheromone has not yet been demonstrated.

Acknowledgement: Research conducted in the author's laboratory was supported by NSF grants BNS 77-03884 and BNS80-02613.

References

Bazire-Benazet, M., Zylberberg, L., 1979.-- An integumentary gland secreting a territorial marking pheromone in <u>Atta</u> sp.:
        detailed structure and histochemistry. J. Insect Physiol. <u>25</u>,
751-765.

Blum, M.S., 1977.-- Behavioral responses of Hymenoptera to pheromones and allomones. pp. 149-169 in Shorey, H.H., McKelvey, J.J.(ed): Chemical control of insect Behavior. John Wiley & Sons, New York.

Blum, M.S., Hermann, H.R. 1978.-- Venoms and venom apparatuses of the Formicidae: Dolichoderinae and Aneuretinae. In: Handbook of Experimental Pharmacology pp. 801-869, Springer-Verlag, Heidelberg-New York.

Hölldobler, B., 1978.-- Ethological aspects of chemical communication in ants. Advances in the Study of Behavior 8, 75-115.

Hölldobler, B., 1982a.-- Trail communication in the dacetine ant Orectognathus versicolor. Psyche, in press.

Hölldobler, B., 1982b.-- Communication, raiding behavior and prey storage in Cerapachys (Hymenoptera, Formicidae). Psyche in press.

Hölldobler, B., 1982c.-- The cloacal gland, a new pheromone gland in ants. Naturwissenschaften in press.

Hölldobler, B., Engel, H., 1978. -- Tergal and sternal glands in ants. Psyche 85, 285-330.

Hölldobler, B., Engel, H., Taylor, R.W., 1982.-- A new sternal gland in ants and its function in chemical communication. Naturwissenschaften in press.

Hölldobler, B., Haskins, C.P., 1977.-- Sexual calling behavior in primitive ants. Science 195, 793-794.

Hölldobler, B., Stanton, R., Engel, H., 1976.-- A new exocrine gland in Novomessor (Hymenoptera; Formicidae) and its possible significance as a taxonomic character. Psyche 83, 32-41.

Hölldobler, B., Traniello, J.F.A. 1980 -- The pygidial gland and chemical recruitment communication in Pachycondyla (=Termitopone) laevigata. J. Chem.Ecol. 6, 883-893.

Hölldobler, B., Traniello, J.F.A., 1980.-- Tandem running pheromone in ponerine ants. Naturwissenschaften 67, 360.

Hölldobler, B., Wilson, E.O., 1977. -- Weaver ants: Social establishment and maintenance of territory. Science 195, 900-902.

Hölldobler, B., Wilson, E.O., 1978. -- The multiple recruitment systems in the African weaver ant Oecophylla longinoda (Latreille) (Hymenoptera: Formicidae). Behav.Ecol.Sociobiol. 3, 19-60.

Janet, Ch., 1898. -- Etudes sur les Fourmis, les Guépes et les Abeilles. Note 17: Système glandulaire tégumentaires de la Myrmica rubra. Observations diverses sur les Fourmis. Paris, Georges Carré et C. Naud, Editeurs.

Jessen, K., Maschwitz, U., Hahn, M., 1979.-- Neue Abdominaldrusen bei Ameisen. I. Ponerini (Formicidae: Ponerinae). Zoomorphologie 94, 49-66.

Kugler, Ch., 1978. -- Pygidial glands in the myrmicine ants (Hymenoptera, Formicidae). Insectes sociaux, 25, 267-274.

Kugler, Ch., 1979. -- Alarm and defense; a function for the pygidial gland in the myrmicine ant, Pheidole biconstricta. Ann. Entomol. Soc. America 72, 532-536.

Maschwitz, U., Schonegge, P., 1977. -- Recruitment gland of Leptogenys chinensis. Naturwissenschaften, 64, 589-590.

Miradoli Zatti, M.A., Pavan, M., 1957. -- Studi sui Formicidae III. Nuovi reperti dell'organo ventrale nei Dolichoderinae. Boll.Soc.Entomol. Ital. 87, 82-87.

Taylor, R.W., 1978. -- Nothomyrmecia macrops: A living fossil ant rediscovered. Science 201, 979-985.

Traniello, J.F.A., Jayasuriya, A.K., 1982. -- Chemical communication in the primitive ant Aneuretus simoni: The role of the sternal and pygidial gland. J.Chem.Ecol. in press.

Wilson, E.O., 1971. -- The insect societies. The Belknap Press of Harvard University Press, Cambridge, Mass.

Compositional Variability: The Key to the Social Signals Produced by Honeybee Mandibular Glands

Robin M. Crewe,
The University of the Witwatersrand

Following the discovery by Ruttner et al. (1976) that worker
Apis mellifera capensis were capable of producing (E)-9-oxo-2-decen-
oic acid (9ODA), and the subsequent generalization of this discovery
to workers of other honeybee 'races' (Crewe, 1977 & Crewe & Velthuis,
1980), the idea that (E)-10-hydroxy-2-decenoic acid (10HDA) is a
unique product of workers while 9ODA is a unique product of queens
(Pain et al., 1960) was shown to be wrong. These results demonstrated
that worker mandibular gland secretions were capable of substantial
variations in fatty acid composition, depending on the age and social
milieu of the bees being investigated. Whether queen mandibular gland
signals could exhibit similar variability, was a question that had
been obliquely investigated by Pain's research group. Thus they showed 1) age dependance in the production of 9ODA by honeybee queens, 2)
variations in 9ODA production between virgin and mated queens (Pain
et al., 1967) and between queens attended by various numbers of workers, and 3) an annual cycle in the quantity of mandibular gland secretion produced. More detailed evidence of this variability was given
(Pain et al., 1974), but its nature was not clearly defined, nor were
the variables that might be producing it. Thus, the nature and significance of the variability in the composition of the mandibular
gland 'sociopheromone' (Velthuis, 1977) of honeybee queens was studied using secretions produced by A. m. mellifera, A. m. adansonii and
A. m. capensis of different ages and mated states.

MATERIALS AND METHODS

The queens investigated were as follows: 15 mated and 3 virgin
queens of A. m. adansonii; 8 mated, 4 swarming, 6 virgin and 16 inbred mated queens of mellifera; and 6 mated queens of capensis. The
mandibular gland secretions were extracted either by placing whole
heads or dissected mandibular glands in dichloromethane. The extracts
were prepared according to a method of Crewe and Bissels (Unpub. data
). The acids in the extracts were derivatized and quantitated according to the method of Gehrke and Leimer (1971). The extracts were analyzed on a Carlo Erba 2150 gas chromatograph equipped with a flame
ionization detector, Grob injector and a Hewlett-Packard fused silica
capillary column coated with methyl silicone gum. Quantitation of the

areas under the peaks on the chromatograms was achieved with a Spectra-Physics Minigrator. Identification of the components in the secretions was by comparison of retention times with authentic standard compounds and by GC-MS analysis.

Multivariate statistical analysis of the results was carried out using BMDP programs from the package developed by the Health Sciences Computing Facility of the University of California, Los Angeles.

RESULTS

1. The secretions of virgin queens

Analyses of virgin queen secretions are given in Table 1, and indicate that young virgins have secretions in which 10HDA predominates (a 'worker-like' signal). This component becomes progressively less important as the bees age, and even though they are not mated, a 'queen-like' signal in which either 90DA or a combination of 90DA & 9HDA comes to predominate, is developed. This appears to be true in both groups of queens investigated, although adansonii queens may produce a signal in which 90DA predominates after 36 hours (Jackson, pers. comm.).

Table 1. -- The percentage composition of the major components of, and the total amount of acids in, the mandibular gland secretions of virgin queens of A. m. mellifera (MEL) and adansonii (ADA)of various ages, and in a group of 4 mellifera (SWA) queens found in a swarm.

Queen & Age	% of Compounds Present					Total acids(ug)
	8HOA	90DA	9HDA	10HDAA	10HDA	
MEL 1 day	1.55	23.85	5.55	2.13	66.22	114.37
MEL 1 day	1.41	28.41	6.40	2.09	61.37	128.38
MEL 10 days	7.06	24.35	21.30	0.64	46.65	132.38
MEL 10 days	7.75	35.93	20.42	0.31	35.58	262.18
MEL 10 days	7.49	26.57	25.34	1.57	39.03	167.93
MEL 23 days	12.19	48.04	28.48		10.68	73.81
ADA 1 day	6.40	12.70	7.30	7.20	64.00	49.60
ADA 2 days	5.10	33.30	3.00	4.30	54.30	20.60
ADA 40 days	11.30	49.30	22.80	1.60	12.40	926.80
SWR 1	3.32	27.49	7.46	2.90	59.15	167.06
SWR 2		48.99	11.26		39.83	11.91
SWR 3	2.60	25.15	6.33	2.24	62.37	115.51
SWR 4	2.73	28.52	11.52	0.13	56.45	188.59

8HOA=8-hydroxyoctanoic acid, 9HDA=(E)-9-hydroxy-2-decenoic acid, 10HDAA= 10-hydroxydecanoic acid.

2. Secretions of swarming queens

An after-swarm which contained 4 queens was captured. Each of the queens was isolated in a hair-curler queen cage, and each was tested for attractiveness to the now queenless swarm. In three replicates, only queen (SWR 2) was attractive to and caused the swarm to cluster, the other three queens did not arouse the interest of the swarming bees. Subsequent to this test for attractiveness, the mandibular gland secretions of all four bees were analyzed. The results are presented in Table 1. Only the queen with the secretion in which 9ODA predominated, was attractive. All the other queens, although producing 9ODA, had secretions in which 10HDA predominated.

3) The secretions of the mated queens

The mated laying queens produced secretions each of which was different from that of other individuals, i.e. the particular mixture of components was unique to each individual. In addition, there were consistent differences between mellifera, adansonii and capensis in the relative proportions of the major components; these are given in Table 2. Stepwise discriminant function analysis of this data reveal-

Table 2. -- The percentage composition of the major components of the mandibular gland secretions of A. m. mellifera(MEL), adansonii(ADA) and capensis(CAP). The individuals were of various ages, but all had been mated and were laying queens in colonies.

Queen	n		% Composition of Major Components Present				
			8HOA	9ODA	9HDA	10HDAA	10HDA
MEL	8	mean	8.11	37.25	32.25	5.11	15.45
		S.D.	1.86	9.16	6.21	3.68	4.80
ADA	15	mean	4.78	65.39	14.43	3.64	8.05
		S.D.	2.54	13.78	5.61	2.74	4.06
CAP	6	mean	3.00	84.83	9.78	0.51	1.01
		S.D.	3.01	11.48	7.38	0.49	1.10

ed that the variable that best discriminates between the groups is 9-HDA, followed by 10HDA and then 9ODA. The classification matrix for the limited number of cases investigated, indicated that all capensis queens were correctly classified, as were all mellifera, while 3 of the 15 adansonii queens were incorrectly classified as capensis. This analysis indicates that the paricular blend of components produced by each of the 'races' of honeybees is characteristic, with the group means ranging from mellifera on the one extreme, through adansonii to capensis on the other (Fig. 1). The proximity of the group means of adansonii and capensis accounts for the misclassification of some of the adansonii queens.

4. Secretions of inbred mellifera queens

The results of a stepwise discriminant function analysis of the mandibular gland secretions of the queens of three inbred strains are shown in Fig. 2. The variable that separated the three strains best, was 9HDA. Only individuals of strain Ka were all correctly classified. Individuals of strains Yd and Gk produced more variable secretions which produced misclassifications of individuals. This was particularly true of strain Gk.

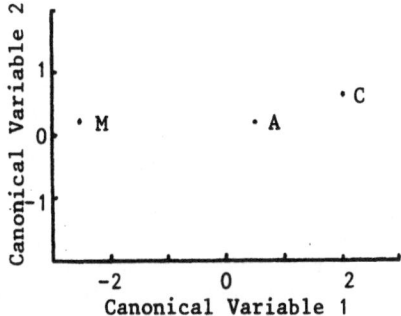

Figure 1. -- Positions of the group means on a plot of the first two canonical variables of a stepwise discriminant function analysis performed on the major components of the secretions of the 29 mated queens. M = mellifera; A = adansonii; and C = capensis.

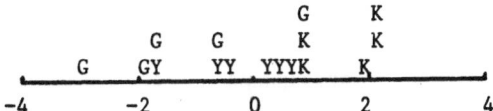

Figure 2. -- Histogram of the canonical variable of the major components of the secretions of the three inbred strains, G=Gk Y=Yd K=Ka, of mellifera queens. Each letter in the histogram represents the position of a single queen.

DISCUSSION

The investigations of chemical signals reported here, have been undertaken assuming that Velthuis' (1977) chemosensory model for their mode of action is correct. Hence the particular blend of chemical constituents determines what the signal will be. In young virgin queens (Table 1), the mixture of constituents has a 'worker-like' character, since 10HDA is the major component. As the virgins age, a more 'queen like' signal develops in which 10HDA no longer predominates. These changes in the relative concentration of the constituents of the mixture, should change its bouquet, and hence change the nature of the signal for the audience (workers). This seems to be bourne out by the study with the swarming queens, where only SWR2 was attractive. The workers in the swarm appeared to recognize a 'queen-like' signal. Thus young queens may not be recognized as queens by workers, since they remain cryptic by producing 'worker-like' signals.

If a particular blend of components present in the mandibular glands is to function as an effective chemosensory signal for workers, then one would expect there to be differences between the races of honeybees which correlate with the ease with which laying workers develop. Thus in capensis, where laying workers develop rapidly, queen signals contain the greatest quantities of 90DA, while in adansonii and mellifera where laying worker development is less facile, there is a progressive decline in the proportion of 90DA (Table 2). The mandibular gland secretions appear to be relatively 'race' specific (Fig. 1).

The inbred queen signals were studied in order to determine the degree of similarity of the signals of sister queens, since evidence is accumulating (Breed, 1981) that workers recognize their queens individually on the basis of odors that are genetically determined and unique. The mated queens each produced a unique blend of components which may be the signal by which the workers recognize them (once they have mated and their signals have stabilized). The signals of the inbred queens were less variable than these, but still showed considerable variability. Breed (1981) found that even in inbred lines, acceptance of transferred queens was only 35%. These results could be explained (Fig. 2) by the degree to which the signals varied. What needs to be determined carefully, is the extent of the difference between blends that can be discriminated by workers.

These chemosensory signals for workers, appear to fulfill the requirements for discriminators or recognition pheromones proposed by Hölldobler and Michener (1980). The signals differ among individuals of a population, they are not of extrinsic origin and they must be learned. Furthermore, they can be varied according to age and social position of the indivdual.

REFERENCES

Breed M.D. 1981. -- Individual recognition and learning of queen odors by worker honeybees. Proc. Natl. Acad. Sci., 78, 2635-2637.

Crewe R.M. 1977. -- Pheromones and colonial defensive behaviour of A. m. adansonii. Proc. Apimondia Symp. on the African Bee., 77-83.

Crewe R.M., Velthuis H.H.W. 1980. -- False queens: a consequence of mandibular gland signals in worker honeybees. Naturwiss., 67, 467-469.

Gehrke C.W., Leimer K., 1971. -- Trimethylsilylation of amino acids, derivatization and chromatography. J. Chrom., 57, 219-238.

Hölldobler B., Michener C.D., 1980. -- Mechanisms of identification and discrimination in social Hymenoptera. Dahlem Konferenzem: Evolution of Social Behavior., 35-58.

Pain J., Hügel M-F., Barbier M., 1960. -- Sur les constituants du mélange attractif des glandes mandibulaires des reines d'Abeilles. C.R. Acad. Sci. Paris, 251, 1046-1048

Pain J., Barbier M., Roger B., 1967. -- Dosages individuels des acides céto-9 décène-2 oïque et hydroxy-10 décène-2 oïque dans les têtes des reines et des ouvrières d'abeilles. Ann. Abeille, 10, 45-52.

Pain J., Roger B., Theurkauff J., 1974. -- Mise en evidence d'un cycle saisonnier de la teneur en acides céto-9 et hydroxy-9 décène-2 oïque des têtes de reines vierges d'Abeille. Apidologie, 5, 319-355.

Ruttner F., Koeniger N., Veith H.J., 1976. -- Queen substance bei eierlegenden Arbeiterinnen der Honigbiene (Apis mellifica L.). Naturwiss., 63, 434.

Velthuis H.H.W., 1977. -- The evolution of honeybee queen pheromones. Proc. VIII Int. Congr. IUSSI, Wageningen, 220-222.

Communication and the Swarming Process in Honeybees

H.H.W. Velthuis,
State University of Utrecht

Honeybee colonies can multiply in two ways: either by producing males that may try to inseminate young queens of other colonies, or by fission (swarming). The latter is by far the most risky way. Its occurrence therefore will depend on rather specific conditions inside and outside the colony, whereas male production is much less dependent upon such factors. Indeed Allen (1958, 1965a) reported the absence of a correlation between drone production and queen rearing or swarming.

At least two kinds of swarms may be distinguished: prime swarms, in which the old queen of the colony and about half of the worker population leave the nest, and afterswarms or secondary swarms, in which one or more young queens depart with the swarm. The old queen has been producing her specific queen pheromones and the bees are familiar with her and discriminate her from other queens (Velthuis & van Es, 1964; Morse, 1972; Boch & Morse, 1974, 1979; Ambrose et al., 1979). The young queens have just started phenomone production and therefore worker familiarity with their specific properties is less likely.

Reproductive swarming is distinguished from the swarms in which the whole colony leaves the nest due to deterioration of the nest site or due to food conditions in the environment. Such swarming is termed absconding. It is related to survival rather than to reproduction. However, in its causation and progress absconding may be similar to reproductive swarming.

The most important factors favoring swarming are crowding (size of the colony in relation to hive space; Simpson, 1958), climatic conditions (both the long-term effects and the immediate effect on emergence) and the age of the queen (probability of swarming increases with the age of the queen; Allen, 1965b).

Preparations prior to swarming start already 2-3 weeks before the issue of the prime swarm. The number of feedings the queen receives from the workers decrease, workers refuse more and more to feed her (Allen, 1956). As a consequence she looses weight (Morse et al., 1966) and reduces her rate of egg laying (Allen, 1956).

Simultaneously, the workers in the colony increase both the volume and the sugar concentration of the contents of their honey stomach (Combs, 1972). In this period the queen starts to oviposit in the special queen cups from which the young queens will emerge. Once there are sealed queen cells, the prime swarm may leave the hive.

The act of swarming is preceded by another phenomenon. Several days earlier scout bees have been inspecting suitable nesting sites. Such scout bees are experienced field bees (Lindauer, 1955), that communicate their discoveries by means of dances, both before and after the departure of the swarm. Eskov (1971) reports that such scout bees determine when the colony will swarm. They emit sounds of 400-450 Hz and of about ¼ sec duration. The number of these sound pulses increases rapidly, stimulating the workers on the combs to move rapidly, and once all the bees are in motion they move rapidly out of the hive, taking the queen with them. Only the very young bees stay behind.

Once the bees are in flight in front of the hive each individual has to make the decision either to return to the hive or to join the swarm. Generally, about 50% of the bees, irrespective of their ages, will return to the hive (Meyer, 1956). This decision is influenced by the queen. If we prevent her from flying with the bees they will all return. She communicates her presence in the air by means of the pheromones from her mandibular glands (Velthuis and van Es, 1964). This stimulates part of the workers to form small clusters on branches etc., a few meters away from the hive, where they are joined by the queen and part of the workers. The presence of the queen in the cluster causes the workers to increase the emission of pheromones from their Nassanoff glands, which makes the cluster more attractive (Velthuis & van Es, 1964). This joint action of pheromones from different sources has been aptly named pheromone concert by Morse & Boch (1971). Several research workers attempted to identify the chemical nature of the queens' pheromone (Velthuis & van Es, 1964; Butler & Simpson, 1967; Morse & Boch, 1971; Boch et al., 1975). However, only one substance from the mandibular gland, the 9 ODA (9-oxo-decenoic acid), has been reported to attract queenless swarm bees. Additional effects of 9 HDA (9-hydroxydecenoic acid) are a matter of controversy.

In the meantime a number of workers also settle on the front of the hive. By emitting Nassanoff pheromone as well, they compete with the swarm in attracting the remaining airborne bees. The swarm will stay on its temporary support until its members have reached agreement about the future nesting site. The bees that returned to the hive will wait for the new queens to mature.

If the remaining colony is still strong enough it will swarm for at least a second time. Now some new components in the decision making are present: the young queens, the emerged one and those that remain in their cells communicate to each other and probably to the workers as well by means of sounds: queen tooting and quacking resp. (Simpson & Cherry, 1969). Tooting by an emerged queen prevents or delays the emergence of the others who respond by quacking (Bruinsma et al., 1982). If two queens meet on the comb fighting occurs and one queen will be killed. Tooting is correlated to the occurrence of vibratory dances (DVAV; Milum, 1955) performed by worker bees on the queen. Also the decision whether an unemerged queen will be destroyed or will have the opportunity to eclose during the take-off of the secondary swarm is correlated to the occurrence of those dances (Fletcher, 1978).

The presence of pheromones from the young queens is essential for the formation of a cluster by the secondary swarm. If the queens are still too young the swarm will return. The next day, when the mandibular glands of the young queens are more mature, clustering may occur in the new swarming attempt. This implies that other queens not having yet the ability to produce the pheromones may also be present in the swarm.

Apart from the importance of vibrations (including those perceived by us as sounds) pheromonal communication seems essential. The simultaneous presence of Nassanoff pheromone and of queen pheromone is necessary for successful clustering and for migration to the new nest site. If queens from clusters are imprisoned in cages their swarms may become airborne but will not depart. If one takes the queen in its cage into the air, the swarm may be led at this moment in any direction (Morse, 1963). If some worker bees of the swarm are provided with some 9 ODA on their body the swarm will depart without its queen. The substance makes the bees responding to the Nassanoff pheromone emitted by the scout bees that guide the swarm to its new home (Avitabile et al., 1975).

Altogether, this provides us with a rather detailed and satis-fying picture of the process of swarming. At this point I like to report on some observations on the pheromonal and sequential aspects which do not fit into this picture.

Several years ago I studied the role of 9 ODA and 9 HDA in attracting swarms. Like all colleagues I brought a captive swarm to the test area, away from its site of origin. The swarm was shaken on a piece of cloth, the queen was removed and a wooden cross bearing two dead and impregnated workers was erected. A steadily increasing number of bees became airborne and searched the environ-ment. Occasionally, a bee landed on or near the two sources of substances but did not start emittance of Nassanoff pheromone. Instead they either departed sooner or later or started biting and pulling the objects, sometimes dismembering them. This behavior was independent of the substances being separate or mixed. Only after a considerable time, being often far more than an hour, some bees started fanning on the cross. Their release of Nassanoff pheromone initiated the formation of the cluster on the cross. Control experiments with a caged queen on the cross led to cluster-ing in far less time, and if I gave a queen the opportunity to fly about 15 minutes after being separated from the swarm, cluster formation started somewhere in the trees within about 5 minutes.

If I left the swarm queenless for about 15 minutes and then approached the bees that were still on the ground with the caged queen they immediately reacted, sometimes while still at a distance of 1-2 m. They oriented towards the cage, started fanning and walked in slow, massive streams in the direction of the queen. As soon as the queen was taken away fanning and locomotion stopped. The perception of the queen is necessary for the occurrence and maintenance of fanning. The queen apparently produces a rather volatile pheromone enabling a quick localization even when she is relatively distant. This substance is produced by the mandibular glands, but is contained only in very fresh preparations of that gland.

In other experiments I held an insect pin bearing a dead bee impregnated with queen extract in my hand. Searching bees landed soon on my hand and, without touching the extract, started fanning. As soon as about 20 bees were fanning I could drop the pin without disturbing the cluster being formed on my hand. The object in the grass was neglected by the bees. Once the bees stimulate each other and the direct information from the queen is apparently not necessary their fanning is regulated in a different way.

Such observations convinced me that we have to differentiate among the worker bees. Some have a leading role, others just follow. Probably the queen also reacts to the leading workers. Most of the experiments on swarming have been rather unnatural in that we studied the attraction of a swarm by a fixed queen instead of the reverse. We have learned a lot about the signals, but not enough about the sequence of them. To really understand the process we should be able to replace the queen with artificial stimuli, producing the right signals at the right time. We tried to do so with colonies whose queens had clipped wings: at the departure of the swarm she fell down and we moved objects with extracts or pure substances in the air. But it was all in vain. Even their own queen, imprisoned in a cage, could not prevent the return of considerable numbers of bees to the hive within 5-10 minutes. Possibly the queen actively produces the secretions of the mandibular gland; if not in flight she refuses to emit this signal.

What we noticed, however, was that in the meantime several miniature clusters had been formed in which a few bees fanned occasionally. This induced the following experiment: In addition to clipping the wings we deprived the queens of their mandibular glands several weeks before swarming was expected. At the moment of emergence of the prime swarm the queen was taken and placed amidst such a group of clustering bees, as soon as we found them. The direct contact with their own queen induced very active Nassanoff pheromone dispersal in the bees leading to the formation of a stable swarm cluster like a normal swarm. Timing aspects were very important; if locating the queen took us too much time the tendency to return to the hive became very strong and only small clusters were formed.

In conclusion I would say that we apparently know a lot, but not all, about the chemical stimuli involved in swarms. The patterns of the signals and their relations in time are, however, insufficiently known. It is through these patterns that we need to understand the functions of the pheromones, because the pattern makes the communication.

References

Allen M.D., 1956. -- The behavior of honeybees preparing to swarm. Brit. J. Anim. Behav., 4, 14-22.

Allen M.D., 1958. -- Drone brood in honey bee colonies. J. econ. Entomol., 51, 46-48.

Allen M.D., 1965a. -- The effect of plentiful supply of drone comb on colonies of honeybees. J. apicult. Res., 4, 109-119.

Allen M.D., 1965b. -- The production of queen cups and queen cells in relation to the general development of honeybee colonies, and its connection with swarming and supercedure. J. apicult. Res., 4, 121-141.

Ambrose, J.T., Morse R. A., Boch R., 1979. -- Queen discrimination by honey bee swarms. Ann. Entomol. Soc. Amer., 72, 673-675.

Avitabile A., Morse R.A., Boch R., 1975. -- Swarming honey bees guided by pheromones. Ann. Entomol. Soc. Amer., 68, 1079-1082.

Boch R., Morse R.A., 1974. -- Discrimination of familiar and foreign queens by honey bee swarms. Ann. Entomol. Soc. Amer., 67, 709.

Boch R., Shearer D.A., Young J.C., 1975. -- Honey bee pheromones: field tests of natural and artificial queen substance. J. Chem. Ecol., 1, 133-148.

Boch R., Morse R.A., 1979. -- Individual recognition of queens by honey bee swarms. Ann. Entomol. Soc. Amer., 72, 51-53.

Bruinsma O., Kruijt J.P., Dusseldorp W. van, 1982. -- Delay of emergence of honey bee queens in response to tooting sounds. Proc. Akad. Wet. Amst., Ser. C.

Butler C.G., Simpson J., 1967. -- Pheromones of the queen honeybee (Apis mellifera L.) which enable her workers to follow her when swarming. Proc. R. Entomol. Soc. (A), 42, 149-154.

Combs, G.F., 1972. -- The engorgement of swarming worker honeybees. J. apicult. Res., 11, 121-128.

Eskov E.K., 1971. -- Sound signals produced by honeybees during the process of swarming. Zool. Zh., 50, 704-712 (in Russian)

Fletcher D.J.C., 1978. -- The influence of vibratory dances by worker honeybees on the activity of virgin queens. J. apicult. Res., 17, 3-13.

Lindauer M., 1955. -- Schwarmbienen auf Wohnungssuche. Z. vergl. Physiol., 37, 203-234.

Meyer W., 1956. -- Arbeitsteilung im Bienenschwarm. Insectes Sociaux, 3, 303-324.

Milum V.G., 1955. -- Honey bee communication. Am. Bee J., 95, 97-104.

Morse R.A., 1963. -- Swarm orientation in honeybees. Science, 141, 357-358.

Morse R. A., 1972. -- Honey bee alarm pheromone: another function. Ann. Entomol. Soc. Amer., 65, 1430.

Morse R.A., Dyce E.J., Young R.G., 1966. -- Weight changes by the queen honey bee during swarming. Ann. Entomol. Soc. Amer., 59, 772-774.

Morse R. A., Boch R., 1971. -- Pheromone concert in swarming honey bees (Hymenoptera: Apidae). Ann. Entomol. Soc. Amer., 64, 1414.

Simpson, J., 1958. -- The factors which cause colonies of Apis mellifera to swarm. Insectes Sociaux, 5, 77-95.

Simpson, J., Cherry S.M., 1969. -- Queen confinement, queen piping and swarming in A. mellifera colonies. Anim. Behav., 17, 271.

Velthuis H.H.W., Es J. van, 1964. -- Some functional aspects of the mandibular glands of the queen honeybee. J. apicult. Res., 3, 11-16.

Abstracts

Ultrastructure of the Dufour Gland in Ants (Hymenoptera, Formicidae)

Johan Billen
Limburgs Universitair Centrum

The ultrastructure of the Dufour gland has been studied in several species of formicine and myrmicine ants. The lumen is lined with a monolayered epithelium covered with a cuticula of approximately 0.3 μm in all species studied. Its basement membrane varies from 0.05-0.1 μm in the myrmicine species to 2 μm in the genus Formica.
The high number of multilamellar bodies is the most conspicuous feature of the cytoplasm of the gland cells. A well developed smooth endoplasmatic reticulum, glycogen deposits and microtubules are also found. Rough endoplasmatic reticulum is very sparse. The mitochondria are scattered over the cell in the Myrmicinae, while in the Formicinae they form a distinct subcuticular layer. Modifications of the cell membrane are restricted to basal infoldings in the formicine Dufour gland. Apical invaginations or microvilli do not occur. Intercellular contacts consist of septate junctions in the apical region.
In the duct of the gland, a different muscular closing apparatus is found in the Formicinae and Myrmicinae.

Different Acceptance of Foreign Queens by Colonies of
Leptothorax acervorum

Dorothea Bruckner
Zoologisches Institut, LMU München

Monogynous, polygynous and pseudopolygynous colonies of Leptothorax acervorum were kept in the laboratory. For the experiments they were made queenless; 24 h later each colony received a foreign queen of the same species. These queens were either accepted (they later started egglaying) or not accepted (often killed) by the workers. In the latter case the test was repeated until the colony accepted a queen. Four days later a second queen was additionally introduced into the colonies.

It was found that virgin queens were never accepted by any colony. Inseminated queens were accepted to a high percentage (70%) as first queens but to a significantly lower percentage (28%) as second queens. Workers of originally monogynous and pseudopolygynous colonies accepted second queens less frequently (9%) than workers of originally polygynous colonies (30%). These results are interpreted in terms of pheromonal labeling systems of individual queens.

Pre-swarming behaviour of workers and virgin queens of Apis mellifera

O. Bruinsma, H. Grooters, and J.P. Kruijt
State University of Groningen, The Netherlands

Behaviour of workers and virgin queens is being studied prior to and during swarming of de-queened laboratory colonies of Apis mellifera. A delay in time of emergence of queens was found in the majority of hives subjected to recorded tooting. The delay occurs only in hives in which worker bees perform vibratory dances at a relatively high rate prior to the emergence of queen(s) (Bruinsma et al., 1981). It is investigated whether individual differences in quacking activity of unemerged queens are correlated to the sequence in which queens emerge, influencing their survival.

Although the tooting queen ususally joins the worker bees during swarming it has been established several times that instead a quacking queens joins the swarm, while the tooting queens remains in the hive. This phenomenon is so far not fully explained by the acoustical communication between queens and workers, and among queens themselves.

Aggression and Territoriality in Ants

Claudine De Vroey
The University of Brussels (Belgium)

A territory is defined as a portion of space actively defended against intrusion of competitors. The territoriality of several Myrmica species has been tested by measuring their aggressiveness on differently marked areas. M. sabuleti and M. scabrinodis are more aggressive when fighting on their own territory. The possible sources of territorial pheromones have been investigated; several substances originating from different parts of the ants' bodies seem to be involved. In M. rubra and M. rugulosa, the workers' aggressiveness is not influenced by territorial marks. These two species are however able to beat off would-be invaders by tagging efficiently explorers of alien societies with an attractive sticker.

Social Organization of the Alarm-Evacuation
Defense in the Ant Genus Pheidole

Robert Droual
City College of CUNY

The ants Pheidole desertorum and P. hyatti use a defense against
the army ant Neivamyrmex nigrescens which involves nest evacuation.
However, unlike a panic alarm defense, their defense occurs in two
phases. In the first phase, workers carrying brood mass at the nest
entrance. In the second phase, an actual evacuation occurs. Also,
during an evacuation these ants follow any recently laid chemical
trails. This defense is called an "alarm-evacuation" defense to
distinguish it from a panic alarm defense. The minor subcastes in
both species make a proportionately greater contribution to the
evacuation of brood during an evacuation than the major subcastes.
However, the majors of P. hyatti carry proportionately less brood
than the majors of P. desertorum. Also, analysis of filmed
evacuations indicate that P. hyatti majors leave the nest in greater
numbers early in the evacuation. This is not true for P. desertorum.
These differences reflect the greater combat role of the P. hyatti
majors. The significance of this with respect to differences in
nest structure, worker size and combat ability is discussed.

The Koschewnikow Gland as the Principal Organ Secreting the Sting
Alarm Pheromone in the Honeybee Worker (Apis mellifera L.)

D. Grandperrin* and B. Mauchamp**
*University of Paris, France - ** C.N.R.A. Versailles, France

Our recent studies have shown the existence of functional
relations between Koschewnikow glands and the setaceous membrane of
the sting apparatus. Glandular secretions reach this membrane via
quadrate and oblongue plates. Gas chromatography and mass
spectrometry analyses show that the same volatile compounds,
especially isoamyl-acetate, are present in the gland and on the
setaceous membrane. Isoamyl-acetate and other volatile compounds
(butyl-acetate, isoamyl-alcohol, hexyl-acetate, 2-nonanol, decyl-
acetate, benzyl-acetate, benzyl-alcohol) have been blended to
prepare an artificial alarm pheromone and its efficiency was
evaluated in a bioassay. The recruiting power of the artificial
pheromone is greater than that of isoamyl-acetate alone. To
achieve the full recruiting power, all volatile compounds secreted
by the Koschewnikow gland have to be present in the artificial
blend tested.

Nestmate Recognition Systems in the Formicidae (Hym.)

Klaus Jaffé
Universidad Simón Bolívar, Caracas

The nestmate recognition system in Atta cephalotes and Solenopsis geminata (Myrmicinae), Camponotus rufipes (Formicinae), Odontomachus bauri (Ponerinae) and Dolichoderus spp. (Dolichoderinae) were studied. Nestmate-recognition in the Myrmicinae and Formicinae is probably based on colony differences in the relative proportion of the various components of the alarm pheromone. In the Ponerinae, colony-differences in the composition of various pheromones, inclusive the alarm pheromone, are used for nestmate recognition.

Absorption of the volatile components of the pheromones on the cuticular waxes seem to be the mechanism by which species-recognition, and sometimes intraspecific colony-recognition is achieved. Intraspecific colony-recognition, i.e. recognition of ants from different colonies of the same species, depends more on the detection of the volatiles when the pheromone is actually released. The implication of these results and its relation to the territorial marking behaviour of the species studied is discussed.

A Learned Component of Sibling Discrimination in Workers of Polistes fuscatus

David W. Pfennig*, Hudson K. Reeve**, and Janet S. Shellman**
*The University of Texas at San Antonio and **Oakland University

We investigated the mechanism whereby workers of Polistes fuscatus discriminate siblings. Nests containing worker-destined larvae were switched between unrelated foundresses before worker eclosion. Subsequently eclosing workers were isolated individually for 20 days and then introduced together in triplets consisting of two sisters and a nonsister each. Preferential association of any two wasps within a triplet was ascertained with two measures of spatial tolerance. We determined that (1) workers can discriminate sisters from nonsisters without prior exposure to adult relatives and (2) sibling discrimination depends upon the learning of acquired cues, presumably odors, which are shared by both interacting wasps.

Treatment of Introduced Foreign Queens by Honey Bee Colonies

Gene E. Robinson
Cornell University

The behavior of workers clustered around in introduced foreign queen was studied in colonies of Apis mellifera. Worker behavior toward these queens was classifed as either attentive, neutral, or aggressive, using individually marked bees of known age. The workers that behaved aggressively constituted a relatively small proportion of the bees in the cluster. Thus, the actions of a small number of bees appear to influence a colony's treatment of an introduced foreign queen.

Reproductive Biology in Halictine Bees

Brian H. Smith
The University of Kansas

Male preference for individual females in the sweat bee
Lasioglossum zephyrum was studied using laboratory reared
individuals. Males were first exposed to a female for a 10 min.
period, and attraction to a second female was then shown to be
negatively correlated to the relatedness between the two females.
Further experiments show this response to result from an odor; work
with animal extracts and synthetic compounds indicates the possible
make-up of the multi-component chemical signal and begins to
explain the ability of the males to discriminate between different
females. This is a learned preference, and it is quite possible
that an important learning period in a male's life is in the nest,
just after emergence, and before leaving. This is the period where
males are exposed to specific females the longest; encounters with
females after leaving the nest last, on the whole, less than one
minute. Since the females in the nest are likely to be close
relatives of the males emerging into that nest, the possibilities
for negative assortative mating in these bees must be considered.

The Trail Pheromone of the Red Imported Fire Ant, Solenopsis invicta
Chemistry, Behavior and Potential for Control

Robert K. Vander Meer, Clifford S. Lofgren, B. Michael Glancey, and
David F. Williams
U.S. Department of Agriculture

Several components of the trail pheromone of Solenopsis invicta
were isolated, identified and synthesized. The component responsible
for trailing activity was active at 0.4 pg/cm, but was inactive in
an attractant-recruitment bioassay. Other components, less active
in the trail bioassay, did elicit a partial recruitment response.
Total activity was achieved by the addition of a Dufour's gland
(source of the trail pheromone) component that by itself was
inactive in both trail and recruitment bioassays. Potential for the
use of the trail pheromone in control of fire ants was demonstrated
on a small scale by significantly enhanced feeding activity on filter
paper disc treated with pheromone plus soybean oil versus soybean
oil alone.

NEUROBIOLOGY AND BEHAVIOR
OF SOCIAL INSECTS
Organized by Randolf Menzel

Introduction

Randolf Menzel,
Freie Universität, Berlin (FRG)

Social insects, in particular the social Hymenoptera,
have an almost inexhaustible behavioral repertoire. Many
of their abilities, such as navigation using patterns of
light polarization in the sky, communication using a
language based on movement symbols, the perception of
color, form and scent, the ability to learn rapidly and
their capacity for long term memory, have continued to
amaze and inspire scientist working in this field. If one
considers that all these behavioral capacities are
controlled by a tiny brain, seldom greater than 1 mm^3 in
size and generally composed of less than 10^6 neurones,
one must assume that during the process of evolution,
extraordinarily economical and effective mechanisms for
the control of the many varied behaviors of these insects
have been developed.

What are the neuronal mechanisms? In 1965, Bullock
and Horridge in their 2 volume work on "Structure and
Function of the Nervous System of Invertebrates" wrote
"There is hardly any physiological data on the neurons
mechanisms which make possible these varied responses".
The situation has changed considerably over the last
decade. Substancial progress has been made possible as a
result of a concentration of interest by neuroanatomists,
neurophysiologists and behavioralists on an animal
initially used under great success by Karl von Frisch
(1966), namely the honey bee <u>Apis</u> <u>mellifera</u>.

The structure and organization of many parts of the
bee brain (retina, first optic ganglion, mushroom bodies)
are already well understood, and with the help of the
new neuroanatomical methods and intracellular marking
techniques now available, the knowledge available concerning these and other regions of the brain continues to
increase. A crucial contribution to this symposium is the
presentation and discussion of the latest neuroanatomical
findings, and the application of these findings to the
derivation of models to explain brain function. For the
first time, it is possible to formulate hypotheses

concerning the flow of information and the strategies of
neuronal integration in the brain of the bee based on
neuroanatomical data, and to compare them with results
obtained using neurophysiological and behavioral methods.
Over the last 10 years, electrophysiological studies
have provided considerable information about the brain
of the insect. In the bee brain, research involved, above
all with visual integration and multisensory assimilation
of information in the brain, has been particularly
successful. Mechanisms involved in color and movement
coding, the modulatory effect of sensory inputs on one
another, and the role played by particular neuropiles such
as the mushroom bodies and the central body, have been
investigated with the help of intracellular recording of
neuronal responses in identified neurones. Although such
analyses are still a long way from providing a full
understanding the mechanisms controlling behavior in
insects, the combination of many such investigations does
allow increasingly precise models to be constructed.
Learning and memory play a central role in the control
of behavior in social Hymenoptera. Since the fundamental
experiments of Forel (1910) and von Frisch (1914), conditioning techniques have not only been applied for many
decades and with great success to studies of sensory
physiology, but in the last decade, have lead also to the
accumulation of considerable information concerning the
analysis of events involved in learning itself. There are
incredible and unexpected parallels between the process
of learning in bees and in mammals including men. The
question of the neural basis of memory function represents
another central point of presentations and discussions
of the symposium. The particular suitability of the honey
bee as an experimental animal, the combination of behavioral studies, electrophysiological and neuropharmacological methods have enabled initial insights into these
questions.
Neurobiology is an exciting research area with rapid
develoment of new concepts and methods. Neurobiologists
working with social Hymenoptera contribute considerably
to the excitement because their experimental animals
provide both behavioral complexity and experimental accessibility to the neural circuitry. In the near future it is
to be expected that studies particularly on the honey bee
will become even of greater importance for neurobiology
in general.

Bullock,T.H., Horridge,G.A.: Structure and Function in
    the Nervous System of Inverbrates. San Francisco,
    London, Freeman & Co. 165

Forel,A.: Das Sinnesleben der Insekten. Eine Sammlung von
    experimentellen und kritischen Studien über Insekten-
    physiologie. E. Reinhardt, München 1910

Frisch,K. von: Der Farbensinn und Formensinn der Bienen.
Zool. Jb. Abt. Allgem.Zool.Physiol. <u>35</u>, 1-182, 1914

Frisch,K. von: The dance language and orientation of bees.
pp 566, Cambridge University Press 1967

Rapid Changes in Dendritic Spine Morphology During the Honeybee's First Orientation Flight

Richard G. Coss and John G. Brandon,
University of California, Davis

Our present research on calycal interneurons in the worker honeybee (<u>Apis</u> <u>mellifera</u> L.) was prompted by the initial discovery in jewel fish that social stimulation shortened dendritic spine stems on tectal interneurons by elongated dilation of spine heads (Coss & Globus, 1978). Honeybees are attractive subjects for this kind of study because previous research has already pinpointed the mushroom bodies as an important memory storage site (see Menzel & Erber, 1978). Our initial study of honeybees attempted to determine if the spine geometry changes seen in jewel fish would appear in honeybees with different levels of cumulative experience (Coss, Brandon & Globus, 1980). Measurements were made of spine density and geometry on Kenyon cells in the mushroom body calyces using rapid Golgi neuronal impregnation and light microscopy. Comparisons were made of newly emerged, 1-2 week-old nurse, and 3-5 week-old forager bees in distinct stages of behavioral development.

The results showed that spine density and overall spine length did not differ appreciably between groups. Spine stems, on the other hand, were markedly shorter in foragers due to elongated expansion of spine heads, a phenomenon which also characterized socially experienced jewel fish. Moreover, stem shortening was progressive throughout development, exhibiting different average rates in shorter and longer spines depending on the level of cumulative experience. For example, the average rate of stem shortening among the shorter spines accelerated between the newly emerged and nurse stages, whereas the stem shortening rate of longer spines accelerated between the nurse and forager stages. These differences in stem shortening rates may reflect both functional differences in spine geometry and the types of afferent projections synapsing on shorter and longer spines.

Theoretical modeling using cable and core conductor theory (Rall, 1978) has provided insight into the possible functional differences of spines with stems of different electrotonic lengths and diameters. According to electrotonic theory, spine stem shortening may increase synaptic efficacy by lowering the input resistance into the parent dendrite. Long stems on long spines would be expected to attenuate synaptically generated current more than short stems on short spines. In further comparisons with long spines, short spines clearly have less stem shortening potential and they tend to show less experimentally induced stem shortening in both jewel fish and honeybees (Coss & Globus, 1978; Coss, Brandon & Globus, 1980).

Taken together, these findings led to the intriguing speculation that changes in synaptic efficacy might vary as a function of overall spine length. Such a constraint on synaptic plasticity might be adaptive, providing changes in neuronal programing commensurate with the ecological demands on behavioral variability. As suggested by our initial findings, shorter spines with less stem plasticity potential might be more involved in the neuronal mediation of less plastic hive maintenance and brood care behaviors. In contrast, spines with greater stem shortening potential might play a more important role in the mediation of highly variable behaviors, such as foraging, which require ongoing adjustments to changing environmental conditions. This speculation, based on our developmental honeybee and jewel fish research, could be considered a morphological analogue to the notion of "preparedness" (Seligman & Hager, 1972), which treats instinct and learning as a graded continuum of behavioral plasticity. Additionally, this speculation provided the rationale for a cross-phyletic program of study in which rapid and long-lasting changes in spine morphology were examined under experimental conditions of varying behavioral plasticity. Here, we report the effects of the honeybee's first orientation flight, a one-trial place learning event.

MATERIALS AND METHODS

1. Subjects and rearing conditions

In September 1979, one hundred fifty six newly emerged bees were gathered from several hives and placed in a small, broodless hive (14 X 16 X 16 cm) with a virgin queen. The hive contained bee candy, water, 1 frame of pollen mixed with dextrose, and 2 frames of honey. A small screened window provided a view of different sun positions and skylight polarization.

2. The first orientation flight

Four bees took orientation flights at 6 days of age and four more flew at 8 days of age to increase the sample size. These bees were examined histologically with age-matched nonflying controls. The flight condition comprised the gentle aspiration of a nonfeeding bee randomly selected from the comb (retinue bees were not sampled). The bee was placed on an unique spot on the hive's top or sides. Typically, the bee took off a few seconds later and flew in a large spiral around the hive. After landing on the take-off spot a few minutes later, the bee was quickly retrieved by aspiration and placed in a tubular wire bee cage covered with white paper. After the last consecutive flight, nonflying controls were placed in similar bee cages and the groups were quickly carried to the histology lab a short distance away. Among the flyers selected for quantification, orientation flights averaged 5 min, 34 sec; SD = 4 min. Prior to histology, the time spent in the bee cages was counterbalanced for both groups; the maximum time interval between flight onset and histology was 30 min.

3. Tissue staining

The rapid Golgi method was used to impregnate neurons. Caged bees were cooled for 90 sec and whole brains were rapidly extracted (under 3 min) from respiring bees and treated by the rapid Golgi protocol described in Coss, Brandon & Globus (1980). Sections were mounted on slides assigned a random number code which was not broken until quantification was completed.

4. Spine quantification

The groups comprised 5 flyers and 5 nonflyers, each contributing 6 or more well-impregnated spiny Kenyon cells (Fig. 1A) for a total of 71 interneurons. The flyer group had four 6-day-old bees and one 8-day-old bee; the nonflyer group had three 6-day-old bees and two 8-day-old bees. Frequency distributions of cells in the inner and outer calycal cups were not significantly different between groups. For examining tissue, we used an Olympus microscope equipped with a Zeiss 100X neofluar objective and modified drawing tube. Spines were inspected throughout the dendritic tree at 1875X and selected for quantification if they lay parallel to the object plane in clear unobstructed focus. Spines and small portions of dendrites were traced in very large scale (1 μm = 6 mm) using a 35X drawing tube lens designed by R.G.C. to shrink the camera lucida image relative to the specimen image. Drawing accuracy was confined by the microscope's 0.2 μm resolution limit. Flyers and nonflyers provided 1073 and 993 spine tracings, respectively, which were measured for overall spine length and spine stem length using a Zeiss MOP-3 digitizing tablet. Overall spine length was measured from the dendrite (or where the stem expanded wider than 0.25 μm near the dendrite) to the spine tip. Stem length was measured from the dendrite, or as described above, to where the spine head expanded wider than 0.25 μm. Lengths of stem varicosities wider than 0.25 μm were added to spine head lengths. Maximum spine head width, profile area, and perimeter were also measured by scanning solid black duplicates with a Quantimet 720 image analyzing computer.

RESULTS

1. Spine stem length

Spine stem lengths contributed by each bee were averaged within 0.25 μm overall spine length categories and a coefficient of variation (CV) was calculated to show stem length variability. Nine overall spine length categories were examined, ranging from 0.75-2.99 μm, to minimize error variance due to small sample sizes at both ends of the flyer and nonflyer distributions. Bartlett's test showed that 6 and 8 day-old flyer and nonflyer groups had homogenous variances. Groups were examined for differences in stem length means and CV's as a function of overall spine length by mixed analyses of variance (ANOVA's) and tests of simple main effects. The ANOVA on mean stem length revealed a significant interaction between groups and categories of shorter and longer spines (F = 2.12, df = 8/64, P < 0.05). Tests of simple main effects showed that the source of this interaction was limited to the longest spine length category of 2.75-2.99 μm (see Fig. 1B).

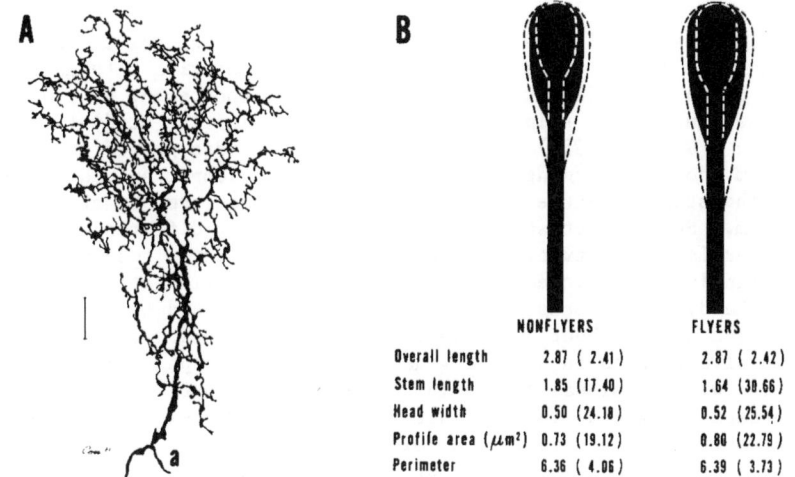

Fig. 1. (A) Camera lucida drawing of spiny Kenyon cell. Axon is labeled by the letter a; calibration bar = 10 μm. (B) Results of grouped data for spines 2.75-2.99 μm long. Means (μm) are shown by a tabular and scale drawing format. Except for overall spine length, dashed contours depict equilibrated intrabee standard deviations. Average CV's (%) are shown in parentheses.

In the 2.75-2.99 μm overall spine length category, the flyer group had a significantly shorter mean stem length (F = 8.61, df = 1/64, P < 0.005) as well as a significantly higher CV than the nonflyer group did (F = 11.18, df = 1/64, P < 0.005). In contrast, inverse F-ratios of simple main effects tests revealed that stem lengths were significantly similar between groups for spines 1.75-1.99 μm long (1/F = 1567, df = 64/1, P < 0.025).

2. Overall spine length
A Chi-square test examined the frequency distributions of flyer and nonflyer spines in all 0.25 μm overall spine length categories derived from equilibrated bee contributions. Flyer and nonflyer group distributions were significantly similar ($\underline{X}^2$ = 4.23, df = 12, P > 0.975), which indicates that flight had no effect on overall spine length.

3. Spine head width, profile area, and perimeter
Using one-tailed $\underline{t}$-tests of means and CV's, group comparisons were made of maximum spine head width, profile area, and perimeter, restricting the analyses to spines 2.75-2.99 μm long which showed flight-induced stem shortening (see Fig. 1B). Compared with the nonflyer group, the flyer group exhibited a significant increase in only the mean profile area ($\underline{t}$ = 2.12, df = 8, P < 0.05). The perimeter means and CV's did not differ appreciably between groups.

DISCUSSION

The results of this study of spine geometry changes in honey-bee calycal interneurons revealed a strong association between the first orientation flight and rapid spine stem shortening by elongated swelling of the spine head. Based on our initial study (Coss, Brandon & Globus, 1980), we expected to find a more graded increase in stem shortening among several categories of longer spines rather than the stepwise increase for the longest spine category of 2.75-2.99 μm. Such abrupt stem shortening may reflect differences in flight-related innervation in addition to providing a record of differential synaptic activity during the first orientation flight.

The dynamic processes underlying rapid elongated swelling of the spine head without appreciable perimeter expansion are likely to involve changes in osmotic gradients during synaptic activity coupled with tensional shifts in the membrane anchoring structure. Long-term changes do include incremental perimeter expansion and progressive stem shortening, albeit, the initial bout of stem shortening averages about half that of flight-experienced foragers. With increasing flight experience and correlated synaptic activity, progressive stem shortening may reach an asymptote in some of the longer spines, transforming their electrotonic properties and stem plasticity to that of much shorter spines. These changes could conceivably decrease neuronal programing potential and concomitant behavioral variation by converting an "open program" (Mayr, 1974) into a more "closed" one. (Supported by NSF Grant BNS-7906843.)

References

Coss R.G., Globus, A., 1978. - Spine stems on tectal interneurons in jewel fish are shortened by social stimulation. Science, 200, 787-790.

Coss R.G., Brandon J.G., Globus A., 1980. - Changes in morphology of dendritic spines on honeybee calycal interneurons associated with cumulative nursing and foraging experiences. Brain Res., 192, 49-59.

Mayr E., 1974. - Behavior programs and evolutionary strategies. Amer. Sci., 62, 650-659.

Menzel R., Erber J., 1978. - Learning and memory in bees. Sci. Amer., 239, 102-110.

Rall W., 1978. - Dendritic spines and synaptic potency. In R. Porter (Ed.), Studies in Neurophysiology. Cambridge Univ. Press, pp. 203-209.

Seligman M.E.P., Hager J.L., 1972. - Biological Boundaries of Learning. Appleton Century Crofts, N.Y., pp. 1-6.

Electrophysiological Analysis of Central Neurons in the Bee and Correlations with Behavior

Joachim Erber,
Freie Universität Berlin (FRG)

In the last years it has become possible to record the activity
of fine neurons intracellularly and to mark these cells with various
dyes. In the bee this electrophysiological technique has been
applied to characterize the cells involved in processing sensory
information in the brain (Erber, 1978; Hertel, 1980). Parallel to
these experiments several specific types of behavior have been
analysed in the bee under laboratory conditions. These behaviors
can be elicited during electrophysiological recordings and correlated with the responses of single neurons.

MATERIALS AND METHODS

The experiments were made with honey bees (<u>Apis</u> <u>mellifera</u> L.)
which were caught at the hive entrance. High impedance electrodes
(approximately 80 - 150 M) were used for the recordings; the cells
were marked with Lucifer Yellow (Erber, 1981).

RESULTS

1. General characteristics of neurons in the median protocerebrum.

Based on neuroanatomical analyses one can expect to find convergence of sensory information in the median protocerebrum. Neurons
from the optic lobes, the antennal lobes, and the suboesophageal
ganglion project into the central part of the brain.

Electrophysiological recordings from single cells support the
neuroanatomical data. The majority of these cells respond to more
than one modality. The reaction patterns of the cells can be very
complex, revealing excitation and inhibition for different temporal
components of the response (Erber, 1978). Using only the response
properties of the central neurons, it is not possible to classify
them in defined groups when the cells belong to the "unstructured"
neuropil. With neurons projecting in defined tracts it is much
easier to classify them as "mainly visual" or "mainly olfactory".
Obviously it is difficult, if not impossible, to determine the
functional properties of these central neurons by applying various
sequential stimuli. These cells could reveal specific response

properties if the adequate stimuli were applied in the right temporal order. Relevant sequences and combinations of stimuli can only
be defined by behavioral experiments. Therefore, an electrophysiological analysis of central neurons in the bee brain can only be
successful when it examines first the general response properties of
single neurons and then, in a second step, tests stimulus combinations which are relevant for behavior. The behavioral experiments,
on the other hand, have to be performed under the same experimental
conditions as the electrophysiological analyses.

2. Inputs to the mushroom bodies.

The mushroom bodies receive their main inputs in the calyx area.
From neuroanatomical findings one can conclude that the major inputs
originate in the antennal lobes, that there are tracts connecting
the optic lobes with the calyces and that the mushroom bodies
receive input from the suboesophageal ganglion also (Mobbs, this
volume).

Neurons of the antenno-glomerular-tract (AGT) were analysed
and identified with electrophysiological techniques by Homberg
(1981 and personal communication). These cells can be classified
in different groups:
(1) neurons responding to all antennal stimuli;
(2) neurons which differentiate between mechanical and olfactory
    antennal stimuli;
(3) neurons responding only to mechanical stimuli;
(4) neurons with mechanosensitive input from the thorax and abdomen;
(5) neurons responding to olfactory and sugar water stimuli applied
    to the antenna.
Obviously there is already a remarkable convergence of sensory
information onto first order interneurons of the AGT.

Neurons connecting the calyces with the lobula were identified.
These cells also respond to more than one modality; interestingly
the visual responses were rather weak. Further experiments are
needed to prove whether this is a characteristic of these neurons
or whether they are responsive to highly specific visual stimuli
which were not presented in the experiments.

The neuroanatomical and electrophysiological results concerning
the input of the mushroom bodies are in good agreement. It can be
concluded that the mushroom bodies receive multisensory information.

3. Information processing in the mushroom bodies.

Due to the enormous density of neurons of minute diameters in
the mushroom bodies, it is difficult to record the activity of
intrinsic cells intracellularly. Measurement of extracellular field
potentials with calculation of current source densities is a very
effective method for gaining some insight into signal processing in
this neuropil.

Application of this electrophysiological technique reveals, in
accordance with the single cell measurements of calycal inputs, that
the calyces receive multimodal information. In the alpha-lobes
information processing is also characterized by multimodal interactions. The intrinsic neurons of the pedunculus can be divided

physiologically into two groups. The extracellular analysis also reveals that the physiological responses in the mushroom bodies are characterized by after effects (Kaulen, Erber, and Mobbs, in preparation).

Intracellular recordings of identified intrinsic neurons are in good agreement with the extracellular measurements. Single intrinsic neurons do not reveal action potentials, they show after effects and multimodal response patterns. The physiological data support the assumption that different modalities are processed in the mushroom bodies of the bee.

4. Multimodal movement-sensitive neurons.

Neurons with large dendritic arborization in the lobula project into the median protocerebrum. Many of these cells are movement sensitive, a large proportion of them is direction selective. The majority of these cells respond to various other stimulus modalities. In this respect movement-sensitive neurons have the same response properties as other central neurons of the bee brain.

5. Behavioral modifications.

Bees show various behavioral responses in the laboratory situation; many of these reactions have the properties of reflexes. A relatively simple response, like a reflex, does not help directly with the analysis of central neurons. Usually the reflexes can be elicited by one stimulus modality alone. The modification of specific reflexes, on the other hand, is very useful in the search for neural correlates of behavior.

There are various reflexes which can be modified in the bee when certain stimulus sequences are presented. One reflex which has been analysed also on the neuronal level is the visual antennal reflex in response to moving patterns. Bees move their antennae towards a moving stripe pattern. When the pattern moves e.g., upward, the antennae are moved downward. If the bee is stimulated with sugarwater this reflex is modified in a following test. After sugarwater stimuli to the antennae and the proboscis, bees show a much stronger response for upward moving patterns (Erber, 1981). Sensitization of this reflex is direction specific. Electrophysiological recordings can help to find the site of neuronal change underlying this behavior. Another type of behavioral change is associative odor learning in the laboratory. Bees learn very rapidly to extend their proboscides in response to a conditioning odor. Electrophysiological recordings while the bee is learning can help to define the role of single neurons during conditioning.

6. Neural correlates of behavior.

Two behavioral modifications were analysed on the neuronal level: sensitization of the antennal reflex and associative olfactory learning. For the visual reflex one can assume that movement sensitive neurons change their direction specific response after sugarwater stimuli. To test this hypothesis, intracellular recordings were made from movement sensitive neurons. The response

pattern of these cells was measured, then the animals were stimulated several times with sugarwater and the response of the neuron
was tested again. At the end of the experiment Lucifer Yellow was
injected into the cell.

These experiments revealed that movement sensitive neurons can
change their direction specificity after sugarwater stimulation. The
changes on the neuronal level compare directly with the changes
observed in behavior. The neurons analysed so far have had projections in the lobula and the median protocerebrum. They responded to
the movement stimulus and also to sugarwater stimuli. Movement
sensitive cells which did not respond to sugarwater did not reveal
any response changes during the experiments.

A similar rule applies to olfactory interneurons whose response
properties change during olfactory conditioning. After conditioning
these cells show an increase of the olfactory response, whereas the
responses to other modalities remain unmodified. The electrophysiological recordings establish direct correlates of behavioral change
on the neuronal level.

SUMMARY AND DISCUSSION

Central neurons in the bee brain are characterized by a high
degree of sensory convergence. First order interneurons of the
antennal system already show this feature. The mushroom bodies
receive multisensory information. Processing of sensory inputs in
this system involves visual, olfactory, mechanical, and gustatory
stimuli. A general characteristic of mushroom body neurons is the
pronounced after effect which can last for several seconds. A
possible function of this maintained activity is a comparison of
previous and present sensory stimuli. Behavioral modifications
during electrophysiological recordings show that multisensory convergence in central neurons can result in modulations of sensitivity
and responsiveness. The experiments done so far suggest that plasticity of the response is a feature of higher order sensory interneurons and that the response changes are due to postsynaptic interactions.

REFERENCES

Erber J., 1978. -- Response characteristics and after-effects of
    multimodal neurons in the mushroom body area of the honey bee.
    Physiol. Entomol., 3, 77-89.
Erber J., 1981. -- Neural correlates of learning in the honey bee.
    Trends in Neuro Sciences, 4, 270-273.
Hertel H., 1980. -- Chromatic properties of identified interneurons
    in the optic lobe of the bee. J. Comp. Physiol., 137, 215-231.
Homberg O., 1981. -- Recordings and Lucifer Yellow stainings of
    neurons from the tractus olfacto globularis in the bee brain.
    Dt. Zool. Ges., in press.

Neuronal (?) Control of Heart Rate in Honey Bees

Harald Esch, University of Notre Dame

The dorsal vessel in <u>Apis</u> <u>mellifera</u> is an important component of the circulatory system. The part lying in the abdomen is called the heart. It pumps hemolymph from the hemocoele of the abdomen through the aorta, and empties it into the head. The hemolymph returns to the abdomen through the body cavity.

The rhythmic beating of the heart can be observed optically. Beat frequencies between 60 and 120 bpm (beats per minute) have been reported (Freudenstein, 1928). There is no agreement whether the frequency is controlled by the nervous system or by myogenic action (Snodgras, 1956).

We attempted to record the electrocardiograms of honey bees and tried to manipulate their heart beat frequencies. We hoped to determine the frequency range, and to find the beat control mechanisms.

MATERIALS AND METHODS

Worker bees were caught at the entrance of a hive during the summer and immobilized in a refrigerator at 8°C. A holder was glued to the dorsal side of the thorax.

Tungsten wire electrodes, insulated except at the tip, were inserted at the dorsal midline of the 6th abdominal segment, and laterally into the middle of the 5th segment. Both electrodes were held in place by a small drop of a wax-resin mixture. The electrodes were connected to the differential input of a Tektronix Model 122 preamplifier. The preamplifier output was connected to a cardio-tachometer coupler in a Beckman strip chart recorder (Model R411), and to an oscilloscope.

Thorax and environmental temperatures were measured with the Beckman recorder using two micro-thermistors (Fenwal BC32-11).

RESULTS

1. Heart beat frequencies and electrocardiograms
    Animals were suspended from a holder. They held a styrofoam ball (diameter 2cm) with their legs and "walked" on it. They could be observed for several days if they were fed from time to time.

347

Heart beat frequencies stayed between 60 and 120 bpm if the thorax remained at the environmental temperature, and the indirect flight muscles were not active (Esch, 1960). A clean electrocardiogram (Figure 1A) could be optically synchronized to the heart beat. In most animals intervals between beats were fairly constant. Some individuals showed arhythmic beating patterns from time to time. The beat stopped for a second or two, and then resumed with some acceleration. The "resting frequency" appeared only as an average over 20 to 30 seconds (Figure 2).

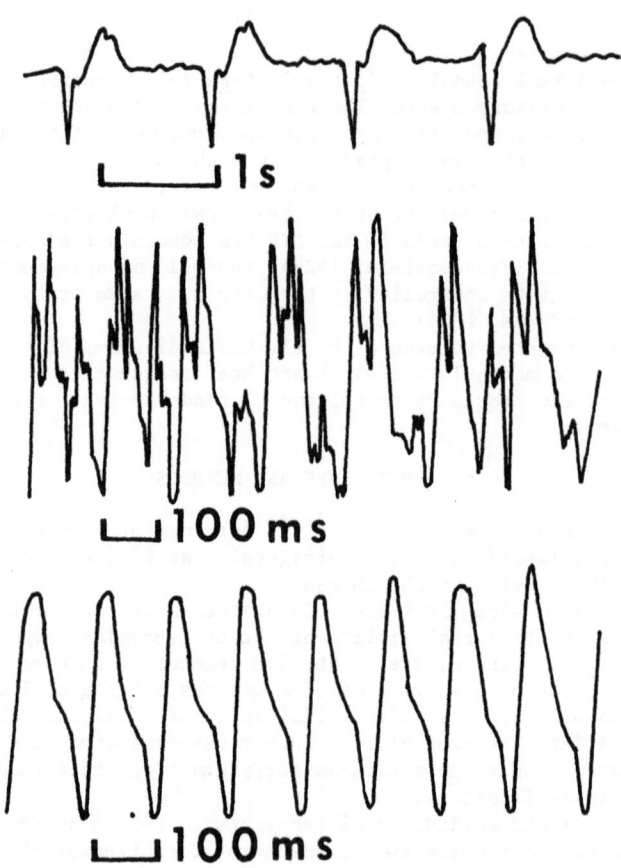

Figure 1. --Electrocardiograms of a honey bee. A.Resting. B.Actively heating the indirect flight muscles. C.Cooling after use of the flight muscles for heating. For further explanations see the text.

The heart beat frequency began to rise as soon as an animal increased the thorax temperature above the environmental temperature by shivering of the indirect flight muscles. Such a heating phase

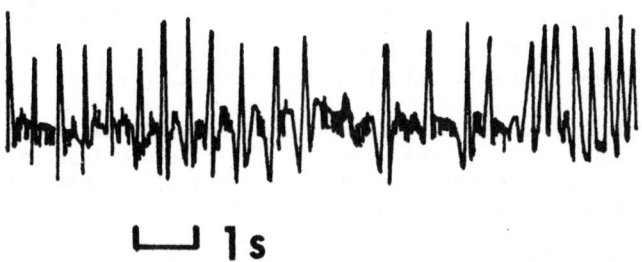

Figure 2. -- Irregular electrocardiogram of a "resting" animal.

is shown in figure 3. The body temperature increased to 34°C at an
environmental temperature of 21°C. The heart beat frequency reached
about 600 bpm. The animal dropped the styrofoam ball and flew (first
arrow) for one minute. It stopped as the styrofoam ball was given
back to it (second arrow). A second heating attempt was made after
it lost the styrofoam ball again (third arrow), but it stopped as
soon as it could reach the ball again. Thorax temperature and heart
beat frequency returned to the resting level within a few minutes.

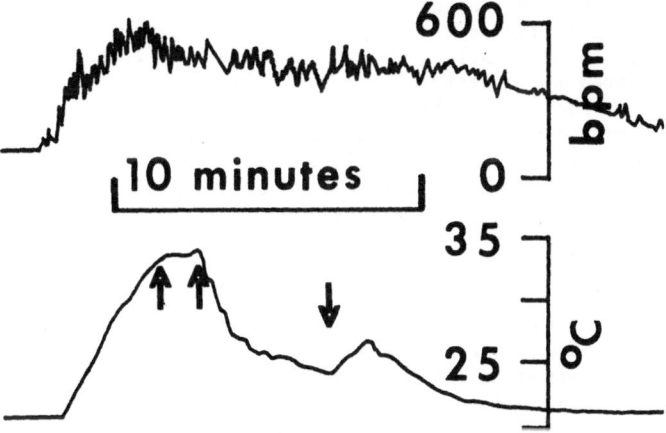

Figure 3. --Heart beat frequency (above) and thorax temperature
(below) during heating, flight, and cooling. See text for
explanation.

The electrocardiogram increased in amplitude during muscle
heating, but it also became very "noisy" (Figure 1B). Oscillations
of a higher frequency were superimposed. The basic beat frequency
could still be determined, but the trigger of the heart rate

coupler was sometimes lost for a few beats. This accounts for the "spikes" in the recording of the heart beat frequency. Heavy pumping movements of the abdomen could be correlated to the disturbances. Dislodging of the recording electrodes or action potentials from other than heart muscles could have been the cause of the "noise". When the active heating stopped, the electrocardiogram amplitude remained high, but the noise disappeared (figure 1C).

2. Heart beat frequency and experimental manipulations

Pinching of the antennae (Norman, 1972), electrical stimulation of the brain or the ventral nerve cord with a wide variety of stimuli (stimulus duration from 0.1 to 1000 milliseconds, voltages between 0.1 and 150 volts) had no effect on the heart beat frequency. Even changes of the environmental temperature from 17° to 36°C, during which the thorax temperature stayed close to the environmental temperature, did not affect the beat frequency.

An amputation of the abdomen revealed exitatory and inhibitory nervous regulation of the heart beat frequency from the thoracic ganglion mass in the house fly (Miller, 1979). Cutting the nerve cord between thorax and abdomen did not change the heart rate in bees.

We cut the petiole carefully, pulled a piece of the aorta from the thorax, and pushed it back into the ventral hemocoele of the abdomen. The hemolymph could circulate, and these preparations survived from hours to days. The heart beat stayed constant between 60 and 120 bpm. Often a Ringer solution was injected from time to time, to prevent a drying of the preparation. A modified Waddington's solution of the following composition was used: 40 mM/1 Na; 1.8 mM/1 $CaCl_2$; 1.3 mM/1 KCl; 2.4 mM/1 NaHCO. Up to several hundred microliters of this solution could be injected without any effect on the heart beat frequency. However, when the sodium concentration of this solution was altered, an injection of a few microliters caused remarkable changes: Concentrations above 40 mM/1 decreased the heart rate. The beat stopped often at 130mM/1. The arrest lasted from a few minutes up to one hour. After this time the normal resting frequency returned slowly. An injection of 100 microliters of an unchanged Waddington's solution brought the normal heart beat back in seconds in such preparations. Solutions with Na concentrations under 40mM/1 increased the beat frequency. A concentration of 20 mM/1 led in many animals to the highest frequencies observed in intact animals (840 bpm). A few microliters of distilled water had a similar effect.

In all experiments in which an injection lead to an increase in the beat frequencies, the electrocardiogram was very similar to the one shown in figure 1C. Altering the concentrations of the other ions in the injected solution did not change the heart beat.

DISCUSSION

We found a much wider range of heart beat frequencies than reported in the literature. Earlier researchers observed the heart beat optically (Freudenstein, 1928). It is easy to miss beat frequencies above 120 bpm with that methode.

Judging by the commonly used criteria, it can be assumed that the heart rate in bees is not under direct nervous control. Electrical stimulation of the nervous system, cutting of all connections between thorax and abdomen, or severing of the neuronal link between ventral nerve cord and heart (Freudenstein, 1928) had no effect.

The sensitive reaction of the beat frequency to the sodium concentration in the body fluid is noteworthy. Sidie (1977) found sodium concentrations between 40 and 55 mM/l in the hemolymph of worker bees, who did not fly before the test. In workers returning from a foraging flight, sodium concentrations amounted to 28 to 34 mM/l. The concentration of all other ions tested did not change. Sidie's results, together with our observations, suggest, that the sodium concentration in the hemolymph might be used to control the heart beat frequency. Heart beat frequencies increased only after an activation of the big, indirect flight muscles. That, however, led to a change in the sodium concentration of the hemolymph.

REFERENCES

Esch H., 1960. --Ueber die Koerpertemperaturen und den Waermehaushalt von Apis mellifica. Z. vergl. Physiol., 43, 305-335.

Freudenstein K., 1928. --Das Herz und das Circulationssystem der Honigbiene. Z. wiss. Zool., 132, 407-475.

Miller Th., 1979. --Nervous versus neurohormonal control of insect heartbeat. Am. Zool., 19, 77-86.

Normann T.C., 1972. --Heart activity and its control in the adult blowfly. J. Insect Physiol., 18, 1793-1810.

Sidie J.M., 1977. --Chemical changes in the hemolymph of foraging honeybees as a result of flight. Amer. Zool., 17, 918.

Snodgras R.E., 1956. --Anatomy of the Honey Bee. Comstock Publishing Associates, 334p.

Circadian and Other Non-Visual Inputs to Identified Visual Interneurones in the Honey Bee

Walter Kaiser,
Technische Hochschule Darmstadt

The optomotor system of insects has been often regarded as inflexible. Long-term recordings from single neurones in the lobula of worker honey bees have provided indications that this is not the case (Kaiser, 1979). The directionally specific, movement sensitive large-field neurones under study here are intimately involved in optomotor behaviour (Kaiser, 1975).

MATERIALS AND METHODS

The neuronal responses to movement of a striped pattern illuminated continuously with constant intensity were recorded extracellularly throughout the preparation's lifetime (up to 4.5 days). The pattern was moved every minute for 5 s in the preferred direction of the neurone. The temperature was held constant at approximately 25° C.

Intracellular recordings made with glass pipettes filled with fluorescent dyes (Procion Yellow or Lucifer Yellow) lasted for up to 40 minutes. Following physiological characterization, dye was injected into the cell in order to subsequently study its morphology and connectivity.

RESULTS

1. Endogenous sensitivity changes.

There are two types of endogenous cyclic sensitivity changes which differ in their periods: circadian and ultradian variations. During the subjective day, the neurones respond strongly to pattern movement; in the night the response is weak. The period length of this endogenous sensitivity oscillation is approximately 22 h; hence they are circadian. The spontaneous activity in the absence of pattern movement as measured immediately before each movement stimulus follows the same course. The periodic, endogenous changes in the responsiveness of the neurones are probably due solely to sensitivity changes, since response-log I functions based on data collected at various times of the day are only displaced along the intensity axis whilst remaining parallel to each other.

Superimposed on this circadian rhythm is an ultradian

rhythm whose period length seems to differ from animal to animal and also to change with the time of day. The period lengths so far found ranged from 5 min to 2 h.

The sensitivity changes are mainly due to hormonal and/or neuronal effects on the neurones themselves or on preceding inter-neurones. This is concluded from the fact that the spontaneous activity of the neurones also shows circadian rhythmicity when the animal is kept in complete darkness. Sign and magnitude of these variations are in agreement with the changes in spontaneous activity recorded during pattern illumination. Only the ultradian oscillations are less obvious in the dark.

2. Exogenous sensitivity changes.

Recently a further non-visual input to these neurones has been discovered: input from mechanoreceptors (Kaiser et al., 1981). The neurones respond to deflection of mechanoreceptive hairs on the compound eye and/or the head. The hairs were bent by carefully-aimed air puffs delivered in complete darkness by a fine pipette. (The antennae had been removed.) If one applies this stimulus to the animal during the subjective night when the neurones are insensitive to visual stimuli or during insensitive phases of the subjective day, the neurones immediately show an increased sensitivity to light; they "wake up". The same effect is elicited by light pulses given to that eye which is not looking at the moving pattern. Apparently the responsiveness of the neurones is dependent on the general stimulus level in the outside world. Thus the responsiveness can adapt appropriately to the environmental situation.

Proprioceptors also influence the sensitivity of the movement detectors: when the neurones are insensitive, spontaneous bursts of nerve impulses sometimes occur. Such a spontaneous burst is always followed by a sudden increase in the sensitivity for pattern movement.

3. Identification of the movement detectors.

There are good reasons to believe that we have identified the neurones from which the above-mentioned extracellular long-term recordings were made: we recorded intracellularly from neurones whose location and physiological properties corresponded to those of the neurones from which I had previously recorded extracellularly (DeVoe et al., in press). The neurones receive excitatory and inhibitory input from both compound eyes. The input from one of the eyes is dominant. The preferred direction of movement for each of the two eyes is such that the neurones respond best to the rotation of the external world which the animal would experience, e.g., when it is turned passively by wind.

Injection of dye into these neurones revealed the following features. Each neurone branches extensively in the lobulae of both optic lobes. The axon which connects both arborizations leaves one lobula through the anterior optic tract, passes through the optic tubercle, runs across the frontal part of the brain and enters the other lobula via its optic tubercle and anterior optic tract. The cell body lies just ventral to the anterior optic tract, near the optic tubercle.

Other movement-sensitive cells were found which cross the brain via a second path, the inferior optic commissure (Kenyon, 1896).

4. Correlation with behaviour.

We expect to obtain further insights into the sources of the sensitivity changes of the movement detectors by comparing their responsiveness with the ongoing simultaneous motor activity of the bee. A first step in this direction has been made. Long-term recordings (up to 5 days) of individual worker bees walking on a tread-mill showed a circadian rhythm as well as ultradian rhythm(s) (Kaiser and Bösebeck, in preparation).

DISCUSSION

The circadian sensitivity changes in the worker honey bee are - to my knowledge - the first ones reported for high-order interneurones which are intimately involved in a specific behavioural reaction. Circadian changes in the sensitivity of single visual neurones have been reported previously only for Procambarus (Aréchiga and Wiersma, 1969) and Limulus (Barlow et al., 1977).

Ultradian sensitivity changes in the visual system have been reported in another invertebrate: the ERG-amplitude in Procambarus fluctuates with a period of approximately 2 h (Sánchez and Fuentes-Pardo, 1977).

In locusts non-rhythmic, short-term sensitivity changes have been demonstrated in certain visual interneurones (Rowell, 1971; Kien, 1976). The sensitivity of these interneurones can be increased by external stimuli. Kien's results are of particular importance: she studied optomotor neurones. These neurones in Locusta are in some respects similar to the interneurones of the bee's optomotor system described above. They show both spontaneous and evoked changes in their activity. However, it is not yet known whether these neurones exhibit rhythmicity.

The interesting feature of our investigation of honey bee visual interneurones is that large-field movement detectors which are part of a reflex-like behaviour (optomotor response) show plasticity. This is expressed in their circadian and ultradian rhythmicity and also in the increase in their sensitivity caused by various external stimuli.

REFERENCES

Arechiga H., Wiersma C. A. G., 1969. -- Circadian rhythm of responsiveness in crayfish visual units. J. Neurobiol., 1, 71-85.
Barlow R. B., Jr., Bolanowski S. J., Jr., Brachman M. L., 1977. -- Efferent optic nerve fibers mediate circadian rhythms in the Limulus eye. Science, 197, 86-89.
DeVoe R. D., Kaiser W., Ohm J., Stone L. S., 1982. -- Horizontal movement detectors of honey bees: Directionally-selective visual neurons,in the lobula and brain. J. Comp. Physiol., in press.

Kaiser W., 1975. -- The relationship between visual movement detection and colour vision in insects, pp. 359-377, in: The Compound Eye and Vision of Insects, Horridge G. A. (ed.). Oxford, Clarendon Press.

Kaiser W., 1979. -- Circadian variations in the sensitivity of single visual interneurones of the bee Apis mellifica carnica. Verh. Dtsch. Zool. Ges., 1979, 211.

Kaiser W., Büsebeck H., in preparation. -- Motor activity of individual worker honey bees walking on a tread-mill.

Kaiser W., DeVoe R., Ohm J., 1981. -- The influence of non-optic stimuli on the sensitivity of identified visual interneurons in the optic lobe of the honey bee. Verh. Dtsch. Zool. Ges., 1981, 173.

Kenyon F. C., 1896. -- The brain of the bee: A preliminary contribution to the morphology of the nervous system of the arthropoda. J. Comp. Neurol., 6, 133-210.   ·

Kien J., 1976. -- Arousal changes in the locust optomotor system. J. Insect Physiol., 22, 393-396.

Rowell C. H. F., 1971. Variable responsiveness of a visual interneurone in the free-moving locust, and its relation to behaviour and arousal. J. exp. Biol., 55, 727-747.

Sánchez J. A., Fuentes-Pardo B., 1977. -- Circadian rhythm in the amplitude of the electroretinogram in the isolated eyestalk of the crayfish. Comp. Biochem. Physiol., 56A, 601-605.

(Supported by the Deutsche Forschungsgemeinschaft under SFB 45: "Vergleichende Neurobiologie des Verhaltens".)

Short-Term Memory in the Honey Bee

Randolf Menzel,
Freie Universität Berlin (FRG)

The time dependence of memory formation provides a powerful tool in psychological studies on learning. Pioneer papers by Müller and Pilzecker late last century demonstrated for human verbal learning an initial state, which is sensitive to interference from new incoming information or from diversion of attention, and a later, consolidated state of firmly established memory. Since then a large literature has accumulated, from studies both in humans and other vertebrates, which describes various psychological and physiological aspects of different temporal memory phases. Whether this time dependence is a continuous process with an even change in the properties of memory or whether memory formation runs through several (three or more) different phases with short transition periods between them and clear distinctions in their physiological properties is still unresolved. Nevertheless, there is no question that memory formation is time dependent and has different qualities, seconds, minutes, hours, days and years following the initial learning experience.

The specific qualities of early and late memories may differ between animal species and between learning tasks. For example, the sensitivity of an early phase to new information or experimental extinction may be high or low, or the time span of an initial memory, which is susceptible to inhibitors of protein synthesis or to electro-convulsive shock (ECS), differs for seemingly small changes in the experimental procedure employed. In most cases, furthermore, it is unknown to what extent the time dependence of memory reflects changes in the substrate of memory storage, or changes in the retrieval of the stored memory. In spite of all these uncertainties a careful analysis of the temporal characteristics should elucidate physiologically relevant features of memory.

In the bee, we have found temporal phases in various associative learning behaviors, which have surprising similarities with memory phases in laboratory mammals and humans. The bee offers a number of advantages for this kind of behavioral study, which are related to its life as a social insect. As we are able to combine electrophysiological, neuropharmacological and neuroanatomical techniques with behavioral tests, we hope to add some insight to the general question of the substrate of the different memory phases.

Honey bees are little learning machines. On their first flight out of the hive they learn quickly the location of their colony with respect to the surrounding landmarks and relative to the sun compass. They learn the color, shape and odor of the hive entrance. On increasing excursions during the following days they learn landmarks further away, use them for relocating their colony and for adjusting their astronomical compass. When a forager bee flies out to collect nectar and pollen from flowers, it learns quickly the direction and distance of the food source, the landmarks, which guide it towards the food source and the signals of the food source itself; color, shape, odor and the distance from immediate surrounding landmarks. These signals of the food source are learned very quickly: natural odors within one trial, colors within one to five trials, black and white patterns within 5-20 trials. The honey bee is able to choose the learned food source with extreme selectivity, because each bee is continuously informed about the food supply in the surroundings by the dances of its hive mates and the food spreading within the colony. This means that the individual bee does not have continuously to explore to find out if it is still working (i.e., food collecting) in the most economical way - as do other flower visiting insects. Both under natural conditions and in behavioral tests, bees choose a food source out of two or more alternatives with more than 90% correctness.

Besides their rapidity of learning and accuracy of choices, bees offer other advantages for behavioral tests: (1) low genetic variation, because all test animals are sisters, (2) test animals are about equal in age and in the same behavioral status (foragers), (3) they arrive at the training station in a highly motivated state, (4) new and naive test animals may be summoned through the dance recruitment of previously exposed test animals, (5) non-reinforcement in the test-exposure has little if any effect on the choice behavior in short-lasting training experiments, (6) one test animal can easily deliver 2,000 choices a day or more than 30,000 in its lifetime.

The learning process is bounded by genetic constraints--a most interesting but little studied field, which will not be explored further in my talk. Its true associative nature has been demonstrated in bees for all aspects commonly thought to define associative learning, including stimulus specificity, acquisition by repetition of learning, necessity of temporal association between the stimuli and the motor action involved, and long duration. For our purposes here it is important to bear in mind that a single learning trial changes the choice between two alternatives from initially a choice of equal probability to 95% correct choices in an odor conditioning experiment and to 75% in a color conditioning experiment when e.g., blue is rewarded and yellow is the alternative. This highly significant change in behavior is true both for free-flying bees and for bees glued to a stage and conditioned through its proboscis reflex using an odor as the conditioned stimulus. Furthermore, the level of correct choices after one reward is almost insensitive to the amount and quality of the rewarding stimulus (sugar water). All the experiments discussed in detail in the talk use this one-trial learning paradigm to study the time-dependence of memory formation in bees.

I will present the results of experiments which demonstrate a sensory memory in bees which lasts for 2-3 secs, and an initial memory phase which is sensitive to experimental extinction, interference with new and contradictory learned information, experimental procedures like ECS, narcosis and cooling. The transition from the initial memory phase, lasting a few minutes, to a stable memory, lasting a lifetime, is seen in a consolidation process, which is represented by an increase of learned performance from about 3 to 15 mins after the single learning trial. Interference by new and contradictory information elucidates an important subdivision of the initial memory phase into an early period (lasting about 30 secs) during which recently stored information is erased very efficiently by a new learning trial.

Next the question will be addressed to how the initial memory phase and the long-lasting memory are interconnected. Erber's 1975 experiments with massed learning trials show that the transition time from the initial to long-lasting memory is not constant but depends on the experimental procedure. Two possible explanations will be given and new experiments outlined.

The susceptibility of the initial memory phase to cooling has been used to study the participation of various brain structures in the transition to long-lasting memory in an olfactory learning task. The time courses are described for the antennal lobes, two lobes of the mushroom body (calyx and α-lobe) and the lateral protocerebrum. The general conclusion is that the initial memory phase is not specifically located in one of the neuropiles tested; rather both the sensory neuropile and the mushroom body are involved, however with clear differences in their time courses. The hypothesis by Vowles of a reverberating circuit between antennal lobes and mushroom bodies as a substrate of a memory trace is rejected.

Storage of learned information is a compound process, the mechanistic basis of which might be expected to act at many levels of the nervous system, starting perhaps with modifiable ion channels, involving the release of modulatory transmitters acting on specific (?) synaptic sites, activating specific (?) elementary networks for the association and maintenance of sensory excitation. Finally the preservation or creation of synaptic contacts may be affected. An intuitive hypothesis correlates the temporal phases of memory formation with the different structural substrates of the engram (Hebb). An experimental approach based upon such a working hypothesis would be incisive if the temporal analysis of learning selectively revealed those factors affecting storage mechanisms, and were less influenced by retrieval mechanisms, motivational effects and other factors of a behavioral method (for example, learning in the test situation, sensory discrimination capacity).

REFERENCES

Erber J., 1975a. -- The dynamics of learning in the honeybee (Apis mellifica carnica). I. The time dependence of the choice reaction. J. Comp. Physiol., 99, 231-242.

Erber J., 1975b. -- The dynamics of learning in the honeybee (Apis mellifica carnica). II. Principles of information processing. J. Comp. Physiol., 99, 243-255.

Erber J., 1976. -- Retrograde amnesia in honeybees (Apis mellifica carnica). J. Comp. Physiol., 90, 41-46.

Erber J., Masuhr T., Menzel R., 1980. -- Localization of short-term memory in the brain of the bee. Physiol. Entomol., 5, 343-358.

Frisch K von, 1965. -- Tanzsprache und Orientierung der Bienen. Springer Verlag Berlin Heidelberg New York.

Lindauer M., 1970. -- Lernen und Gedächtnis - Versuche an der Honigbiene. Naturwiss., 57, 463-467.

Menzel R., 1967. -- Untersuchungen zum Erlernen von Spektralfarben durch die Honigbiene (Apis mellifica). Z. vergl. Physiol., 56, 22-62.

Menzel R., 1968. -- Das Gedächtnis der Honigbiene für Spektralfarben I. Kurzzeitiges und langzeitiges Behalten. Z. vergl. Physiol., Physiol., 60, 82-102.

Menzel R., 1969. -- Das Gedächtnis der Honigbiene für Spektralfarben II. Umlernen und Mehrfachlernen. Z. vergl. Physiol., 63, 290-309.

Menzel R., 1979. -- Behavioral access to short-term memory in bees. Nature, 281, 368-369.

Müller G. E., Pilzecker A., 1900. -- Experimentelle Beiträge zur Lehre vom Gedächtnis. Z. Psychol., 1, 1-288.

The Effects of Amines on Behavior and Neural Activity in the Honey Bee

Alison R. Mercer,
Freie Universität Berlin (FRG)

Biogenic amines have been identified in the CNS of insects, but due to a paucity of information from pharmacological and biochemical studies, the functions of amines in the insect CNS remain obscure. The predominant amines in insect nervous systems are the catecholamine dopamine, the phenolamine octopamine, and the indolalkyamine 5-hydroxytryptamine.

Catecholamine-containing fibres have been identified histochemically in the brain of the honey bee (Elofsson and Klemm, 1972; Klemm, 1974, 1976), evidence of 5-hydroxytryptamine-containing fibres was not found in these studies. The functions of amines in the insect can be investigated using behavioral as well as electrophysiological techniques. The large amount of information available concerning olfactory learning and behavior in the honey bee makes this animal particularly suitable for an investigation of the roles played by biogenic amines in the insect CNS, and their involvement in the control of behavior. Studies are now underway to determine the effects of exogenously applied amines on the behavior and neural activity of the honey bee. A brief account of this work is presented in the following report.

MATERIALS AND METHODS

1. Behavioral studies.

Associative learning in the honey bee can be demonstrated by pairing a reflex, such as proboscis extension in response to sugar water stimulation of the antenna (Kuwabara, 1957), with an olfactory (or visual) stimulus. The proboscis extension reflex was used in this study to investigate the effects of exogenously applied amines on conditioned and unconditioned responses to olfactory stimuli. Fixed honey bees were conditioned to an orange scent in a one-trial learning routine. Amines were injected directly into the brain of the bee as described by Mercer and Menzel (1980).

2. Electrophysiology.

Averaged evoked potentials resulting from stimulation with light, scent or air were recorded in the alpha lobe of the mushroom body of the bee brain, as described by Mercer and Erber (in preparation). The effects of iontophoretic application of dopamine or octopamine on the size of evoked potentials were measured. Amines were injected into either the alpha lobe or calyx of the mushroom body.

RESULTS

1. Behavioral studies.

Responsiveness to unconditioned olfactory stimuli is enhanced by octopamine. 5-hydroxytryptamine and dopamine have little effect on the numbers of bees responding to unconditioned olfactory stimuli. However both 5-hydroxytryptamine and dopamine reduce responses to a conditioned olfactory stimulus. Reduction in the percentage of bees responding to the conditioned stimulus occurs regardless of whether 5-hydroxytryptamine or dopamine is injected before or after one-trial learning. The effects of dopamine are more powerful than those of 5-hydroxytryptamine. Inhibition of responses resulting from dopamine treatment was found to be time dependent. Response levels initially reduced by dopamine show a recovery in bees retested 60 minutes after drug administration.

2. Electrophysiology.

Iontophoretically applied octopamine increases the size of evoked potentials resulting from light stimuli. Potentiation of the light response occurs very rapidly if octopamine is injected into the calyx, the maximum effect being reached 3-5 minutes after drug administration. When octopamine is injected into the alpha lobe, the increase in size of light-evoked potentials takes much longer to become apparent, and reaches a maximum 15 minutes after octopamine is applied.

Evoked potentials resulting from stimulation with air or scent are reduced by octopamine, and the sign of the potentials is often reversed. The effects of octopamine on potentials evoked by scent and air are particularly marked when the amine is injected into the calyx. Injection of octopamine into the alpha lobe produces similar results, but the effects take significantly longer to develop.

In contrast to octopamine, dopamine injected into the calyx or alpha lobe has little effect on evoked potentials resulting from stimulation with light. However, potentials evoked by stimulation with air or scent are immediately reduced by the application of dopamine to either the calyx or the alpha lobe. A recovery in responses is seen 10-15 minutes after dopamine treatment. Dopamine, like octopamine, tends to reverse the sign of potentials resulting from stimulation with scent or air.

DISCUSSION

Exogenously applied amines influence the responsiveness to olfactory stimuli and the retrieval of information during memory recall in the honey bee. Response levels to a conditioned stimulus are dramatically reduced by the primary catecholamine dopamine. However, dopamine does not appear to affect information storage.

Reduction in response levels to a conditioned olfactory stimulus occurs regardless of whether dopamine is applied before or after conditioning, and response levels initially reduced by dopamine recover in later tests. Treatment with octopamine does not result in the reduction of behavioral responses to olfactory stimuli.

Octopamine and dopamine also affect neural activity in the brain of the bee in different ways. The electrophysiological studies provide us with more direct information about the nature and sites of action of these amines in the brain of the honey bee. Receptors mediating the effects of octopamine on light-evoked potentials appear to be in the calyx because potentiation of responses to light becomes apparent immediately after administration of octopamine into this region of the brain. Dopamine appears to have little effect on the octopamine-sensitive receptor system which modulates responses to light stimuli. The delayed onset of the effects of octopamine injected into the alpha lobe on potentials resulting from stimulation with light, suggests that the receptors mediating these effects are located some distance away from the injection site, and are reached, after a certain time interval, by diffusion of octopamine through the brain tissue.

Evoked potentials resulting from stimuation with air or scent are reduced by both octopamine and dopamine. The effects of these amines become apparent immediately if dopamine or octopamine is injected into the calyx. Further studies are required to determine whether both amines activate the same receptor mechanism, or whether the effects of dopamine and octopamine involve different receptor sites. The fact that dopamine has an immediate effect on responses to air and scent when applied to the alpha lobe, but octopamine does not, suggests the latter may be true.

There is good evidence that both dopamine and octopamine are present in the mushroom bodies of the insect brain (Evans, 1980), a region of the brain known to play an important role in the process of olfactory learning in the honey bee (Menzel et al., 1974; Erber et al., 1980). Monoamine-containing fibres have been located in the calyx, pedunculus and alpha and beta lobes of the bee brain (Klemm, 1974, 1976). The globuli cells of the mushroom bodies in the brain do not show aldehyde-induced fluorescence (Klemm, 1974, 1976), but do stain with neutral red (Mercer, unpublished observation), a dye which has been used to selectively stain amine-containing cells (Stuart et al., 1974). Intense staining with neutral red also occurs in the globuli cell layer of the mushroom bodies of the cockroach P. americana, and radioenzymatic assays have been used to show that in the cockroach, the globuli cell bodies, as well as the calyx itself, contain considerable quantities of octopamine (Dymond and Evans, 1979).

Multidisciplinary studies will be necessary to understand fully the functions of biogenic amines in the insect nervous system. In the present investigation, behavioral and electrophysiological techniques have been used to investigate amine functions in the bee brain. At this stage, care must be taken in any attempt to combine the results of these different approaches. However it is interesting to note that both behavioral and neural responses to .

olfactory stimuli are reduced by dopamine. Octopamine on the other hand, reduces potentials evoked by olfactory stimuli, but does not reduce behavioral responses to a conditioned olfactory stimulus. Octopamine, unlike dopamine, enhances potentials resulting from light stimuli. Such differences between the effects of dopamine and octopamine on neural activity in the bee brain may be related to the differences observed between the influence of these two amines on behavioral responses. Studies using pharmacological antagonists may help to differentiate the actions of dopamine and octopamine in the bee brain more clearly.

REFERENCES

Dymond G. R., Evans P. D., 1979. -- Biogenic amines in the nervous system of the cockroach, Periplaneta americana: association of octopamine with mushroom bodies and dorsal unpaired median (DUM) neurones. Insect Biochem., 9, 535-545.

Elofsson R., Klemm N., 1972. -- Monoamine-containing neurones in the optic ganglia of crustaceans and insects. Z. Zellforsch. mikrosk. Anat., 133, 475-499.

Erber J., Masuhr T., Menzel R., 1980. -- Localization of short-term memory in the brain of the bee (Apis mellifera). Physiol. Entomol., 5, 343-358.

Evans P. D., 1980. -- Biogenic amines in the insect nervous system. Adv. Insect Physiol., 15, 317-473.

Klemm N., 1974. -- Vergleichend-histochemische Untersuchungen über die Verteilung monoamin-haltiger Strukturen im Oberschlundganglion von Angehörigen verschiedener Insektenordnungen. Ent. germ., 1, 21-49.

Klemm N., 1976. -- Histochemistry of putative transmitter substances in the insect brain. Prog. Neurobiol., 9, 99-169.

Kuwabara M., 1957. -- Bildung des bedingten Reflexes von Pavlovs Typus bei der Honigbiene (Apis mellifera). J. Fac. Sci. Hokkaido Univ. Ser. VI, Zool., 13, 458-464.

Menzel R., Erber J., Masuhr T., 1974. -- Learning and memory in the honeybee, pp. 195-217, in: Barton-Browne L. (ed.), Experimental Analysis of Insect Behavior. Springer, Berlin Heidelberg New York.

Mercer A. R., Menzel R., 1980. -- The effect of biogenic amines on conditioned and unconditioned responses to olfactory stimuli in the honeybee Apis mellifera. J. Comp. Physiol., in press.

Mercer A. R., Erber J. -- The effects of dopamine and octopamine on average evoked potentials recorded in the alpha lobe of the mushroom bodies of the bee brain. In prep.

Stuart A. E., Hudspeth A. J., Hall Z. W., 1974. -- Vital staining of specific monoamine containing cells in the leech nervous system. Cell Tiss. Res., 153, 55-61.

The Neural Architecture of the Honeybee Mushroom Bodies

Peter G. Mobbs,
Queen Mary College, University of London

The mushroom bodies of the Hymenoptera, perhaps because of their historical associations with the supposed intelligence of social insects, have received considerable attention from behavioural scientists. Wadepuhl and Huber (1979) in a series of systematic stimulation experiments have provided evidence to support the view initially proposed by Huber, on the basis of earlier stimulation and lesioning research, that the mushroom bodies are responsible for the selection and co-ordination of different behaviour patterns. A prominent feature of stimulation experiments is that momentary application of current in the mushroom body area can produce complex sequences of behaviour. Recently Menzel et al (1974) and Erber et al (1980) have demonstrated the association of olfactory memory with the honeybee mushroom bodies.

Despite the fascination of highly patterned neural structures, such as the mushroom bodies, for neuroanatomists, these neuropiles have not received the attention that has been afforded to the crystalline precision of the optic ganglia. The work of Vowles (1955) is the only detailed histological study of the honeybee mushroom bodies. Vowles analysis was, however, based on reduced silver stains and provided no details of the three dimensional organisation in relation to the morphologies of the constituent intrinsic and extrinsic neurons. The present study utilises results obtained from Golgi staining and intraneuropilar injection of cobalt chloride to establish the morphologies and connections of the mushroom bodies.

METHODS

The results described in this study are based upon the parallel analysis of reduced silver, Golgi and cobalt stains. Two reduced silver techniques were routinely employed, the Holmes-Blest, and the Bodian, technique. A variety of Golgi methods were also employed. The application of these techniques to the bee brain are described in detail in a previous publication (Mobbs, 1981). The intraneuropilar introduction of cobalt chloride within the brain utilising broken micro-electrodes was found particularly useful. Strausfeld and Hausen's (1977) technique was adapted as follows. One millimetre glass capillary electrodes were broken to a

twenty-five micron tip diameter and filled with 3 M cobalt chloride.
The capillaries were introduced into the brain through a window cut
in the head capsule by sliding them along a modelling clay platform.
The electrodes were left in the target neuropile for up to one hour
and then withdrawn.   The preparation was then left for a further
hour to allow the cobalt to diffuse.   The heads were then washed
briefly in saline and treated with 0.1% ammonium sulphide.   The
preparations were fixed in alcoholic Bouin and intensified following
the technique of Bacon and Altman (1977).   Cobalt stained material
was embedded in Epon and sectioned at 50 µm.

RESULTS

1.   Intrinsic neurons.
     The mushroom bodies of the bee are paired neuropiles that
dominate the midbrain.   They consist of the compacted arborisations
of approximately 340,000 small intrinsic neurons (Kenyon neurons or
K-cells) all of which have a characteristic lambda shape.   This
morphology is a reflection of the overall form of the mushroom
bodies.   Each K-cell has an arborisation in one or other of the
paired calyces and sends a projection into the pedunculus that
divides entering the alpha and beta lobes.
     The K-cells can be divided into five groups on the basis of the
morphology of their dendritic terminals and the neuropile area of
the calyx in which they arborise (Table 1).   The cell bodies of the
intrinsic neurons are located in one of three positions.   One group
of smaller cell bodies (4-5 µm) forms a conical pile above the centre
of each calyx.   This group is surrounded by another that consists
of larger cells (6-7 µm) that fill the remaining space in the
protocerebrum above the calycal neuropile and spill over into the
space about the lip.   The third group of K-cell bodies form a layer
two to three cells deep beneath each calyx.   The distribution of the
different intrinsic neurons is illustrated in figure 1.
     The K-cells project from the calycal neuropile zone of their
origin into the pedunculus in a regular and ordered fashion.   The
K-cell projections form bundles that represent the peduncular
processes of cells with neighbouring and overlapped dendritic
fields in the same calycal zone.   The two-polar organisation of the
calyces is transformed into a Cartesian one within the pedunculus.
The K-neurons within the basal ring represent the most posterior part
of the pedunculus and beta lobe, those of the collar, the central
region, and those of the lip, the anterior layer.   Within the alpha
lobe the K-cells of the lip form the most ventral layer, and those
of the basal ring the most dorsal, with collar fibres located
centrally.   As a result of the two-polar to Cartesian transform the
calyces can be considered as a series of parallel arrays in which
each calyx represents a continuation of the arrays of its partner.

Type	Dendritic Morphology	Field Geometry	Soma Position	Distribution
K I	Dense spiny	All similar – spheroidal to flattened	Around lip	Lip only
K II	Dense spiny	Some variation but mainly cylindrical to flattened	Outer cell body group above calyx	Collar only
K III	Sparse spiny	Several different geometries, some stratified	Inner cell body group above calyx	Basal ring only
K IV	"Lumpy"	Several different geometries, typically very small field	Inner cell body group above calyx also below basal ring	Basal ring only
K V	Clawed	Several different geometries, some stratified	Inner cell body group above calyx, also below collar and basal ring	Collar and basal ring – some dendrites penetrate lip

Table 1.   Distribution and characteristics of K-cells

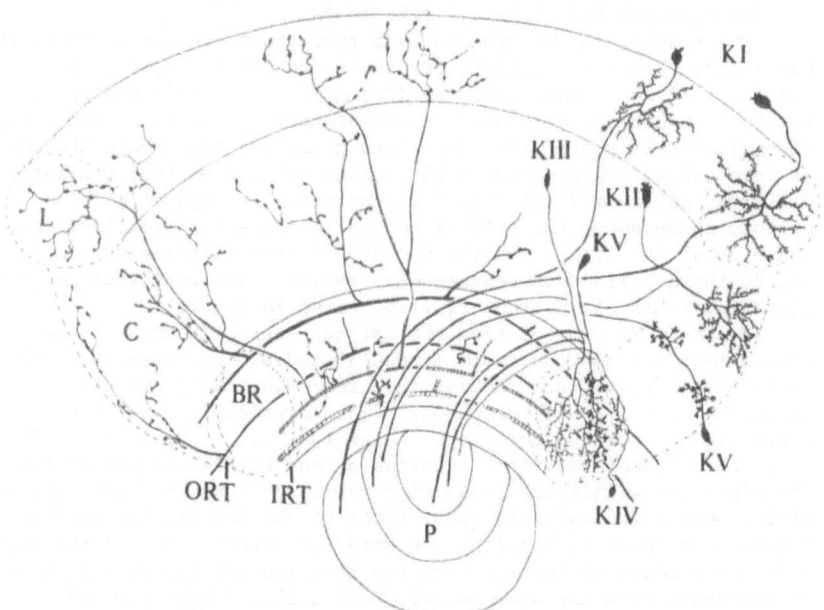

Figure 1   Diagram to show K-cell geometries in relation to the distribution of the primary calycal inputs from the antenno-glomerular tracts running within the inner ring tract (IRT) and optic fibres in the outer ring tract (ORT). (L – lip, C – collar, BR – basal ring).

2.   Extrinsic input neurons.
     There are three major origins for input to the calyces. Fibres
between the antennal lobes and the calyces form three tracts.
Termed the antenno-glomerula tracts (AGTs). The AGT's carry fibres
with a variety of morphologies. The majority, however, arise in the
antennal lobe and terminate in the lip zone of calyx. The second
major input to the calyces arises from the optic lobes and fibres
enter the calyx from three tracts with origins in the medulla and
lobula.   These cells arborise in the basal ring and collar zones.
The fibres of optic origin terminate at different depths within the
collar and there is some evidence to suggest that the collar is a
map, in its depth, of the outer face of the medulla.   The
morphology of some of these primary input neurons are illustrated in
figure 2.   The third main source of calycal input are feedback
neurons running from the ipsilateral alpha and beta lobes.   These
feedback neurons run within the protocerebro-calycal tract
(PCT) (figure 3).

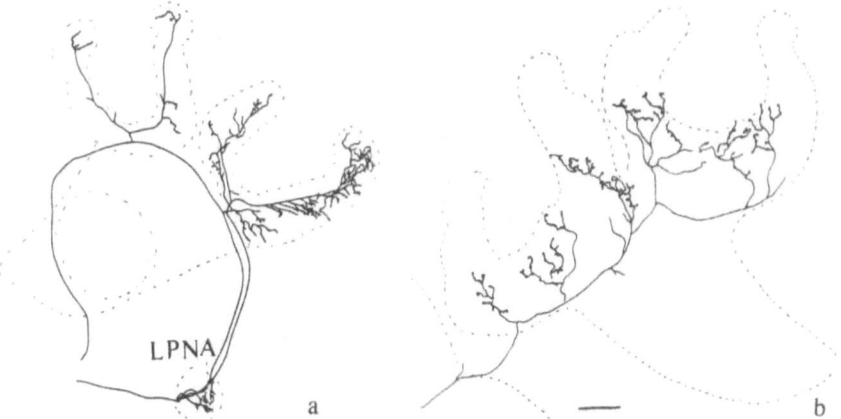

Figure 2   a) Drawings of antenno-glomerular tract neurons
               that arborise solely within the lip of the calyx.
               Note branches in the lateral protocerebral neuropile
               (LPNA).
               b) Drawing of a fibre from the medulla branching in
               the basal ring and collar of the calyx.   (Bars 100
               μm).

3.   Extrinsic output neurons.
     Fibres, with morphologies that suggest they are output neurons,
leave the alpha and beta lobes at two main points.   The first of
these, termed the alpha exit, is located on the lateral margin of
the alpha lobe, carries fibres from four tracts. The PCT carries
feedback fibres from this exit point to the calyx along with fibres
from the anterior dorsal protocerebral commissure (ADPC), anterior
lateral protocerebral tract (ALPT), and alpha lobe to anterior
optic tubercle tract (alpha-AOTT).   The ADPC carries fibres from
the alpha and beta lobes to the neuropile around the contralateral

alpha lobe. Some of the fibres in this tract have processes that penetrate the contralateral alpha and beta lobes. The ALPT carries fibres from the alpha exit point to the ipsilateral lateral protocerebral neuropile. The fourth tract, the alpha-AOTT contains axons that run to the ipsilateral calyx and from the alpha exit.

The second exit point, termed the beta exit, is located medially at the junction of the alpha and beta lobes. Three groups of fibres leave the mushroom bodies at this point. A small group form a tract joining the beta lobes (beta-betaT) Twenty or thirty substantial fibres run toward the undersurface of the lower part of the central body but only one fibre in this group has been Golgi stained, and the extent of the connection with the central body remains unknown. The third tract is a branch of the ADPC which forms a conspicuous X-shaped chiasm in the neuropile between the alpha lobes. This tract carries fibres that connect the alpha and beta lobes on one side of the brain with the alpha lobe and neuropile around it on the other side. Fibres in this tract also have processes that enter the contralateral calyx and pedunculus.

A conspicuous feature of many of the neurons leaving the alpha and beta lobes are their stratified dendrites over narrow windows along the length of the K-cell processes. Many of the extrinsic neurons arborise in both the alpha and beta lobes (figure 3).

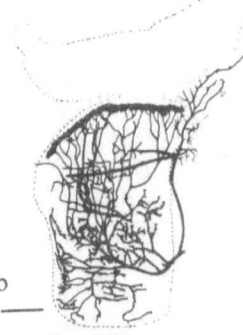

a b

Figure 3   a) Drawing of a feedback fibre in the PCT.
            b) Drawing of an extrinsic neuron running to the
            lateral protocerebrum.   (Bars 100 µm).

DISCUSSION

Vowles (1964) developed a model of mushroom body function in which the K-cells formed part of a series of reafferent loops. The calyces, he argued, receive input from "sensory centres" and the alpha lobes send an output back to them. He suggested that the beta lobe received a copy of the information sent to the alpha lobe and relayed this to "motor centres". He supposed that the purpose of this circuitry was to correct the motor output to accord with the prevailing sensory input. The present study provides no support for specific sensory and motor associations of the alpha and beta lobes. Many of the extrinsic neurons arborise in both lobes and there is an absence of direct connection between the mouthpart and antennal motor neuropiles and the beta lobe.

Schurmann (1974) made an interesting comparison between the mushroom bodies and the vertebrate cerebellum. He suggested that the K-cell projections in the pedunculus and alpha and beta lobes represent a delay line utilised for the transposition of distances into the sequential activation or inhibition of extrinsic fibres. The present study supports Schurmann's hypothesis but also suggests a number of other possibilities.

The timing and delay capabilities of the mushroom bodies may be significantly enhanced by the presence of large numbers of direct and indirect feedback loops. The system need not only operate in the millisecond range of the K-cell delay line as information may loop within it a number of times. The inputs of the calyces clearly show that the mushroom bodies are involved in the analysis of polymodal information. The stratification of extrinsic neurons dentrites within one or more K-cell layers suggests that some of them may act as feature detectors, which because of the looped pathways, respond to sequences of relevant stimuli. The system may be used to sequentially activate pathways involved in the production of the complex sequences of behaviour that have been shown to result from the instantaneous electrical stimulation of the mushroom bodies. Feedback inhibition or excitation may also be relevant in the production of the long lasting after-effects recorded by Erber (1978).

REFERENCES

BACON, J., ALTMAN, J.S., 1977. - A silver intensification method for cobalt filled neurons in wholemount preparations. Brain Res. 138, 359-363.

ERBER, J., 1978. - Response characteristics and after effects of multimodal neurons in the mushroom body area of the honey bee. Physiological Entomol., 27, 37-60.

ERBER, J., MASUHR, TH., MENZEL, R., 1980. - Localisation of short-term memory in the brain of the bee, Apis mellifera. Physiological Entomology, 5, 343-358.

MENZEL, R., ERBER, J., MASUHR, TH., 1974. - Learning and memory in the honeybee, pp. 195-217. In: Insect Behaviour. Barton Browne (Ed.). Springer Verlag.

MOBBS, P.G., 1980. - The brain of the honeybee Apis mellifera - I. Phil. Trans. Roy. Soc. B. In Press.

SCHURMANN, F.W., 1974. - Bemerkungen zur funktion der corpora pedunculata in gehirn der insekten aus morphologischer sicht. Exp. Brain Res., 19, 406-432.

STRAUSFELD, N.J., HAUSEN, K., 1977. - The resolution of neuronal assemblies after cobalt injection into neuropile. Proc. R. Soc. Lond. B., 199, 463-476.

VOWLES, D.M., 1955. - The structure and connections of the corpora pedunculata in bees and ants. Quart. J. Micro. Sci., 96, 239-255.

VOWLES, D.M., 1964. - Models and the insect brain. pp. 377-399. In: Neural Theory and Modelling. Reiss (Ed.). S.U.P.

WADEPUHL, M., HUBER, F., 1979. - Elicitation of signing and courtship movements by electrical stimulation of the brain of the grasshopper. Naturwissenschaften, 66, S.320.

Modulation of Neurosecretion During Caste Determination in *Apis mellifera* Larvae

Heinz Rembold and Gabriele Ulrich,
Max Planck Institute for Biochemistry
Martinsried (Munich, FRG)

The female honey bee larva can develop into two phenotypically different forms, the queen and the worker. The feeding-behaviour of the nurse bees finally decides the emergence of a queen or of a worker. Quality of the nutrient is the first step in the process of caste formation (v. Rhein, 1933, Jung-Hoffmann, 1966, Rembold and Lackner, 1981, Zander and Becker, 1925). Differential nutrition of the young larva entails a different hormonal response, which ends up in different metabolic and morphogenetic activities (Fyg, 1959, Townsend and Shuel, 1962, Weaver, 1966, Dogra et al., 1977). By use of histological and tracer techniques, we have now studied the manifestation of an initial difference in food quality by the larval endocrine system.

MATERIALS AND METHODS

Larvae of the Carnica bee, Apis mellifera, from field hives were used. The age of the larvae was defined according to Rembold et al. (1980). For histological examination paraffin sections were used, stained either with paraldehyde - fuchsin (PAF) or AZAN (Adam and Czihak, 1964). The autoradiographic studies were carried out after Dörmer (1973), using [5-^{3}H]uridine.

RESULTS

The endocrine system of insects consists of the neurosecretory cells of the brain (NSC), the corpora cardiaca (CC), the corpora allata (CA), and the prothoracic glands. This also applies to adult honey bees. In 1st instar larvae of the honey bee, all these organs are present. However, all of them are only in an embryonic or in an early stage of development. They differentiate during further larval life, in queen larvae faster than in the worker.

This caste-specific retardation in the developmental program can well be seen in the development of the neurosecretory cells (NSC). In 1st instar larvae, the NSC are embryonic cells without axons. In queen larvae, outgrowth of the axons begins in the 2nd instar and is completed at the end of 3rd larval instar. In worker larvae, on the contrary, onset (3rd instar) and termination (4th instar) of axonal outgrowth is delayed by one larval instar.

In the NSC cell bodies of queen larvae, transcription of genetic information, i.e., incorporation of ^{3}H-uridine into RNA, begins
near the end of 3rd to beginning of 4th instar. In worker larvae,
however, incorporation of ^{3}H-uridine is first found in the 5th instar. Stainable neurosecretory material is detectable in late 5th
instar queen larvae (Fig. 1).

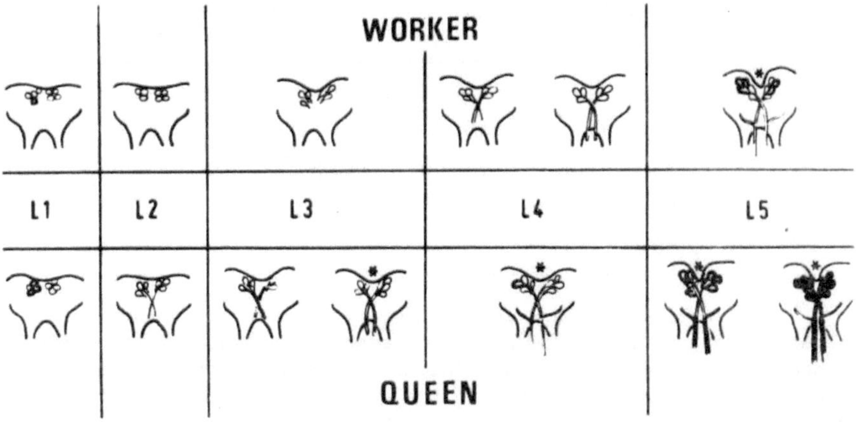

Fig. 1. -- Scheme for the caste specific differentiation of the
neurosecretory cells during larval development of the female honey
bee. Differences between the castes are demonstrated for axon outgrowth, incorporation of ^{3}H-uridine (*), and stainable neurosecretory material.

The corpora cardiaca (CC) are embryonic in both the castes till
end of 3rd instar. In the 4th instar, they begin to grow and to differentiate. At the end of 4th instar, the CC of queen larvae incor-
porate ^{3}H-uridine, whereas in worker larvae, radioactive RNA is
detectable only in the 5th larval stage.

The retardation of worker larvae as compared with queen larvae,
is also shown in the development of the corpora allata (CA). From
3rd instar onward, the queen CA is always bigger in size than that
of the worker. In the maximum (5th larval stage), the volume of
queen CA is twice that of a worker (Tab. 1). During larval development, the nuclei of the CA undergo several stages of endomitosis as
shown by photometric estimation of Feulgen stained DNA. Also in this
case, the queen larva gains a time advantage. Here, the last phase
of endomitosis was found at the end of 4th instar, in workers in the
5th larval instar only. If the two phenomena, caste and growth specific changes in volume and DNA synthesis of CA, are compared, they
seem to be connected with each other. The volume of this hormone
gland is always increased after a phase of condensation and decondensation of the cell nuclei. The effect is more distinct in the CA
of the queen than of the worker larva. ^{3}H-thymidine, a DNA precursor,
is incorporated after decondensation of the cell nuclei, whereas no

radioactivity is present in condensed nuclei. This tracer experiment is another demonstration for several steps of endomitosis in the larval CA.

Tab. 1. -- Volume of CA in larval stages L1-L5 of the female honey bee. Mean value $\pm$ S.E.

Stage	Queen (μm^3)	n	Worker (μm^3)	n
L1	-		0.275	2
L2	0.58	2	0.475	2
L3	1.54 ± 0.442	14	0.96 ± 0.494	5
L4	3.09 ± 0.961	34	1.32 ± 0.730	10
L5	5.48 ± 1.483	10-	3.20 ± 0.490	11

    Incorporation of ^{3}H-uridine into prothoracic glands during different stages of larval development is shown in figure 2. No difference between queen and worker larvae can be demonstrated. Silver grains were primarily located on the cell nuclei. According to these data, labelling of the gland is fairly constant and at high level till end of 3rd instar. It then decreases during the following instars and reaches a zero level at the end of 5th larval instar (PP).

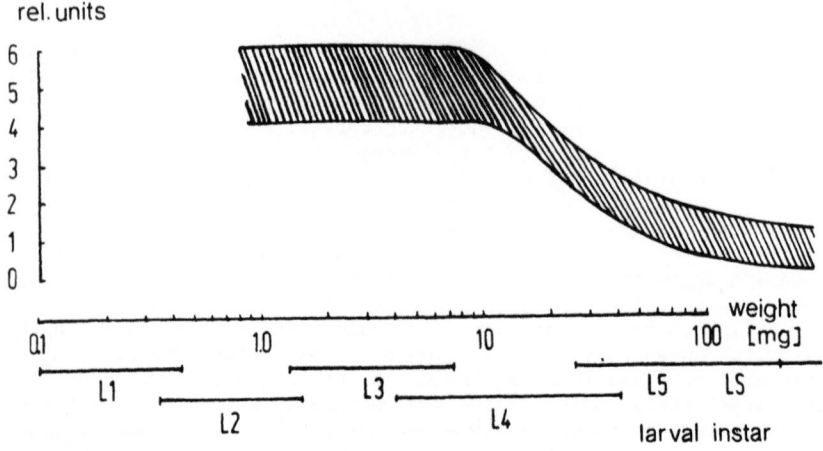

Fig. 2. -- Incorporation of ^{3}H-uridine into honey bee prothoracic glands during larval development. The silver grains located on the cell nuclei in autoradiography were counted and expressed in relative units.

DISCUSSION .

Our data clearly proof the existence of caste specific differences in the development of the endocrine system in the honey bee.
If the larvae are supplied with an optimal diet, queen larvae differentiate and activate their endocrine tissues earlier than the worker larvae. This time shift finally results in differences of hormone
titer (Hagenguth and Rembold, 1978, Rembold and Hagenguth, 1981).
The classical scheme of endocrine regulation of insect morpho-
genesis has been described by many authors (for review, see Novák,
1975). The NSC control, with their neurosecretory products, the
activity of the peripheral hormone gland (CA and prothoracic glands).
As described above, in the queen larvae metabolic activity of the
NSC begins in the 3rd, and in worker larvae in the 5th instar. It
has to be postulated therefore, that a controlling system, which is
different from the NSC, must be active in the young larvae. It may
be speculated, that this regulating system is located in the ventral
ganglions. Some of our data hint at such a possibility, however, no
definite results are available:

Controlling system   Food   Brain with NSC
in young larvae   in elder larvae

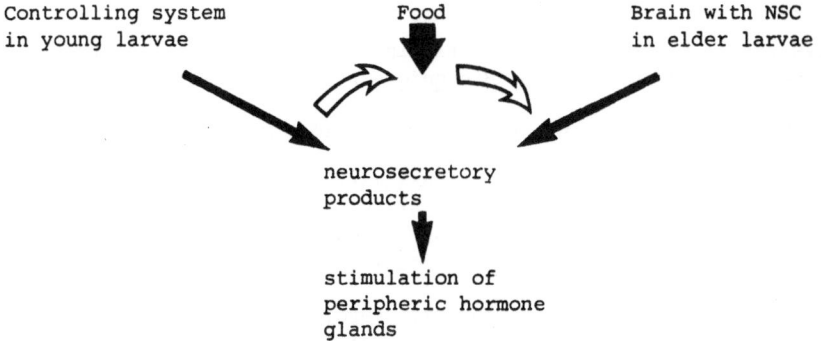

neurosecretory
products

stimulation of
peripheric hormone
glands

In this sense, the queen larvae switch over from the early larval controlling system to the neurosecretory system of the brain
much earlier than the workers. By this time shift, they gain a developmental advantage. This earlier switch-over from one regulatory
system to another one is most likely stimulated by an optimal nutrition of the sensitive 3rd instar larvae. A caste-specific developmental program is switched on as a consequence of different maturation of the neuroendocrine system. This is primarily reflected by a
caste specific modulation of ecdysteroid (Hagenguth and Rembold,
1978) and juvenile hormone (Rembold and Hagenguth, 1981) titres.
Interesting enough, these caste specific events at the hormone level
are also reflected by histological differences, as has been shown
here for the CA (JH synthesis) and prothoracic glands (ecdysteroid
synthesis). We now understand the mechanism of caste determination
in the 3rd instar honey bee larva as a combination of an external
stimulus (food quality) with an endogenous, genetically programmed
event, i.e., the switch on of the neurosecretory system in the brain
of the female larva.

References

Adam H., Czihak G., 1964. -- Arbeitsmethoden der makroskopischen und mikroskopischen Anatomie. Gustav Fischer Verlag, Stuttgart.

Dörmer P., 1973. -- Quantitative autoradiography at the cellular level. In: Molecular Biology, Biochemistry and Biophysics 14, Micromethods in Molecular Biology. Ed. V. Neuhoff, Springer-Verlag, Berlin.

Dogra G.S., Ulrich G.M., Rembold H., 1977. -- A comparative study of the endocrine system of the honey bee larvae under normal and experimental conditions. Z. Naturforsch., 32c, 637-642.

Fyg W., 1959. -- Normal and abnormal development in the honey bee. Bee World, 40, 57-66.

Hagenguth H., Rembold H., 1978. -- Kastenspezifische Modulation des Ecdysteroid-Titers bei der Honigbiene. Mitt. dtsch. Ges. allg. angew. Ent., 1, 296-298.

Jung-Hoffmann I., 1966. -- Die Determination von Königin und Arbeiterin der Honigbiene. Z. Bienenforsch., 8, 296-321.

Novák V.J.A., 1966. -- Insect Hormones. 3. Ed., Methuen, London.

Rembold H., Hagenguth H., 1981. -- Modulation of hormone pools during postembryonic development of the female honey bee castes. Regulation of Insect Development and Behaviour (Sehnal F., ed.). Wroclaw Techn. Univ. Press, Wroclaw pp. 427-440.

Rembold H., Kremer J.-P., Ulrich G.M., 1980. -- Characterization of postembryonic developmental stages of the female castes of the honey bee, Apis mellifera L. Apidologie, 11, 29-38.

Rembold H., Lackner B., 1981. -- Rearing of honey bee larvae in vitro: effect of yeast extract on queen differentiation. J. Apicult. Res., 20, 165-171.

v. Rhein W., 1933. -- Über die Entstehung des weiblichen Dimorphismus im Bienenstaat. Arch. Entwicklungsmech., 129, 601-665.

Townsend G.F., Shuel R.W., 1962. -- Some recent advances in apicultural research. Ann. Rev. Entomol., 7, 481-500.

Weaver N., 1966. -- Physiology of caste determination. Ann. Rev. Entomol., 11, 79-102.

Zander E., Becker F., 1925. -- Die Ausbildung des Geschlechtes bei der Honigbiene II. Erlanger Jb. Bienenk., 3, 163-223.

The Bee's Celestial Map –
A Simplified Model of the Outside World

Rüdiger Wehner, The University of Zurich

One of the most staggering behavioural abilities of some insect species is to use the pattern of polarized light in the daytime sky as a compass. Santschi (1923) had already noticed that ants were able to navigate correctly even if they could not see more than a small patch of the blue sky, and von Frisch (1949) later demonstrated in bees that it was the polarized light in the sky of which the insect made use under such conditions. The question, however, remained: How does the insect derive compass information from the pattern of polarized light in the sky? This first begs the question as to what the skylight patterns look like.

SKYLIGHT PATTERNS

As skylight polarization arises from the scattering of sunlight within the earth's atmosphere, the whole pattern is fixed relative to the sun (Fig.1). Thus, skylight patterns can be described most conveniently in terms of a sun-related system of coordinates where the sun and the antisun form the (unpolarized) poles of the sphere, and the line of maximum polarization runs along the equator. In such a system of coordinates the meridians mark the planes within which sunlight is scattered. Consequently, due to the laws of Rayleigh scattering, the e-vectors are aligned with the parallels of altitude.

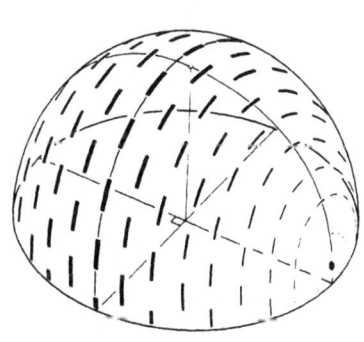

Fig.1: The pattern of polarized light in the sky. The directions and widths of the black bars mark the angles χ and degrees D of polarization, respectively. Elevation of sun (black disc): μ_o = 6°. Two great circles are shown: One includes the solar and antisolar meridian (and forms the symmetry line of the whole pattern), the other symbolizes the line of maximum polarization.

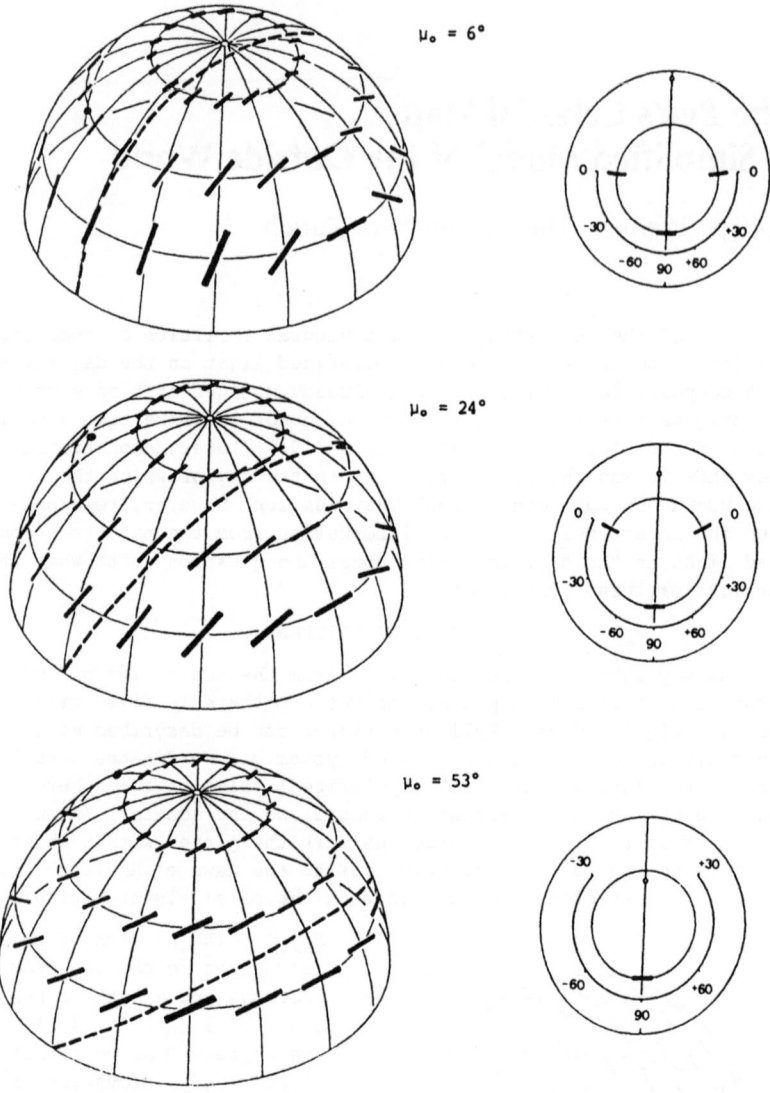

Fig.2: The pattern of polarized light in the sky for three elevations μ_0 of the sun. In contrast to Fig.1, the reader views the antisolar meridian. The dashed line marks the line of maximum polarization. In the right figures, the e-vector distributions are given for an elevation of $\mu = 45°$ and the highly polarized part of the sky.

Within the horizon system of coordinates, the skylight patterns change during the course of the day. As the sun moves up, the line of maximum polarization tilts down (Fig.2). Light is always polarized horizontally along the solar and antisolar meridian, but at any other point in the sky, the angle of polarization varies throughout the day.

COMPASS STRATEGIES

In general, there are two possibilities by which the insect could use the patterns of polarized light in the sky as a compass: (1) The insect performs some kind of spherical geometry (Kirschfeld et al., 1975). (2) The insect uses map information. Either it remembers the pattern last seen, and later tries to match the memorized pattern with the current one (Stockhammer, 1959; von Frisch, 1965: 398), or it is informed about all possible patterns as they occur in the sky during the course of the day.

THE BEE'S CELESTIAL MAP

Even though neither possibility is applied by the bees in the strict sense mentioned above, the second possibility comes closer to the truth. Bees use a simplified master-image of the sky that does not change with the elevation of the sun. Of course, it rotates about the zenith as the sun changes its azimuth (Wehner and Lanfranconi, 1981), but this is not a point to be considered here.

Samuel Rossel and I have reconstructed the bee's celestial map by displaying single beams of artificially polarized ultraviolet light to the bees. While the bees were "dancing" on a horizontal comb, the directions of their waggle runs were recorded by a video camera. Thus, we were able to determine systematically the positions in the sky where the bees expected any particular e-vector to occur. For experimental details see Wehner (1982a,b) and Rossel and Wehner (in prep.).

As a result, bees make mistakes. The bee's internal representation of the celestial canopy, its celestial map (Fig.3), comprises only those e-vectors which lie in the highly polarized part of the sky. This is in accord with the earlier finding that any one e-vector occurring twice at a given elevation is interpreted by the bees as the one lying further away from the sun, irrespective of its degree of polarization (Rossel et al., 1978; Brines and Gould, 1979). Furthermore, it is even in the highly polarized part of the sky that the bee's celestial map does not coincide exactly with any one pattern of polarization present at any time of the day. Instead, the bee's celestial map reflects the mean distribution of e-vectors that lie along the line of maximum polarization (Fig.4).

It is intriguing to think of the antisolar meridian (characterized by horizontal e-vectors) as the main reference line of the bee's e-vector compass. Do the bees prefer this line? In all our experiments performed so far, in which a horizontal and a vertical e-vector were presented simultaneously, the bees did not exhibit any preference for the horizontal e-vector. They seem to use a celestial map in which all e-vectors are of equal importance.

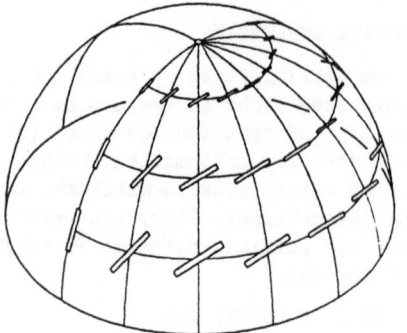 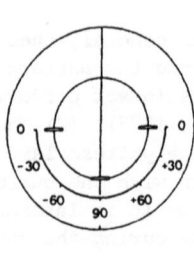

Fig.3: *The bee's internal map of the sky. Irrespective of the elev-ation of the sun, the bee refers invariably to a stereotyped ce-lestial map. This map includes only the highly polarized half of the celestial hemisphere, and is characterized by identical e-vector dis-tributions (see right figure) for all parallels of altitude.*

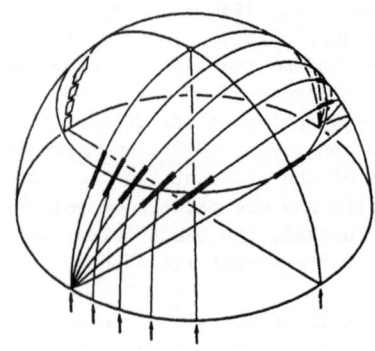

Fig.4: *As the sun moves up (white arrow), the line of maximum polariz-ation tilts down. Consequently, the maximally polarized e-vectors $\chi(D_{max})$ also tilt down. Their distribution is shown for an elevation of $\mu = 45°$, but is very similar for all other elevations $\mu > 30°$. Within the bee's celestial map (Fig.3) the dis-tribution of e-vectors is identical for all elevations (at least for $\mu > 30°$) and coincides with the mean distribution of the maximally polar-ized e-vectors $\chi(D_{max})$ as calculated for all elevations $\mu > 30°$.*

CONCLUSION

What the bee has incorporated into its brain, by either evolutionary or individual experience, is a simplified rather than a correct copy of the patterns of polarized light as they actually occur in the sky.

The strategy to rely on a simplified master-image of the sky is remarkably efficient. The navigational errors, which are necessarily introduced by such a strategy, are usually small, and only show up if the insect can not view large but only small parts of the sky. On the other hand, in relying on such a simplified master-image of the sky, the insect needs not resort to more complicated strategies, which may be profound in mathematical terms but not amenable to an insect's brain.

After all, insects are not astronomers. Neither do they perform spherical geometry in the sky, nor do they dispose of an astronomical almanac where all possible skylight patterns are written down. Instead, they use a stereotyped celestial map based on some principal aspects of skylight polarization. Thus, they sacrify absolute precision for a workable neural strategy.

In addition, to keep navigational errors down, they use some powerful backup systems drawing upon the sun and non-celestial cues such as landmarks.

I thank Dr. S. Rossel for his cooperation. This work was supported by Swiss National Science Foundation Grant 3.313.78.

References

Brines M.L., Gould J.L., 1979. Bees have rules. Science 206, 571-573

Frisch K.v., 1949. Die Polarisation des Himmelslichts als orientierender Faktor bei den Tänzen der Bienen. Experientia 6, 210-222

Frisch K.v., 1965. Tanzsprache und Orientierung der Bienen. Berlin, Heidelberg, New York: Springer

Kirschfeld K., Lindauer M., Martin H., 1975. Problems of menotactic orientation according to the polarized light in the sky. Z. Naturforsch. 30c, 88-90

Rossel S., Wehner R. (in prep.). The bee's e-vector compass.

Rossel S., Wehner R., Lindauer M., 1978. E-vector orientation in bees. J. Comp. Physiol. 125, 1-12

Santschi F., 1923. L'orientation sidéral des fourmis. Mém. Soc. Vaud. Sci. Nat. 4, 137-175

Stockhammer K., 1959. Die Orientierung nach der Schwingungsrichtung linear polarisierten Lichts. Ergeb. Biol. 21, 23-56

Wehner R., 1982a. Himmelsnavigation bei Insekten. Neurophysiologie und Verhalten. Neujbl. Naturf. Ges. Zürich 184, 1-132

Wehner R., 1982b. The perception of polarized light. In: The Biology of Photoreceptors. Symp. Soc. Exp. Biol., eds. Cosens D., Vince'Prue D. Cambridge Univ. Press (in press)

Wehner R., Lanfranconi B., 1981. What do the ants know about the rotation of the sky? Nature (Lond.) 293, 731-733

Basic Mechanisms of Sensory Antennal Information Processing in Insects, with Special Reference to Social Insects

Claudine Masson, Bures Sur Yvette, France

THE ANTENNAL SENSORY SYSTEM: ORGANIZATION AND FUNCTION

Chemical stimuli play an important role in insect behaviour. They are essential to the life and survival of the insect.

Specific chemical signals of particular importance to insects, such as pheromones (e.g. sex attractants or aggregation pheromones) and allelochemicals (e.g. plant attractants or phagostimulants) tend to be complex in structure and are composed of several different chemicals.

They are detected by two different sensory systems which act independently, but in a complementary way, to trigger specific behavioural sequences. The olfactory system is employed in the detection of more volatile substances which are carried by air flow, whereas the gustatory system detects substances by direct contact.

The triggering of a particular behavioural pattern in response to a specific odor depends upon (i) antennal detection by specialized structures, the olfactory sensilla; (ii) the peripheral identification of each component in the mixture of substances which makes up the odor; (iii) integration of the chemical message together with signals of other sensory modalities also acting on behaviour.

On the basis of neuroanatomical and neurophysiological investigations, it is possible to give a very general picture of the insect olfactory system with special reference to social insects and cockroaches.

In the last ten years information about the afferent antennal pathways in insects has been provided by studies on honey bees, ants, cockroaches and moths. The central distribution of antennal afferent axons, the organization of the deutocerebrum, and the connections with other centers in the brain are similar in all species, which suggests that general rules governing neural connectivity exist for the CNS of insects.

1 - <u>Antennal detection and peripheral olfactory coding</u>.
The olfactory sensilla are distributed along the
antennae. This distribution seems not to be at random
(Esslen, Kaissling, 1976; Altner, Sass, Altner, 1977;
Arnold, Masson, 1981). In ants, for example, the sensilla
are arranged according to a chemical gradient with a
maximum number in the distal segment (Masson, 1973; Delabie,
Masson, 1982). The sensilla are modifications of the
antennal cuticule and have a wide variety of shapes, for
example, hairs of different length and thickness as in ants
(Masson, 1973) and cockroaches (Masson, Brossut, 1981) or
pore-plates as in honey bees (Arnold, Masson, 1981) and
bumble bees (Fonta, Masson, 1982). Independently of the
shape, the cuticule is perforated by pores, the "gateways"
for the odorant molecules to reach the underlying
membranes of dendrites of the receptor cells. The olfactory receptor cells are bipolar sensory neurones which
send their axons directly to the CNS; they are responsible
for the selectivity and/or the speci ficity of detection
and for discriminating mechanisms.
Data suggest that functionally defined types of olfactory receptor cells are located in specific morphological types of antennal sensilla (Altner, 1977).
Electrophysiological investigations in different
insects indicate that on the basis of their quality
spectra, two main groups of sensory neurons exist. Firstly,
neurones with large quality spectra in which the
different spectra either have several chemicals in common,
and are therefore overlapping, as in cockroaches, or the
relatively large spectra tend not to overlap, as in the
bee. Such a "generalist system", is characterized by a
very high olfactory discriminating power and furthermore,
is very economic because only a few such neurones are
required to detect very complex and different odorant
mixtures, as food odors. Secondly, sensory neurones with
very narrow spectra - sometimes restricted to only one
odorant - whose spectra are very specific. Although such
a "specialist system" has a reduced olfactory discrimination performance, the response threshold of each cell
is very low and this population of cells compose the
odorant detector system with the highest specificity and
sensitivity. Such a system is certainly the best adapted
to detect very specific stimuli such as sex attractants.
Each insect has representatives of both types of
sensory neurone The different ratios of wide and narrow
spectra neurones in different insect species is a function
of their way of life, some being more generalist insects
(Social Insects, cockroaches...) others being more
specialist insects (moths, pest insects...).

2 - <u>Central antennal pathways and processing</u>.
As each olfactory receptor cell axon reaches the CNS

directly, the question arises as to how the CNS reads the
different peripheral patterns in order to evaluate the
stimuli in quality and in quantity, and transform them to
be used in terms of motor patterns. Combined classical
microscopy, cobalt chloride staining and electrophysiological methods reveal the following features of the
antennal afferent projections in the CNS. Axons of different antennal sensory modalities combined together form the
sensory bundles of the antennal nerve; some are fascicu-
lated in bundles of apparently functionally related fibers
(Masson, 1973; Masson, Strambi, 1977). They run to the
brain without synapses or branching outside the CNS. At
their entrance in the brain they are distributed among
the glomeruli of the sensory part of the ipsilateral
deutocerebrum (so-called antennal lobe). In the glomeruli,
most of the receptor axons are connected to local interneurons (about 2/3) and the rest are connected to neurons
leaving the deutocerebrum. The glomeruli are produced
by the huge convergence of antennal fibers onto only a
few hundred output deutoneurones (ratio - 1/100). In cockroaches histological and computer analysis of the spatial
organization of the glomeruli has demonstrated that the
glomeruli are strictly invariant in number, position and
volume, and are therefore identifiable (Chambille, Rospars, Masson, 1980). Studies now in progress suggest
that a similar situation exists in the bee. A sexual
dimorphism exists in the structure of the glomeruli: in
the male, but not in the female, there is a "macroglome-
rulus" (Chambille, Rospars, Masson, 1980) which cobalt
chloride precipitations and electrophysiological recordings have demonstrated receives mainly the afferents of
specific peripheral receptors to female odors; in the
drone, such staining experiments in progress suggest a
special complex of glomeruli.

The deutocerebral convergence is spatial and multimodal, qualitative and quantitative (Masson, Brossut,
1981). In terms of functioning the deutocerebral convergence is very useful, it works (i) to amplify the specific
signal-to-noise ratio and (ii) to integrate simultaneously the different parameters of a complex odorant which
stimulates the peripheral sensilla.

The antenno-glomerular tract of the tractus olfac-
torio-globularis represents the main efferent bundle of
the sensory antennal lobe and connects it with the corpora
pedunculata, lateral part of the protocerebrum, and suboesophageal ganglia. Recently we have demonstrated a thin
direct tract to the contralateral side of the antennal
lobe in the worker bee.

ONTOGENESIS AND MATURATION

The deutocerebrum is a highly convergent structure

with only partial overlapping of synaptic fields of
adjacent fibers; therefore, few changes can occur in its
connectivity and it looks like a hardwired network with a
more or less constant output to input relationship. The
corpora pedunculata are, on the contrary, a highly
divergent - highly convergent structure, with a densely
interwoven network of fibers resulting in a fully versatile
output to input correspondence provided some device
selects the appropriate combination of synapses among
countless equiprobable possibilities.

Some experimental evidence clearly supports the idea
that neuronal plasticity is relevant to insect olfactory
behaviour, and that it is mainly located in the corpora
pedunculata.

Moreover these data suggest that no true plasticity
exists in the deutocerebrum. It appears that the basic
information storage of olfactory cues in the workerbee
implies a dynamic form and requires simultaneous
integrity of deutocerebrum and corpora pedunculata.

Finally, the plasticity of the olfactory performances must depend on two main sets of neuronal networks,
and correlate with specific features of their connectivity, those of the deutocerebrum and of the corpora
pedunculata.

Consequently it must be of interest to consider the
different steps of their connectivity during insect
development between eclosion to emergence.

The first important step is the connection between
the first order neurones and the secondary ones.

Cobalt sulfid staining of nymphs of the workerbee, day
after day, at different ages, reveals the main features
of the antennal afferent projections in the CNS, ipsi-
and contralateral sides. The first synapses between
primary sensory neurones and local interneurons of the
deutocerebrum arise the third day of the nymphal period.
The total number ("adult number") of glomeruli seems to
be reached three days later (3 days before emergence),
simultaneously, some fibers projected into other centers of
the brain, and at day 6 of the nymphal period, the projec-
ions are very similar to those of the adult the first day
after emergence. The "mature organization" three days
before emergence is undoubtedly only "apparent" and we
expect that synaptic elements are partly "free" and
capable of being stabilized under experimental conditions in the first days of the adult life. In previous
experiments (Arnold, Masson, 1980) at different postembryonic ages and/or following social deprivation,
data suggest that during the postnatal maturation of
the olfactory system in the worker bee different developmental stages occur and that the period between emergence and day 8 may be critical.

384

REFERENCES

ARNOLD G., MASSON C., 1980. - Postnatal maturation of the
olfactory system of worker bee; effects of early
social deprivation. Int. Congr. Olfaction and Taste
VII. H. van der Starre ed. IRL London, p 279.

ARNOLD G., MASSON C., 1981. - Evolution, en fonction de
l'age, de la structure externe des sensilles olfac-
tives de l'antenne de l'ouvrière d'abeille, Apis
mellifera L. C.R. Acad. Sci., Paris, Ser. 3, 292,
681-686.

ALTNER H., 1977. - Insect sensillum specificity and structure = an approach to a new typology. Proc. VI
Internat. Symp. Olfaction and Taste. J. Le Magnen
ed. IRL London, 295-303.

ALTNER H., SASS H., ALTNER I., 1977. - Relationship between structure and function of antennal chemo-,
hygro- and thermoreceptive sensilla in Periplaneta
americana. Cell Tiss. Res. 176, 389-405.

CHAMBILLE I., ROSPARS J.P., MASSON C., 1980. - The deuto-
cerebrum of the cockroach Blaberus craniifer.
Spatial organization of the sensory glomeruli. J.
Neurobiology, 11 (2), 135-157.

DELABIE J., MASSON C., 1982. - Le système antennaire de
la fourmi champignonniste Acromyrmex octospinosus
(Hym. Formicidae Attini). Attini 12.

ESSLEN J., KAISSLING K.E., 1976. - Zahl und Verteilung
antennaler Sensillen bei der Honigbiene. Zoomorphol.
83, 227-251.

FONTA C., MASSON C., 1982. - Analyse de l'équipement
sensoriel antennaire du bourdon Bombus hypnorum L.
submitted for publication. Apidologie.

MASSON C., 1973. - Contribution à l'étude du système
antennaire chez les fourmis. Approche morphologique,
ultrastructure et électrophysiologique du système
sensoriel. Thèse Doctorat d'Etat. Marseille, 332 pp.

MASSON C., STRAMBI C., 1977. - Sensory antennal organization in an ant and a wasp. J. Neurobiology 8,
537-548.

MASSON C., BROSSUT R., 1981. - La communication chimique
chez les insectes. La Recherche 121, 406-416.

Disruption of Ant Recruitment by the Frontal Gland Secretion of a Termite: A Chemical Defense Strategy

Manfred Kaib,
University of Bayreuth (FRG)

Ants are the main predators of termites. The efficiency of this predation depends mainly on the recruitment mechanisms of the predators to the food sources and on the defense strategies used by the prey itself. In most ant species nestmates are easily recruited to termite colonies, which are excellent resources because of the extremely high local population density in the termite hives. This ever present predation has consequently led to the evolution of defense strategies in termites. Schedorhinotermes lamanianus, an arboreal Rhinotermitinae with nests of tremendous dimensions, has evolved defense mechanisms on various levels. Socially organized barriers are firstly, the covering of connecting trails or foraging areas with soil, wood, or fecal material and secondly, the foraging regime of this termite in which initial foraging is done by minor soldiers. Foraging-trails are soldier-generated and induce gallery building by workers which are guarded by minor soldiers (Kaib and Leuthold, in prep.). Thirdly, we find an unproportional concentration of minor soldiers in the connecting galleries and at the termini of established foraging areas and of major soldiers in the hives (Kaib, 1980). Under undisturbed conditions more than 60 % of the termites in the galleries are workers, the rest are soldiers. This proportion reverses during mechanical disturbance or when ants are present (Kaib, unpubl. data). Fourthly, nestmates are recruited to sites of attack (Kaib and Kriston-Meissl, 1979). On an individual level termite workers and soldiers use their biting mandibles to disable small attacking ants on a mechanical basis and their frontal gland secretion to repell or to temporarily disable predators chemically.

MATERIALS AND METHODS

The field studies were made in Kenya at the South Coast and in Shimba Hill forests with freshly collected S. lamanianus on the natural foraging ground of Myrmicaria spec., Oecophylla longinoda, Paltothyreus tarsatus and Camponotus spec., respectively, and with Cubitermes umbratus as a termite control species.

Supported by the DFG (Ka 526).

RESULTS AND DISCUSSION

1. The influence of the termite frontal gland secretion (FGS) on the survival of attacking ants.

The FGS of S. lamanianus is composed of a blend of aliphatic ketones (Prestwich et al., 1975), which is dabbed on the cuticle of the attacking ant during combat or it simply evaporates. Contaminated ants are repelled and show a typical behaviour, which is described in detail later. However, all ants gradually recover.

Myrmicaria ants were caught following an encounter with Schedorhinotermes minor soldiers and were kept for 10 to 14 h in plastic petri dishes of 9.5 cm Ø. Additionally foraging ants were kept in petri dishes which were treated with the FGS of 10 minor soldiers each. The results show clearly that the ketonic secretion of S. lamanianus is not toxic to ants when topically applied by ter-

Table 1. The survival rate of Myrmicaria spec. after combat with S. lamanianus soldiers or after contact with the vapour of minor soldier FGS.

Treatment of ants	duration (hours)	n	normal	disabled	dead
control	10 – 14	80	79	1	0
attacked by soldiers	12 – 13	29	28	0	1
vapour of FGS	10 – 14	80	50	18	12

mite soldiers during naturally occuring combat (Tab. 1). When Myrmicaria are kept for 10 - 14 h in an atmosphere highly enriched with the vapour of the FGS, the locomotory activity of these ants is drastically increased during the period of treatment. The increased mortality observed may be due to this activity (exhaust of energy resources) rather than to a toxic effect of the FGS.

2. The reduction of the vulnerability of termites to ant attack by volatile components of the soldier secretion.

At foraging sites of S. lamanianus minor soldiers explore new areas and lay the pheromone trails which lead to gallery building by workers. Minor soldiers are on guard during this initial stage of building and patrol in the galleries. The presence of the soldier secretion reduces the vulnerability of both soldier castes and of the workers to ant attack. The following experiments support this hypothesis. In petri dishes groups of 20 termite workers (control) and mixed groups of 10 termite workers and 10 minor soldiers, or groups of termite workers treated with FGS of 10 minor soldiers were simultaneously offered to Myrmicaria at the natural foraging areas during high activity periods of the ants. Tab. 2 summarizes the number of termites caught and carried away by ants during the time required to empty the control dishes. This effect depends neither on the density of termites nor on that of the attacking ants. The vapour of the secretion of minor and major soldiers is equally effective to Myrmicaria although the chemical composition of their secretion differs slightly.

Table 2. Rate of attack by <u>Myrmicaria</u> to workers and minor soldiers
of <u>S</u>. <u>lamanianus</u>. * Mean number of termites ± SD.

	n	control	workers*	soldiers*
10 workers + 10 minor soldiers	10	20	5.1 ± 0.8	1.2 ± 0.4
20 workers + FGS	10	20	6.5 ± 1.3	-

3. The behavioural pattern of <u>Myrmicaria</u> after treatment with FGS.

During an 1:1 encounter between attacking ants and <u>S</u>. <u>lamanianus</u>
soldiers the termite soldiers release FGS which flows into a brush
at the tip of the soldier's labrum and is dabbed on the cuticle of
the predator by a rapid jerking movement. As a result the head or
the antennae of the ants are contaminated but usually the ants are
not injured by the biting mandibles of the soldiers. The ants then
immediately release the termite prey and retreat. They move side-
wards and backwards with stiff hind legs in a zigzag course and ro-
tate rapidly while dragging their body on the ground. The antennae
are kept in a stiff backwards position. Dragging gradually declines
and the time spent for cleaning increases. The duration of this behaviour pattern depends on the amount of secretion applied and it
may be more than 15 min until the ants recover fully. Injured or
killed termite soldiers usually remain in the battle area and the
ants move back to the nest without prey.

During the encounter between ants and termites - especially
when termite soldiers are killed - large amounts of the FGS are released which repell additional ants. This repellant effect even
counteracts a gathering of <u>Myrmicaria</u> during combat which is caused
by an ant secretion possibly originating from the gaster.

Similar behaviour patterns can be observed following encounters
between <u>S</u>. <u>lamanianus</u> soldiers and <u>Paltothyreus</u>, <u>Oecophylla</u>, and
<u>Camponotus</u> but not following encounters between these ants and sol-
diers of many other termite species which also have developed a
frontal gland.

4. The influence of the termite FGS on the recruitment of <u>Myrmicaria</u>.

Little is known about the recruitment behaviour and the orientation mechanisms of <u>Myrmicaria</u>. These ants seem to search for prey
individually with palpating movements of their antennae. The number
of ants in the foraging area increases rapidly when prey is located
by single ants. Prey is then carried rapidly to the nest, apparently
by the shortest route. The number of ants capturing one termite depends on the size and the mechanical fortification of the prey and
ranges between 1 ant in an encounter with 1 <u>S</u>. <u>lamanianus</u> worker to
more than 10 ants in the case of 1 <u>Cubitermes</u> alate. However, the
number of ants decreases rapidly in the presence of the vapour of <u>S</u>.
<u>lamanianus</u> FGS. As Fig. 1 demonstrates, ants having caught a termite
worker run directly towards their nest (mean 69.0 ± 4.65 cm/10 sec,
n = 22, black arrow) while ant movement is disorganized after contact

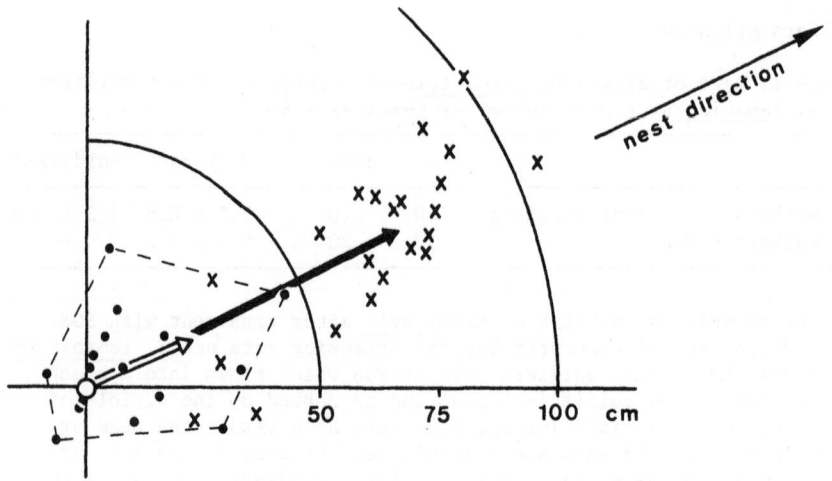

Figure 1. Homing movement of <u>Myrmicaria</u> within 10 sec after capturing termite workers (x) or minor soldiers (●) of <u>S</u>. <u>lamanianus</u>.
O = attack site.

with FGS (mean 15.8 ± 3.25 cm/10 sec, n = 16, white arrow).

CONCLUSIONS

Like many other termite species <u>S</u>. <u>lamanianus</u> uses common defense strategies, e.g. covering of trail or foraging areas, soldier-
generated recruitment (Traniello, 1981; Kaib, unpubl. data) etc., to
prevent attacks by predatory ants. Additionally, volatile components
of the FGS of their minor soldiers, which always accompany workers,
serve in alarm communication similar to that of <u>Nasutitermes</u> <u>exi-
tiosus</u> (Eisner <u>et al</u>., 1976). The FGS of both soldier castes of <u>S</u>.
<u>lamanianus</u> does not immobilize predators mechanically like that of
other Nasutitermitinae, nor is it toxic. It acts chemically on different levels. 1) It repells ants - including scout ants - and thus
reduces the likelihood of massive ant predation. The centres of <u>S</u>.
<u>lamanianus</u> colonies are protected by the vapour of the FGS due to the
high number of major soldiers concentrated there, each carrying
approximately 4 mg of secretion. This serves as a chemical barrier
for intruding ants without physical contact to termites. 2) The FGS
temporarily disables attacking ants during combat which causes a loss
of the possible prey and stops the ant's homing behaviour. Both mechanisms complicate the recruitment of ants to <u>Schedorhinotermes</u> as
prey. This defense strategy seems to be more economical and on a more
highly evolved level than heavy armourment, since it does not cause
a considerable loss of nestmates during combat.

References

EISNER T., KRISTON I., ANESHANSLEY D.J., 1976 - Defensive Behavior of a Termite (Nasutitermes exitiosus). Behav. Ecol. Sociobiol., 1, 83 - 125.

KAIB M., 1980 - Defensive Behaviour in the Termite Schedorhinotermes lamanianus: The Biological Function of Chemical Signals by Soldiers. Olfaction and Taste VII, 415.

KAIB M., KRISTON-MEISSL I., 1979 - Frontaldrüsensekret von Termitensoldaten: chemische Zusammensetzung und biologische Bedeutung. Verh. dtsch. Zool. Ges.1979, 232.

KAIB M., LEUTHOLD R.H. - Gallery building of Schedorhinotermes lamanianus, a defense mechanism. in preparation.

PRESTWICH G.D., KAIB M., WOOD W.F., MEINWALD J., 1975 - 1,13-Tetradecadien-3-one and Homologs: New Natural Products isolated from Schedorhinotermes soldiers. Tetrahedron letters, 52, 4701 - 4704.

TRANIELLO J.F.A., 1981 - Enemy deterrence in the recruitment strategy of a termite: Soldier-organized foraging in Nasutitermes costalis. Proc. Natl. Acad. Sci. USA, 78, 1976 - 1979.

On Synaptic Connexions, Tracts and Compartments in the Brain of the Honeybee

F. W. Schürmann,
The University of Göttingen (FRG)

The brain of the bee and of other social insects have attracted particular interest since Dujardin (1850) first pointed out that the volumes of the paired mushroom bodies in the protocerebrum are correlated with behavioral complexity. Many anatomists have shown that the size of the two main types of neuropile (corpora pedunculata, central body, bridge, the three optic neuropiles, antennal lobes) and the non-glomerular, "non-structured" neuropile varies considerably in the insect groups (Howse, 1975). Some investigators have made contributions to fill these compartments with cell types and groups of neurons (s.Schürmann, 1974, Strausfeld and Nässel, 1980, Ribi and Scheel, 1981, Mobbs, 1982). Despite these efforts the anatomical organization is far from being understood.

The obvious order, size and geometry of neuronal arrays in brain compartments has led many neurophysiologists to look upon brain and neuron functions as being attributed to distinct neuropile areas or cells which are thought to map the compartments shown by anatomy (s.Howse, 1975).

The aim of this study is to increase our knowledge of ascending and descending fibres in the brain of the honeybee and to draw some outlines of the brain glomerular and non-glomerular neuropile connexions.

MATERIALS AND METHODS

For the histological studies of 400 brains of foraging bees (Apis mellifera L.) the following methods were used (for details s.Strausfeld and Miller, 1980) : Golgi impregnations, Protargol-stained serial sections, cobalt-chlorid-backfilling and silver intensification, experimental degeneration in the connectives, electron microscopy.

RESULTS

1. Cobalt backfilling

Cobalt backfilling via the connectives between suboesophageal and thoracic ganglia reveals a mixture of ascending and descending

nervous elements connecting the brain and the suboesophageal ganglion with the ventral nerve cord. We report here the distribution of
neuron parts considered as primary fillings (non-transneuronal
stained elements).

Three pairs of neuronal somata groups in the supraoesophageal
and two pairs in the suboesophageal ganglion (Fig.1) can be distinguished by their cell body position and in parts by their fibre projections. They are considered to be descending neurons.

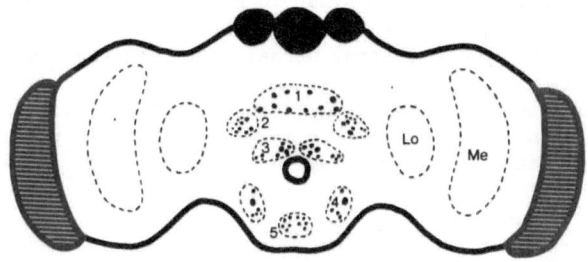

Fig.1 Distribution of somata groups 1-5, Lo lobula, Me medulla

Group 1 consists of at least 6 pairs of somata (diameters 30-45 μm)
in the upper pars intercerebralis at the rim of the inner mushroom
body calyces and at the level of the bridge and upper central body.
Group 2, at least 15-20 pairs of somata (diameters 20-45 μm), join
group 1 somata ventro-laterally, caudad from the central body down
to the level of the ß-lobes of the mushroom bodies. Group 3 contains
at least 25 pairs of somata and lies at the frontal margin of the
lower protocerebrum at the level of the optic tubercles medially
and the upper parts of the deutocerebral antennal lobes. Groups 4
and 5 comprise lateral and medial pairs of 10 and 20 neurons, the
fibres of which could not be analyzed. The total minimum of stained
somata in the brain is about 100, in the suboesophageal ganglion
about 30.

Group 1 somata belong to the giant ocellar fibres which contribute to a flattened protocerebral synaptic plexus ventral to the
bridge neuropile (Pan and Goodman, 1977). Though the bridge is partially penetrated, there is no micromorphological evidence for synaptic contacts. Group 2 neurons send two bundles of axons ventrally.
Their arborizations form a second synaptic neuropile area overlapping partially with the described ocellar plexus and extending to
the ventral ipsi- and contralateral parts of the protocerebrum
(neuropile beneath the ß-lobes and lateral protocerebral lobes).
Group 3 neurons can be subdivided into two clusters of cells with
different fibre projections. A dorsal symmetrical set of neurons
(10 pairs of cells) gives off axons ascending laterally, twisting
around the α-lobe and then descending medially from the peduncles to
the caudal protocerebral neuropile where they form horizontally
arranged dendritic layers. The ventral set of 30 somata sends axons
posteriorly forming layered dendrites in the medial and lateral

parts of the lower protocerebrum and in the dorsal lobes of the
deutocerebrum.

The dendritic-like synaptic domains of the brain somata groups
are all found in the caudal and ventral non-glomerular neuropile
areas partially overlapping in the proto- and deutocerebrum. The
descending axons of all these somata join in the caudal brain.

We describe as ascending fibres those elements without somata
in the brain and suboesophageal ganglion. They belong to a large
mass of small diameter fibres, some medium sized and "giant" fibre
elements. Obviously their main synaptic areas are the lower caudal
protocerebral brain parts (especially parts of the protocerebral
lateral lobes , the deutocerebral dorsal lobes and the small trito-
cerebral areas).

Both ascending and descending fibres do not normally project
into glomerular neuropiles, but are mainly confined to parts of the
non-glomerular neuropile where they partially overlap.

2. Connective organization

The large cobalt stained fibres can be traced in whole mounts
down from the brain to the dorsal parts of the connective between
suboesophageal and prothoracic ganglion. Transverse sections reveal
symmetrically arranged fibre groups in the dorsal half of the connective (130 fibres with diameters greater than 5 μm , 14 fibres
with diameters of 10-15 μm) and a majority of fine fibres in the
ventral half (diameters 0.05-5 μm, about 10 000 fibres as calculated from measurements and counts in electron micrographs).

Degeneration experiments show that parts of the small fibre
areas belong to multisegmental interneurons or sensory cells of unknown position in the thoracic and abdominal system. Other small
fibres are axons of sensory cells of the head (antennal and mandibular nerves). A clear cut determination of all ascending and
descending fibres cannot be made from the present degeneration experiments. Obviously only a small number of fibres can be attributed
to descending brain neurons.

3. Glomerular neuropile connecting elements

Synaptic connexions of the most prominent tracts of the central
brain, the antenno-glomerular tracts in the non-glomerular neuropile
are seldom reported. These tracts contain fibres branching into the
dorsal deutocerebral lobes and include projections to the lateral
protocerebrum at the base of the lobula (Mobbs, 1982). Fibres entering the mushroom body stalks and lobes with unknown somata and
central body neurons show synaptic branches in the non-glomerular
neuropile around these mushroom body parts. Optic lobe fibres (connexions to the lobula and medulla) synapse in the lateral lobes, in
the optic tubercles, in caudal parts of the protocerebrum and the
deutocerebral dorsal lobe. Though the texture of connexions cannot
be completely described from silver stains and Golgi impregnations,
it is evident that extrinsic elements of the glomerular neuropile
can meet the projection areas of the cobalt stained ascending and
descending fibres.

DISCUSSION

A minimum of 100 neurons is found in the brain of the bee by cobalt-chlorid-backfilling experiments. They connect the head ganglia with the ventral nerve cord. There is good evidence that they represent descending nerve cells (from an anatomical point of view), repeatedly found at the same position with a characteristic branching pattern in the non-glomerular neuropile. Fibres from these cells were traced to the connectives where they belong to the large-diameter fibres including some "giant"axons.

The non-glomerular neuropile of the·brain is mainly made up of 53500 neurons (Witthöft, 1967), including the descending nerve cells. Among these 53000 cells we have to expect the unknown numbers of interneurons linking the ascending and descending neurons with the large glomerular neuropile masses mainly formed by small local interneurons (globule cells, Kenyon cells). These small interneurons represent the great majority of the cells in the brain of the worker bee (433000 in the optic lobes, 340000 in the mushroom bodies, Witthöft, 1967). From these counts and our cobalt stains we derive a ratio of 7730 glomerular globule cells : 535 brain interneurons : 1 descending brain neuron. A high convergence of brain interneurons to descending nerve cells can be predicted and has been shown for mushroom body cells to output-interneurons (Schürmann, 1974).

As ascending and descending elements show partial overlap of their synaptic projections in the non-glomerular neuropile, a mono-synaptic or short circuit of these two nerve cell types might exist in the brain. Glomerular brain neuropiles exerting influence on ascending and descending bi- and plurisegmental neurons do this by means of output-interneurons which project to non-glomerular neuropile areas of these connecting elements. Some of these brain interneurons have been recorded from and stained by micropipettes in bees (Erber, 1981) and in crickets (Schildberger, 1981). They project to the lateral protocerebral lobes and dorsal lobes of the deutocerebrum. Connexions of the optic lobes extend to the same areas.

Large parts of the non-glomerular neuropile do not exhibit ramifications of the cobalt-stained fibres (parts of the lateral lobes , the anterior neuropile around and between the mushroom bodies, optic tubercles). These areas contain neurons linked to the glomerular neuropile representing nerve cells restricted to the brain. Though tracts stained with silver elucidate connexions between brain compartments they do by no means reflect the potential of interactions between single neurons and sets of neurons forming suborders of the neuropile. Functional contacts might be established largely by local circuits.

Some of the descending interneurons have been identified and recorded (s.Guy et al., 1979, Strausfeld and Nässel, 1980). If we accept a small number of brain descending interneurons (a hundred in the bee), we have to consider how the brain uses such a limited number of channels tó influence and induce so many behavioral patterns of differing complexity with the thorax and abdomen. The anatomical results presented here support the concept of "command" neurons (Wiersma and Ikeda, 1964). All the large fibres in the connectives can be investigated by intracellular electrophysiological recordings.

REFERENCES

DUJARDIN F., 1850. - Mémoire sur le système nerveux des insects. Ann. Sci. Natur. Zool., 14, 195-206.

ERBER J., 1981. - Neural correlates of learning in the honeybee. Trends in Neurosci., 4, 270-273.

GUY G.R., GOODMAN L.J., MOBBS P.G.,1979. - Visual interneurons in the bee brain: synaptic organization and transmission by graded potentials.J. Comp. Physiol., 134, 253-264.

HOWSE P.E., 1975. - Brain structure and behavior in insects. Ann. Rev. Entomol., 20, 359-379.

MOBBS P.G., 1982. - The brain of the honeybee Apis mellifera - I. The connections and spatial organization of the mushroom bodies. In the press .

PAN K.C., GOODMAN L.J., 1977. - Ocellar projections within the central nervous system of the worker honeybee, Apis mellifera. Cell. Tiss. Res., 176, 505-527.

RIBI W.A., SCHEEL M., 1981. - The second and third optic ganglia of the worker bee. Cell Tiss. Res., 221, 17-43.

SCHILDBERGER K., 1981. - Some physiological features of mushroom-body linked fibres in the house cricket brain. Naturwiss., 68, 623-624.

SCHÜRMANN F.W., 1974.- Bemerkungen zur Funktion der Corpora pedunculata im Gehirn der Insekten aus morphologischer Sicht. Exp. Brain Res., 20, 406-432.

STRAUSFELD N.J., MILLER T., 1980. - Neuroanatomical Techniques. Springer New York, 496 p.

STRAUSFELD N.J., NÄSSEL D.R., 1980. - Neuroarchitecture of brain regions that subserve the compound eyes of crustacea and insects. Handbook of Sensory Physiology. Springer Berlin, 132 p.

WIERSMA C.A.G., IKEDA K., 1964. - Interneurons commanding swimmeret movements in the crayfish Procambarus clarki. Comp. Biochem. Physiol., 12, 509-525.

WITTHÖFT W., 1967. - Absolute Anzahl und Verteilung der Zellen im Hirn der Honigbiene. Z.Morph. Tiere, 61, 160-184.

Abstracts

Male Territoriality in the Carpenter Bee <u>Xylocopa</u> <u>virginica</u> <u>virginica</u>

Edward M. Barrows

Georgetown University

Males of <u>X</u>. <u>v</u>. <u>virginica</u> choose more types of territorial sites
than are known to be used by any other species of bee. Their
territories are established at nest sites, near food plants, and
near landmarks away from nests and food. They have been seen
copulating in the first two sites, only. At nest sites, males
defend 33,000 cm^3 "hover spaces" surrounded by hemispherical "attack
spaces" with 20-m radii. They drive away other males and
heterospecific insects. Based on mid-spring time budgets, males
spend significantly more time interacting with each other than
interacting with females or reacting to other airborne objects.
Probably due to their limited vision, males leave territories in
pursuit of birds and in flying towards distant airplanes. Their
response to artificial territorial intruders (black spheroids)
relates to size of intruders. Interloping males invade nest-site
territories; other males without territories patrol wildlflowers
and attempt mating with foraging females.

The Central Projections of the Hairplate Receptors of Episternal
Cone, Petiole, and Mesocoxa, of the Honeybee

William Fletcher and Lesley Goodman

Queen Mary College, London

The sensory neurons innervating the hairplates of the episternal
cone, petiole, and mesocoxa, of the worker honey bee <u>Apis</u> <u>mellifera</u>
have been stained by axonal filling with cobalt chloride. Following
intensification with silver the central projections of the mechano-
sensory hairs were examined using serial sections and wholemounts.
The hairplate receptors of the episternal cone enter the prothoracic
ganglion through nerve I N 1 and arborize ipsilaterally and
dorsoventrally in the prothoracic ganglion, anterodorsally, in the
mesometathoracic ganglion, and dorsally in the suboesophageal
ganglion. In contrast the hairplate fibres of the mesocoxa

terminate in the mesothoracic neuromere alone, both ipsilaterally and contralaterally. The hairplate neurons of the petiole terminate exclusively in the mesometathoracic ganglion. These fibres enter via nerve II N 13 and arborize dorso-ventrally in the abdominal neuromeres, and ventrally in the mesothoracic and metathoracic neuromeres. The distribution of the hairplate fibres is considered in relation to their possible function in graviception.

Descending Neurons in the Brain and Thorax of the Honeybee

Lesley Goodman, Peter Mobbs, and Richard Guy
Queen Mary College and City University

The descending multisegmental interneurons of the bee brain arborise in the posterior slope neuropile of the protocerebrum. The axons of these neurons run in the most dorsal part of the ventral connective and represent the major part of the median dorsal longitudinal tract (MDT) within the thoracic ganglia. Electron microscope studies show that there are 34 neurons of over 8 μm in diameter within the MDT as it enters the prothoracic ganglion. Colbalt staining of these neurons by intracellular injection during recording experiments, and by intra-neuropilar and axonal diffusion, has enabled the mapping of many of the larger descending cells into the neuropiles of the brain and thorax. The patterns of arborisation revealed are related to the electrophysiology of individual cells and their possible inputs as shown by other histological techniques. Recording experiments suggest that many of these neurons are involved in flight stabilisation and other behaviours requiring rapid re-orientation.

Social Context: If Ant Behavior Depends on It, Should our Methods Be Leaving it Out?

Deborah M. Gordon
Duke University

Many investigations of social insect behavior measure the reactions of individuals to some experimental manipulation. But these reactions may be affected by the social context, or behavioral state, prevailing in the colony at the time of the expriment. Two experiments using harvester ants (Pogonomyrmex spp.) demonstrated the effect of social context on ant behavior. Social context was specified as the number of ants engaged in each of five activities. The first experiment tested the response of P. badius to oleic acid, a midden pheromone. Objects treated with oleic acid are taken to the midden, as previously reported, when a large percentage of the colony is doing midden work or nest maintenance. But when a large percentage is foraging or convening, treated objects are taken into the nest as if they were food items. The second experiment was a field study of P. rugosus. Recruitment to bait was high when large numbers of ants were patrolling or doing nest maintenance. But when large numbers were foraging on a trail to a distant food source, ants would not recruit to newly offered bait. These results suggest that social context may be a significant factor to be considered when designing experiments to investigate ant behavior.

Identification of Motion Sensitive Descending Neurons in the Bee
Apis mellifera

R. G. Guy*, P. G. Mobbs**, and L. J. Goodman**
*The City University and **Queen Mary College, London, England

In the honeybee Apis mellifera visual information plays a major role in flight stabilisation. During flight the bee is free to move in roll, yaw and pitch planes. In this study we have examined the responses of descending neurons in the brain and suboesophageal ganglion to targets suggesting these movements of the animal. The intracellular injection of cobalt dye has enabled the identification of these neurons. The parallel staining of groups of intrinsic neurons by the intraneuropilar injection of cobalt has revealed the possible inputs of some of these movement sensitive units. Of particular interest are neurons with tonic responses to movement in preferred directions. In order to derive unambiguously both the direction and velocity of the target the thoracic motor neuropiles must carry out a cross-fibre analysis of the information carried by many such fibres.

Pattern analysis and simulation in the study of social insects

P. Hogeweg & B. Hesper
Bioinformatica, Padualaan 8, Utrecht

Research on social insects should study the ways in which behaviour patterns of individuals can generate behaviour patterns of societies. In this paper we examine how a combination of pattern analysis and discrete event, individual-oriented simulation (hogeweg & Hesper, 1979, 1981a,b,c) can be used to this end. Pattern analysis is used initially to detect patterns in the behaviour of individuals (van Honk & Hogeweg 1981), and subsequently to compare observed and simulated behaviour of individuals and societies. Individual oriented simulation is used to implement the behaviour of individuals recognised so far and to generate the societal patterns generated by this behaviour of individuals. The pattern analysis and simulation steps are interated as our knowledge increases.
The procedure is illustrated by its use in the study of the ontogeny of bumble bee colonies.

Hogeweg, P. & Hesper, B. (1979) in: B. P. Zeigler et al. (eds.) Methodology of systems modelling & simulation. North Holland.
-----1981a in: T.I. Oren (ed.) Proc. Cybersoft Symposium, Int. Society for Cybernetics, Namur.
-----1981b in: UK CS Conference on Computer Simulation, Westbury H.
-----1981c in: J. Theor. Biol. Vol. 93, in press.
van Honk, C.G.J. & Hogeweg, P., Beh. Ecol. Sociobiol. 9: 111-119.

Pupal and Post-emergence Development of the Mushroom Bodies of the
Brain of the Honeybee

Michael Noble and Peter Mobbs
Queen Mary College and St. Mary's Hospital Medical School

The mushroom bodies of the adult bee consist of 340,000 similar intrinsic neurons arranged in a series of highly ordered parallel arrays. The precision of the organisation of these neuropiles makes them an ideal subject for developmental studies. Brains from pre-pupal and pupal bees of known age have been serially sectioned in a variety of planes to establish the changes in the number of instrinsic neurons & the volume and spatial organisation of the mushroom bodies. The mushroom bodies are prominent neuropiles within the late larval, and pupal, bee brain. In larval bees the pedunculus, and the alpha- and beta-lobes can be readily distinguished even though their volumes are greatly reduced. The calyces, however, are poorly differentiated as the lip, collar, and basal ring neuropiles, that are characteristic of the adult structure, can barely be distinguished. The groups of Kenyon cell-bodies responsible for the formation of these three calycal zones can be identified in early pre-pupal development. The changes that occur in the gross disposition, number, and dendritic complexity, of these cells has been charted. This development has been related to the formation of the tracts from the antennal lobes and compound eyes that represent the major inputs to the calyces.

The Development of the Brain of the Honeybee

Michael Noble and Peter Mobbs
Queen Mary College and St. Mary's Hospital Medical School

A method based on that of Cameron Jay (1962, Bee World 43, 119-122) that enables the aging of workerbee pupae upon morphological criteria, including progressive changes in eye and cuticular colour, was developed. Brains from pre-pupal, pupal and newly emerged bees were aged to within six hours using this technique. Serial sections stained with reduced silver and haematoxylin and eosin were used to investigate the development of the major brain neuropiles and their connections. Changes in the architecture of individual neurons within the compound eyes, antennal lobes and mushroom bodies were examined in Golgi stained preparations. Although from early in pupal development all the gross compartments of the brain are apparent, there are a number of spectacular changes in some neuropiles. The ocelli for example, which in the adult are fused into the dorso-posterior protecerebrum, are connected to the same region in the pupal brain by long periperhal nerves composed of receptor axons. The glomerular structure of the antennal lobes, prominent in the adult, is absent in $1\frac{1}{2}$ day old pupae.

The Release of Sexual Behavior in Males of Polistes fuscatus

David C. Post and Robert L. Jeanne
The University of Wisconsin-Madison

Males of Polistes fuscatus perch in territories located in
female hibernation areas, where they intercept and attempt to mate
with females. Males respond to females with the following ordered
chain of behaviors: (1) orient, (2) pursue, (3) grasp, and (4)
attempt copulation. Various tethered objects were presented to
territorial males in order to determine the stimuli which attract
and release sexual behavior in males. Wasp-shaped, black and yellow
banded objects were found to elicit pursuit and grasping behavior.
A female sex pheromone was found to attract males over short
distances and release male copulatory behavior. A homosexual
response of caged males to the female odor was used as a bioassay
to show that the source of the sex pheromone was the venom gland
and sac.

Investigations of the Mating Behavior of the Eastern Yellowjacket, Vespula maculifrons

Kenneth G. Ross
Cornell University

The mating behavior of V. maculifrons was studied in a
laboratory setting. The courtship and mating behavior was described
and quantified. Males were found to be capable of inseminating up
to 3 queens, while queens were also capable of repeated matings.
Newly emerged queens were assayed for ability to mate and were found
capable of being inseminated at 2-3 days of age. Newly emerged
queens could not fly properly until 7-9 days of age. These results
suggest the possibility of sibling matings in the natal colony
during the period when virgin queens are receptive but not yet able
to undertake mating flights. Choice experiments revealed that
virgin queens did not discriminate against male siblings in mating
trials. However, matings did not readily occur in the absence of
light suggesting that this may serve as a proximate mechanism
preventing sib mating in the parental nest cavity. Dissections of
series of fall queens from 7 nests support this suggestion.
Inseminated queens were found in only one nest which had been
disturbed so that light penetrated the nest structure.

The Sense Organs on the Flagella of Ants (Formicoidea)

Joachim R. Walther
Free University of Berlin/Fed. Rep. of Germany

The morphology and ultrastructure of all nine types of sensilla
on the flagella of the male, female and worker of Formica rufa L.
have been investigated using light and electron microscopy (SEM and
TEM). Additionally, the antennal sensilla of several other species

from eight families of the Formicoidea have been studied. Though
some differences between the patterns of sensilla of the different
families were observed, it was clear that a special pattern of
sensilla of the Formicoidea exists. Three olfactory sensilla, one
gustatory + mechanosensitive sensillum, the mechanosensitive Sensilla
campaniformia and the tactile hairs, the Sensilla ampullacea and
coeloconica and one poreless single-walled sensillum belong usually
to the pattern of sensilla of the ants.

Abstracts: General Topics*

Micro-electronic identity tags for bumble bees,
to monitor arrivals and departures from nests

Tracy Allen
University of California--Berkeley

Bees wear micro-electronic identity tags, which are decoded automatically as the bees pass to and fro through the entrance of their nest in the course of foraging. Identities of bees and times and directions of passage are recorded by a computer system for subsequent analysis of individual and aggregate foraging patterns.

The tags are inexpensive, hand-wound coils of wire, which are 2 mm in diameter and weigh 4 mg. Each tag has a unique resonant frequency (like a cow-bell). A tag is glued to the notum of each of up to 20 bees in a nest. The detection apparatus reads the resonant frequencies of tags that come within a 2 cm range of a detection antenna, which is usually positioned above the nest entrance tunnel. The prototype detection apparatus (minus the data logger) costs about $600. This system is tailored to bumble bees, but may be adapted for studies of social behavior of other animals.

In Bombus edwardsii and B. vosnesenskii hives studied, major shifts in individual and aggregate foraging patterns were associated with changing weather and with changing conditions within the nest. Bee-to-Bee variation in foraging behavior was great. For example, a few "elite" bees did most of the foraging, and nests had many inactive bees.

Trees from glue: chemical biogeography
of nasute termites

Barbara L. Bentley, Glenn D. Prestwich, & Thomas J. Gush
State University of New York at Stony Brook

This chemical structure of nasute defense compounds ("glue"), as determined by gas chromatography, can differ among populations within species as well as among species. In this study, we used these differences to determine the degree of similarity among three populations of Nasutitermes in Costa Rica: 1) from the La

---

*(These abstracts were not easily assigned to symposium topics
or arrived too late for inclusion under those topics.)

Selva rainforest on the eastern Atlantic slopes, 2) from a tropical dry deciduous forest in the northwestern province of Guanacaste, and 3) from a rainforest on the Osa penninsula in southwestern Costa Rica. The presence or absence and the relative concentrations of both monoterpenes and diterpenes were used in a cluster analysis to create a dendrogram of similarities among and within populations. These data can be used to establish if the Osa populations are derived from western, dry forest populations or from eastern, rainforest populations.

In addition, we have analyzed secretions from individual termite heads, and thus we can determine variation within a colony as well as among colonies. These data are the first of their kind and are currently being investigated for use as taxonomic discriminators.

Population Genetic Structure of Dolichoderine Ants

Rudi Carol Berkelhamer
University of California, Berkeley

Genetic structures of populations of three species of Dolichoderine ants were analyzed and compared. The predicition that polygynous species are relatively inbred was tested. This very important prediction is based on W. D. Hamilton's haplodiploidy/kin selection model of the evolution of eusociality. If the prediction fails to be upheld, then the relatively common occurrence of polygyny is a major argument against Hamilton's genetical theory.

Analyses of electrophoretic data using Wright's F-statistics show that populations of both polygynous species studied, Iridomyrmex pruinosum and Conomyrma bicolor, are relatively inbred. By contrast, populations of a related monogynous species studied, Conomyrma insana, are not inbred. Further analyses of population structure using a graphical clustering technique indicate that populations of Iridomyrmex pruinosum, one of the polygynous species, are especially inbred and may exhibit extremely local breeding. These findings support the predicitions of the kin selection model as well as those of an alternative trait group selection model proposed by the author.

Termites of Southern India and their Affinities to those of Neighbouring Regions

Geeta Bose
Zoological Survey of India, Calcutta

Termite fauna of southern India is extremely rich and is represented by 95 species belonging to 33 genera of the families Kalotermitidae (5 genera, 18spp.), Hodotermitidae (one genus, one sp.), Rhinotermitidae (3 genera, 5 spp.), Stylotermitidae (one genus, one sp.) and Termitidae (23 genera, 70 spp.).

Four termitid genera (Termitinae, 1; Nasutitermitinae, 3) are endemic to southern India and nine others endemic to the oriental region are present here, the remaining 20 being common to those from other parts of the Indian subcontinent and other zoogeographical regions.

The Energetics of Wax Production in Apis mellifera and its
Importance for the Evolution of Eusociality in Apis

Stephen L. Buchmann, Justin O. Schmidt, and Robert L. Schmalzel
U.S. Dept. of Agriculture,
Carl Hayden Bee Research Center, Tucson, AZ

The caloric costs of comb production by Apis are not known.
Wax was produced rapidly using 12-18 day-old caged bees with a
queen. Late in the year (Oct. - Nov.) wax production ceased apparently in response to an unknown circannual rhythm. Bees consume
about 7 grams of sucrose (27,600 calories) to produce 1 gram of
beeswax (10,000 calories).

When constructing new comb, Apis does not adulterate wax as do
the meliponines. Adulteration requires additional procurement costs
(flight time and energy). Instead, honey bees remain relatively
quiescent while secreting wax scales and manipulating the scales
into the architecture of the comb. If we assume a conversion
efficiency of .35 of honey into wax and that Apis uses about 4.2 g
wax to store 100 g honey in sealed honey comb, then the amount of
honey transformed into wax to sequester an additional 100 g of
honey can be determined. These values are preliminary but suggest
that at least one-fourth of the honey required by a colony is for-
feited in securing honey in a comb.

Chemical Evolution of an Exocrine Secretion
from Bees (Hymenoptera: Apoidea)

James H. Cane
The University of Kansas

The intra- and interfamilial distributions of Dufour's gland
lipids and waxes secreted by female bees, for purposes of sting
lubrication, nest cell water regulation, and larval nutrition, can
be predicted by evolutionary considerations. Biosynthetic pathways
limit the spectrum of biochemical possibilities. Experimental
ecological chemistry indicates the biologically relevant contexts of
secretion, and so aids in hypothesizing means of persistence of
certain biochemical novelties in the face of selection and metabolic
expense over evolutionary time. Thirdly, cladogenesis, the branching patterns of genetically-related lineages, predicts phylogenetic
relationships as a function of shared, uniquely derived chemical
characters which indicate common ancestry. Equally parsimonious
trees can be tested by a priori biosynthetic predictions. Biosynthetic simplicity and strong directional selection can yield convergences which are predictable by this approach.

Examples are presented demonstrating 1)a unique common ancestor
shared by the Colletidae, Oxaeidae, and Halictidae, 2)three clades
within the Andrenidae, and 3)the potential for this method in
deducing phylogenetic relationships among the social bees.

404

Distribution and Zoogeography of Oriental
Termitidae (Isoptera)

Om Bahadur Chhotani
Zoological Survey of India, Calcutta

Family Termitidae, in the oriental region is represented by
482 species belonging to 50 genera of the subfamilies Apicotermitinae, Termitinae, Macrotermitinae and Nasutitermitinae;
Apicotermitinae by 4 genera and 29 species (all endemic to the
region), Termitinae by 16 genera (10 endemic) and 143 species,
Macrotermitinae by 5 genera (2 endemic) and 115 species and
Nasutitermitinae by 25 genera (21 endemic) and 195 species. Of the
13 non-endemic genera, two each are cosmopolitan and cosmotropical
and 9 are common to two or three zoogeographical regions of which
8 are, however, common to oriental and ethiopian regions.

The distributional pattern and zoogeography of the oriental
genera is discussed with a view to their origin.

Genetic Structure of Natural Populations in the European
Species of the Genus Reticulitermes (Isoptera)

Jean-Luc Clement
The University of Paris (France)

Enzymatic polymorphism in the european populations is very
marked (50% of loci are polymorphic) and it gives us informations
for each population on the number and quality of reproductors in
each society. A complementary study of aggression using an ethological test between nests in each population gives us informations
on colonial individuality and possible colonial fusion. In humid
and oceanic climate there is no agonism in summer between nests and
societies can exchange reproductors and workers during colonial
fusions. Workers always build galleries linking tree trunks in
large wet pine forest and neotenics outbreeding is favored. Allelic
frequencies of many colonies are in harmony with the Hardy-Weinberg
law, so that we could consider each colony as a part of a population.
Genetic uniformity is high. In dry and semi-arid climate, colony
formation being due to the swarming of winged individuals and
colonial individuality is very high. Aggressive behaviour always
prevent colony fusion. Workers are generally F1, allelic
frequencies approximate theoretical Mendelian frequencies. There
are high genetical differences between neighbouring nests.

Heat Balance in Foraging Honey Bees
(Apis mellifera caucasica) in the Desert

Paul D. Cooper[1], Steven L. Buchmann[2], William M. Schaffer[1]
[1]University of Arizona and [2]Carl Hayden Bee Research Laboratory

Foraging honey bees often fly during periods of high ambient temperature, and intense solar radiation in the desert. Additionally, bees gain heat as a result of metabolism. Previous studies have indicated that bees cool evaporatively when ambient temperature exceeds 38 C. By measuring convection coefficients of bees at flight speed, and balancing this avenue of heat loss with metabolic heat, solar heat gain, and long wave radiation input, the amount of heat lost by evaporation under field conditions is estimated. The ability of honey bees to forage under extremely hot conditions is suggested to be constrained by a combination of availability of water and/or nectar for cooling the bee, while the necessity of foraging at high temperatures may result from colony water balance requirements.

Adaptation of Alate and Worker Mandibles of Neotropical Nasute
Termites to Soil-feeding Habit

Luis Roberto Fontes
Universidade de São Paulo
Instituto de Biociências, Depto. de Zoologia - Brasil

The mandibles of the eight genera of soil-feeding nasutes occuring in the Neotropics are studied and compared to the six genera of Ethiopian nasutes with similar diet. The former are divided into 3 groups, according to the development of the mandibles of the alate and worker castes. Group 1 (apical teeth usually large; index of left mandible 0.66-3.59 and second marginal tooth smaller than in non-soil feeding nasutes) includes 2 types of workers, differing in the width of the gap between the third marginal tooth and the molar prominence of the right mandible; in the same colony are found workers "with broad gap" and "with narrow gap";these may originate from different sexes.

Reclassification of the Nearctic Species
in the Ant Genus Myrmica (Formicidae, Hymenoptera)

André Francoeur
Université du Québec à Chicoutimi

The nature of the Myrmica species actually recognized in the nearctic region has been reevaluated through a comparative analysis of their morphology, ecology, and biogeography. Examination of all the specimens of the type series revealed numerous misleading conceptions and assumptions among these species, as well as in the species groups, and their relationships: many type series included more than one taxon. These facts and abundant recent material brought out new species and species groups. The number of different

forms and phylogenetic lines from the western half of North America
has been underestimated, particularly in the Rockies. The proposed
reclassification of the species exhibits a better ecological correlation with the presence of the genus in the biomes of the nearctic
region.

Patrilineal Group Dynamics in the Honeybee Apis mellifera

Wayne M. Getz and Katherine Smith
University of California at Berkeley

Using a recessive mutant for cuticle color (cordovan) and
instrumental insemination, hives have been set up in which two
patrilineal lines can be visually distinguished. Data obtained
from these hives on kin grouping during swarming, preferential
brood care, kin recognition and egg laying patterns is reported.

Isopteran Sex-ratio Strategies

Susan C. Jones,[1] Jeffery P. La Fage,[2] and Ralph W. Howard[1]
[1]Forest Service-USDA, Gulfport, Mississippi
[2]Louisiana State University

Isopteran populations conform to an overall 50:50 alate sex
ratio, although there is variation within individual colonies.
Populations with equal numbers of males and females in a significant
proportion of the colonies are characterized by an extended flight
season, nonsynchronous flights from different colonies, and small
numbers of alates in each swarm. Populations having a significant
proportion of colonies with skewed sex ratios are characterized by a
brief flight season, synchronous flights from many colonies, and
large numbers of termites in each swarm. All available data on
termite sex ratios are compiled and discussed with reference to
genetic implications.

Taxonomy of Ants (Fam. Formicidae) in A. R. Egypt

A. H. Kaschef and A. H. Mohamed
Department of Entomology, Faculty of Science, Ain Shams
University, Cairo.

The Egyptian ants are economically indifferent and a considerable number of species are injurous to Man. Most important
among these are the house ants while other species are well-known
garden and field pests in both the Nile-Valley and desertic areas.
The work included survey, taxonomy supplied with keys of the known
species and field observations on their habitats. Great attention
has been given to mountanious areas along the Red Sea shore and
oasis as well as daily and seasonal activities.
    22 genera belonging to 5 subfamilies (Ponerinae, Dorylinae,
Dolichoderinae, Formicinae & Myrmicinae) were considered. 141
species, subspecies and varieties were recorded in our country among
which 55 species are intercepted in A.R. Egypt.

A Comparative Biochemical Study of the Contents
of the Eggs of the Ant Pheidole pallidula Nyl.

B. E. Lorber[++], L. Passera[+], and B. Colas[++]
+ Laboratoire de Biologie des Insectes
118, route de Narbonne, F-31062 TOULOUSE, FRANCE
++ Laboratoire de Biochimie, IBMC-CNRS
15, rue R. Descartes, F-67084 STRASBOURG, FRANCE

A comparative analysis of the contents of queen biassed eggs
(laid by mated queens after hibernation), worker biassed eggs (laid
one month later and during the activity season) and trophic eggs
(produced by virgin queens prior to swarming) of the ant Pheidole
pallidula Nyl, has been carried out. We have studied the following
molecules; proteins, nucleic acids, lipids and sugars.
   The U.V. absorption of the soluble crude extracts of reproductive eggs is always higher than that derived from trophic eggs.
However no difference can be observed in protein distribution or
quantity by gel electrophoresis. Only small differences can be
found in the pattern of lipids and the concentration of sugars is
very low. Taken together our results indicate that the reproductive
and trophic eggs (0-1 day old) have a similar biochemical composition with the exception of a $\geqslant$10-fold excess of high molecular
weight nucleic acids which is found in the former species of eggs
relative to the latter.

Geographic Patterns of Genetic Variation
in a Termite

Peter Luykx
The University of Miami
(Coral Gables, Florida)

Patterns of allozymic variation were studied in natural populations of Incisitermes schwarzi (Kalotermitidae) by means of starch
gel electrophoresis. X-linked alleles in different populations in
the Miami area appear to be in Hardy-Weinberg equilibrium, but are
in marked disequilibrium with respect to X- vs. Y-linkage. Preliminary findings indicate that X-linked alleles are fixed in populations
where inbreeding is highest. Geographic variation in allele frequencies appears not be be correlated with geographic differences in
chromosome arrangements (translocations) in the Miami area.

Effects of Environmental Conditions on Hygienic Behavior
of Africanized Bees (Apis mellifera)

Dejair Message[1] and Lionel S. Gonçalves[2]
1 - Universidade Federal de Vicosa (MG-Brasil)
2 - F.F.C.L. Ribeirão Preto - USP (SP-BRASIL)

The hygienic behavior of honey bees provides a mechanism of
resistance to brood diseases, consisting of the uncapping of cells
containing dead brood and removing the bodies. Our results show
that the behavior of bees varies within the same colony in different
tests, or for different colonies in the same test, due mainly to environmental factors. Colony conditions greatly influence hygienic
behavior. The age of dead brood also has a marked influence. The
maximum efficiency is obtained using combs containing dead brood
aged 9 to 10 days or 19 to 20 days and the minimum efficiency is
obtained when dead brood with 17 to 18 days of age is used.
(PIG/CNPq).

Multivariate Generic Comparison of Ant Flights

Elwood S. McCluskey
Loma Linda University

Treated as a taxonomic character the daily hour of mating
flight indicates a difference in timing from one genus to another.
But is is insufficient by itself to assign species to genera
correctly. Would the addition of other behavioral characters make
this possible? Genera where there are flight records for at least
five species each were compared. Flight hour alone segregates all
the species of genera 1 and 2 from genera 3, 4 and 5, and most of
the species of 1 from 2. Flight data alone segregates all the
species of 3 from 4 and 5. Thus 80% of the species of 1, 2 and 3
are assigned to their respective genera by these two characters
(verified by discriminant analysis). Genera 4 and 5 are similar
morphologically and were at one time considered congeneric.
Additional data available for several species of each of these two
genera were tested: temperature at flight onset correctly sorts
the species to genus, and light intensity at onset nearly does so.
Accumulating enough data from sporadic phenomena such as ant flights
to make a multivariate comparison is most difficult, and the analysis
here was possible mainly because of Talbot's many years of
observations.

Effect of Ionizing Radiation and Temperature on Apis mellifera L.
(Hymenoptera Apoidea): Analyses of morphometric characters, acid
gland morphology and physical abnormalites.

Luiza Nakayama and Catarina Satie Takahashi
University of São Paulo

Temperature shocks administrated to *Apis* pre-pupae and young
pupae caused changes in some morphological trait dimensions of the
head and of the wings, decreasing it (Nakayama & Takahashi, 1981).
But when the temperature treatment was combined with 2000R of gamma
radiation, the effects were not as homogenous as those obtained
with temperature alone. The combination of hypo-(10-20-30°C) or
hyperthermia (40-42°C) and radiation or vice-versa sometimes increased and sometimes decreased the dimension of the morphological
characters. In addition, normal frequency of the bifurcated acid
gland ($\pm$ 75%) in the workers was considerably altered. Several
malformed workers were observed after this kind of treatment such
as: alterations or lack of wings, one cyclopic female with altered
genital characters, proximity and fusion of ocelli and swollen
antennae without any change in the number of antennae segments.

Genetic structure of *Formica* populations

Pekka Pamilo
University of Helsinki, Finland

Three components of population structure have been examined
in *Formica* ants: interaction structure, mating structure, and
production of sexuals. The main problems studied are:
Interaction structure: What is the level of polygyny? Is
polygyny functional? How related are the coexistent gynes?
Mating structure: Are the species polyandrous? Are there
departures from random mating? How large is a breeding unit?
Production of sexuals: Do the workers produce males? How
does the population structure affect sex ratios? Who controls the
sex ratio in the population, gynes or workers?
The material comes from 14 species and the results are mainly
based on genotype distributions among nests in natural populations.
There are great interspecific differences in population structure
and also clear intraspecific variation. The populations seem to
be subdivided into small breeding units, leading to important genic
differentiation within the populations. The coexistent gynes are
generally related to each other, they are chiefly monoandrous, and
local inbreeding is rare. The reproductives are mainly produced
by the gynes, and it is possible that the gynes also control the
sex ratio, at least in the *Formica rufa* group species.

410

A Method for Hand Rearing Vespula Larvae
(Hymenoptera: Vespidae)

Mark Parrish
Rutgers, The State University of New Jersey (Cook College)

Yellowjacket larvae were reared on isolated combs without attendant adults. Larvae were incubated at 30°C and 100% humidity. Strained baby food, because of its high protein content and pasty texture was fed to the larvae through a B-D ^{TM} V46 hypodermic needle. Rapid and efficient collection of salivary secretion was attained with a B-D ^{TM} I-41-4 hypodermic needle attached to an aspirator. Larvae pupated and eclosed as active adults. This method permits bioassay of physiologically active chemicals and observations of caste differentiation.

Chromosome analyses in some wasp species
(Polistinae : Hymenoptera).

Silvia das Gracas Pompolo and Catarina Satie Takahashi
University of São Paulo

Few cytogenetics studies have been done with wasps, due to the difficulty of chromosome visualization. With conventional techniques they show up as simple dots. By integration of some techniques already known, it was possible to get more conclusive data about the chromosome number in some species of the subfamily Polistinae found in the region of Ribeirão Preto (state of São Paulo-Brazil). In this study it had obtained the following results: *Polybia paulista* n=17 (male) and 2n=34 (female): *Stelopolybia multipicta* 2n=64 (female); *Stelopolybia pallipes* n=32 (male); *Polistes versicolor* n=31 (male) and 2n=62 (female); *Brachygastra lecheguana* 2n=56 (female and *Pseudopolybia vespiceps* n=8 (male) chromosomes.

On some Biotic Factors Limiting the Evolution of Bumblebee Colonies

A. Pouvreau
Station de Recherche sur l'Abeille et les Insectes Sociaux

At every stage, in the overwintering of queens, in nestbuilding, and in the rearing of colonies, adverse factors limiting population growth are at work. Bad weather, insufficient food, especially in the spring, diseases, and enemies such as the larvae of the wax moth, Aphomia sociella, all tend to reduce bumble-bee populations. Even man inadvertently becomes an enemy of the bumble-bee in its natural habitat, when he uses herbicides and insecticides.

The author gives a brief outline of the main parasites and predators affecting Bumble-bees, and investigates the nature of the interactions between bumble-bees and their principal enemies.

Ecological Consequences of Regulation of Nest
Temperatures at High T_a in the Honey Bee

William M. Schaffer, Paul D. Cooper, and Stephen L. Buchmann*
The University of Arizona and
(*) USDA Bee Research Center, Tucson

Honey bee activity (bees/inflorescence) at Schott's agave in
Southern Arizona is strongly correlated with the standing crop (SC)
of available nectar. On hot days ($\bar{T} > 30^{\circ}C$), the bees quit the
flowers even though nectar still remains $[SC_q = 1.5 \pm .3 \mu l$ (N=1034)$]$
On cooler days ($\bar{T} < 30^{\circ}C$), the bees continue to forage until virtually all of the nectar has been consumed $[SC_q = 0.2 \pm .1 \mu l$
(N=408)$]$. Since honey bees are capable of flight at T_a up to $46^{\circ}C$,
we hypothesize that the changes in foraging behavior observed in
the field reflect variation in colony demand for liquid for nest
temperature regulation. The key assumption of this hypothesis is
that under most circumstances the rate at which liquid can be
collected in the form of water far exceeds the rate at which it can
be collected in the form of nectar. Observational and experimental evidence from artificially heated and cooled hives supporting this point of view is reviewed.

Dynamics of Food Flow in the Imported Fire Ant,
Solenopsis invicta Buren

A. Ann Sorensen and S. Bradleigh Vinson
Texas A & M University

The dynamics of food flow were studied in laboratory colonies
of Solenopsis invicta Buren. Methods included the use of radio-
iodinated protein mixed with food, marked worker ants, and quantitative measurements of specific digestive enzymes. Food flow was
influenced by nutritional needs, distribution of digestive enzymes, colony hunger, and division of labor. The distribution of
proteins, oils, and carbohydrates was related to nutritional requirements and availability of digestive enzymes in caste members.
Starvation levels in foragers and nurses influenced respectively
the quantity of carbohydrates and proteins brought into the colony.
Queens and larvae did not directly influence the amount or type of
food brought back. However, partially digested proteins regurgitated by well-fed larvae provided a positive feedback to the
nurses. The amount, turnover time, and internal distribution of
food in each worker correlated with its current "career" as either
a food gatherer, food storer, food relayer, or brood tender. The
amount of food per caste was dependent on the total amount of that
food in the colony, with turnover rates differing between castes
as the amount of food increased. The brood represented a form of
food storage for the colony and during stress cannibalism increased
significantly.

Classic Metabolism Physiology as
Applied to Social Insect Colonies

Edward E. Southwick
State University of New York
College at Brockport

The fundamental features of intermediary metabolic pathways of
energy transformations of foodstuffs underlie all living systems.
Any organism derives about 5 cal for each ml oxygen consumed when
metabolizing sugar. Energy metabolism integrates a number of
physiological factors into an overall performance of the organism.
Colonies of social insects can be treated as individual super-
organisms and classic metabolic physiological analyses utilized.
Standard methods were applied to intact colonies of wintering
honey bees (Apis mellifera) and metabolic cost of thermoregulation
determined. Below 10° C, metabolic rate, determined from oxygen
consumption measurements, increased as a function of decreasing
environmental temperature following the equation:
MR (Wkg) = 7.96 - 0.24T(C)

Honey Bee Drinking Behavior Within The Hive

Hayward G. Spangler
U.S. Department of Agriculture, Carl Hayden Bee Research Center

Electronic methods measured water consumption by a honey bee
colony within a single story hive, thermal input into the hive,
relative humidity of the outside air, and number of bees leaving
and entering the hive. During early morning flight, bees consumed
water at a rate of usually less than 2 g/hr. As external ambient
air temperature rose to 40°C, water consumption rose sharply to as
high as 15 g/hr. A screen placed over the hive entrance blocked
bees from leaving the colony and caused water consumption to increase 4-5 fold when temperatures were high. When the input line
to the watering device was cut off, the total available water
dropped to a minimal level of 19.1 g. Accordingly, numbers of bees
returning to the hive increased 4-9%. When bees were deprived of
the customary water supply within the hive, more of them foraged
for water.

Cytological mechanism of syngamy in the automictic laying
workers of Apis mellifera capensis

Savitri Verma, F. Ruttner and L.R. Verma
Institut fur Bienenkunde der universitat Frankfurt
am Main, F.R.G.

Cytological investigations revealed that impaternate (laying
workers) females of Apis mellifera capensis were normal diploid
with 32 as the chromosome number. During first maturation division chromosome pairing, synapsis and diakinesis were observed
during prophase I and chromosome number was reduced to 16.

as a result of second meiotic division 4 haploid nuclei were formed
in the form of a chromatin mass with no surrounding cytoplasm and
a membrane. These nuclei were arranged in a sagittal plane and
did not lie perpendicular to the surface of the egg as reported
previously. Later, two central nuclei i.e. egg pro-nucleus and
the descendant of the first polar body (second division nonsister nuclei) fused to form the zygote nucleus. Genetic consequences of such mechanism of automixis are also discussed.

Ecotype studies on Apis cerana indica F. of Northwestern Himalayan region

L.R. Verma & V.K. Mattu
Himachal Pradesh University, Simla (India)

Comprehensive morphometric studies were made on worker bees
of Apis cerana indica F. collected from different altitudes of
North western Himalayan region (comprising Himachal Pardesh and
Kashmir) in order to identify the different ecotypes of this
species. Statistical analysis of the data obtained on the fifty
five different morphological parts revealed a significant positive
correlation with altitude in certain characters in both the regions.
Bee samples collected from mountainous zones were bigger in size
and darker in colour as compared to those from submountainous zone.
Worker bee samples collected from different altitudes of Kashmir
region showed significantly higher values for several morphological
characters than those found in different altitudes of Himachal
Pradesh. Thus bees from Kashmir region were bigger in size and
darker in colour than bees of Himachal Pradesh representing the two
different ecotypes. The great phenotypical variability observed
in these two populations of honeybees provide excellent opportunity
for improving its quality by selection and breeding.

Colony structure and genetic relatedness in a species complex of ponerine ants

Philip S. Ward
The University of California at Davis

In Australian ants of the Rhytidoponera impressa group two
distinctive types of colonies can be found in most species: (i)
queenright, monogynous colonies with sterile workers (Type A), and
(ii) worker-reproductive colonies, in which one or more mated
workers occurs in lieu of a queen (Type B). Mated workers never
cohabit with a functional queen, but the two colony types are often
found in the same population. Allozyme markers confirm that worker
offspring in Type A nests are full sibs and that all males are
produced by the queen. The mean relatedness among unmated workers
in Type B colonies is estimated by regression analysis of allele
frequencies to be $b \approx 0.30$, and about 5% of the male alates are
produced by uninseminated workers. Type A colonies have larger,
more aggressive workers and more female-biased sex ratios of
investment than Type B colonies. Both colonies produce males,

but Type A colonies also release numerous colony-founding queens, while Type B colonies reproduce by fissioning (budding off daughter colonies with one or more mated workers). Selective forces favoring the evolution of worker-reproductive colonies are discussed.

Field Studies of the Lesser Attine Ant Genera

Neal A. Weber
Florida State University

The growing interest in the mutualistic relations between attine ants and their fungi has prompted this demonstration of the obscure minor genera. Seven genera contrast greatly with Acromyrmex and Atta, the well-known leaf-cutters. The minor genera have much smaller populations of a few hundred workers in the mature nest. The nest may have one cell to a few and the external indication of each genus and species tends often to be characteristic. Indications are that the ant fungi may have unusual properties and the fungi associated with these attines have received little attention. The only known source for the fungus cultures of minor genera is the Weber Ant Fungi Collection in the New York Botanical Garden. The present illustrations may assist the investigator in searching for additional cultures from the minor ant genera.

Genera illustrated by nests and gardens are Cyphomyrmex, Mycetophylax, Mycocepurus, Myrmicocrypta, Apterostigma, Sericomyrmex and Trachymyrmex.

Index to Authors, Genera, and Higher Taxa

(Reference is ordinarily made to the first but not to subsequent uses of a name in the same paper.)

About the Book and Editors

<u>The Biology of Social Insects</u>

edited by Michael D. Breed, Charles D. Michener,
and Howard E. Evans

   In this book internationally known experts provide a comprehensive view of current knowledge of social insect biology including much previously unpublished information. Particular emphasis is given to the relationships between social insects and humans; sections are devoted to economically important social insects, pollination, foraging, and the role of insects in ecosystems and agroecosystems. The authors also discuss communication, behavior and caste within insect colonies. A special section focuses on the neurobiology of social insects.
   A series of papers considers the presocial insects, which live in family groups but without caste differences. Also well represented are the fields of sociobiology and the origins and evolution of social behavior. The book will be valuable to agricultural scientists as well as to entomologists, sociobiologists, ecologists, ethologists, and natural historians. Endocrinologists and neurobiologists will also find important new material.

   Michael D. Breed is assistant professor in the Department of Environmental, Population, and Organismic Biology at the University of Colorado, Boulder. *Charles D. Michener* is Watkins Professor of Entomology and of Systematics and Ecology, the University of Kansas. *Howard E. Evans* is professor of entomology at Colorado State University.